# The Mathematical Principles of Quantum Mechanics

# The Mathematical Principles of Quantum Mechanics

D. F. Lawden

*Emeritus Professor*

*University of Aston in Birmingham, U.K.*

Dover Publications, Inc.

Mineola, New York

*Bibliographical Note*

This Dover edition, first published in 2005, is an unabridged republication of the work first published by Methuen & Co., Ltd., London, in 1967. Readers who would like solutions to the exercises may send an e-mail request to: editors@doverpublications.com.

*Library of Congress Cataloging-in-Publication Data*

Lawden, Derek F.
The mathematical principles of quantum mechanics / D.F. Lawden.
p. cm.
Originally published: London : Methuen, 1967.
Includes bibliographical references and index.
ISBN 0-486-44223-3 (pbk.)
1. Quantum theory—Mathematics. I. Title.

QC174.12.L393 2005
530.12—dc22

2004059119

Manufactured in the United States of America
Dover Publications, Inc., 31 East 2nd Street, Mineola, N.Y. 11501

# Contents

# Preface

As the most powerful and widely used tool of modern mathematical physics, quantum mechanics has now established its right to inclusion in every university course intended for mathematicians who wish to proceed to an Honours degree with special qualifications in applied mathematics and natural philosophy. Thus, whereas some years ago, instruction in the subject at the undergraduate level was restricted to lectures delivered in university physics departments, most universities in the British Commonwealth now arrange courses on this topic which are specifically designed to be taken by specialist mathematicians. Such students will already have been presented with developments of the classical theories of mechanics and electromagnetism which, while paying proper heed to the physical interpretation of the mathematical formulae, have nonetheless proceeded to establish the fundamental ideas from the basic principles or laws in a systematic, natural and logical manner. They therefore expect, and it is desirable that they should be offered, a similar development of the theory of quantum mechanics. However, the approach to the subject preferred by physicists when lecturing to their students on this topic is not usually of this type; very naturally, their emphasis is placed upon the results and their physical interpretation, rather than upon the details of the argument and, on account of the central importance of quantum theory for their subject, physics students must perforce be introduced to the main ideas involved at a relatively early stage in their studies, before they are sufficiently mature mathematically to follow an argument which makes some claims to rigor and generality. For these reasons, the form taken by a textbook of quantum mechanics directed towards applied mathematicians following a first course in the subject, usually in their third university year, should be different in many respects from that of a text intended to be read by students who are specialising in physics. However, the majority of texts seem to have been written with physicists in mind, so that I felt that a book which attempts to adopt the more leisurely and precise attitude towards the subject, characteristic of the natural philosopher, might be welcomed by students of applied mathematics.

Of all the mathematical models which are employed in contemporary physics, the quantum mechanical one is certainly the most difficult to construct *ab initio* in a manner which will appear natural and convincing to the student who is not already familiar with the concepts involved. One reason for this is that the principles which form the most convenient starting point for the argument and which constitute the natural axioms upon which to base the logical structure, are not immediately suggested by the relevant experimental data, but have, as it were, been precipitated by theoreticians out of the mixture of observations only after treating it with numerous reagents of a non-logical kind. The principles can certainly be shown to lead quite unambiguously to results which are in agreement with the experimental evidence, but it is by no means so clear that, if other reagents had been brought into action, a quite different set of principles, giving every appearance of being equally as reasonable as those which have in fact been accepted, might not have been discovered. However, A. Landé has been successful in discovering reasons why this physical model must possess those features which are usually regarded as being most characteristic of quantum mechanics. A full account of his approach will be found in *New Foundations of Quantum Mechanics*, but I have made use of ideas from this work in the introductory stages of my own development and it is hoped that this has resulted in an argument which will yield every evidence of springing from our common experience of the physical world in a natural and satisfying way.

Another reason for the difficulty which is generally experienced in reaching a clear understanding of the concepts and principles of the subject, is that these differ quite radically from those which were employed in the classical models of mechanics and electromagnetism. Only by presenting the student with numerous particular instances of the application of the general principles to physical problems and the proper interpretation of the results obtained, will he eventually become so familiar with the landmarks that he will feel at home in the microphysical world. To assist him to effect this change of outlook, I have been careful to illustrate the theory by many problems which are soluble in fairly simple terms and to provide sets of exercises containing further such examples at the end of the book (page 249). However, it should be made clear at this point, that this is intended to be a textbook of applied mathematics, not a textbook of physics. The emphasis has everywhere been placed upon the principles and structure of the theoretical model and the solutions of particular physical problems and explanations of physical phenomena have only been given where it was felt that these would contribute to a clarification of the general principles of the subject. The physicist will note the omission of many topics which are normally treated in a textbook of quantum mechanics intended for reading by his own students; for example, the very important applications of the theory in the field of physical chemistry are not referred to at all, since it was decided that these would occupy a great deal of space without assisting the reader's understanding of fundamentals to any degree. Nevertheless, it is hoped that any student who masters the ideas and arguments presented in this book, will experience no difficulty in following the applications of the theory by other authors to the solution of the manifold special problems of atomic physics.

The mode of development of the subject is as follows: in Chapter 1, the concept of a pure quantum mechanical state is described and its representation by a vector in a complex vector space is explained. It is assumed that the number of distinct eigenvalues for the observables being considered is finite, the generalisation to the case when the number is infinite (and even not enumerable) being postponed until Chapter 3. The representation of observables by matrices and the relationship between representations erected upon two different bases is then treated and the chapter concludes with an examination of the mathematical implications of the compatibility of observables and the representation of a system which is separable into two subsystems. The general principles elucidated in Chapter 1 are exemplified in Chapter 2 by application to the simplest possible system to exhibit quantum mechanical features, viz. a system possessing only two distinct eigenstates, such as a particle with spin $\frac{1}{2}\hbar$ whose translatory motion is ignored. A more complex system comprising two such particles is then examined and this leads, for the case of indistinguishable particles, to a formulation of the Exclusion Principle. The generalisation to observables possessing continuous spectra is made in Chapter 3, the case of the coordinate and momentum observables being examined in detail. It is demonstrated that the mathematical relationship between the coordinate and momentum representations can be derived by application of the special principle of relativity as was suggested by Landé. The concept of a Hamiltonian operator determining the manner in which the state of a system varies with the time is introduced in Chapter 4 and this leads quite naturally to an account of the Heisenberg type of representation. Chapter 5 is devoted to a detailed examination of the properties of the angular momentum operators and to the solution of the problem of an electron moving in the radial field of an atomic nucleus. Chapter 6 provides an introduction to the vast field of perturbation techniques and touches incidentally on a number of topics such as scattering and the induced emission of radiation from an atom, but it makes no attempt to give a complete account of these phenomena. A conventional account of Dirac's relativistic theory of the electron concludes the book.

A few notes regarding the notation I have adopted will be helpful. Scalar observables are denoted by italic symbols and vector observables by German Fraktur symbols. A matrix representing an observable $x$ is denoted by the corresponding bold roman symbol $\mathbf{x}$ and its conjugate transpose by $\mathbf{x}^\dagger$. An operator representing the same observable is shown thus, $\hat{x}$; occasionally, the representative of an observable $x$ proves to be a matrix having operators as its elements; this is then denoted by a symbol $\hat{\mathbf{x}}$. The expected value of an observable $x$ is denoted by $\bar{x}$ and the conjugate of a complex number $z$ has been shown as $z^*$. Vectors representing states of a system are denoted by Greek letters, e.g. $\alpha$ and the corresponding column matrices whose elements are the components of the vector relative to some reference frame, by bold Greek letters, e.g. $\boldsymbol{\alpha}$. The scalar product of two vectors $\alpha$, $\beta$ is shown as $\langle\alpha|\beta\rangle$ and is defined so that it is anti-linear with respect to the first factor and linear with respect to the second. The square root of minus one is denoted by $\iota$.

I should like to place on record my indebtedness to the authors of the large number of books I have consulted on various points; those which have proved to be most helpful have been listed in the Bibliography at the end of the book. I also wish to thank my colleague, Professor Andress, who has read the entire manuscript and offered a number of useful suggestions which have resulted in a clarification of the argument in many places. Lastly, I would like to assure my secretaries, Lee Trenberth and Pam Lovell, who together completed the arduous and skilful task of preparing the typescript, that their efforts in this regard were very much appreciated.

D. F. LAWDEN

*Mathematics Department,*
*University of Canterbury,*
*Christchurch, N.Z.,*
*February,* 1966.

# The Mathematical Principles of Quantum Mechanics

CHAPTER 1

# Vector representation of states

## 1.1. The Leibniz Principle

According to Leibniz, the physical world is so constituted that a small change in any cause will generate only a small change in its effect, i.e. the cause-effect relationship is continuous. Our everyday experience makes this principle appear eminently reasonable; a small adjustment of the throttle setting causes a small change in the speed of a motor car and, similarly, the range attained by a missile fired from a gun is a continuous function of the muzzle velocity and the angle of projection. There are, however, apparent exceptions to this rule. Thus, if the potential difference between two electrodes separated by air is raised steadily, no current will flow until a certain critical level is reached, when any further small increase in the P.D. causes a breakdown in the insulating property of the air and a relatively large current then commences to flow. However, even in circumstances such as these it is not difficult to interpret the principle in such a way that it is seen to be still valid.

Thus, suppose steel balls are rolled down an adjustable chute so that they impinge upon a vertical knife edge. For a certain range of positions of the chute, the balls will all be deflected to one side of the knife edge and for another range of positions, they will all be deflected to the other side. Since each ball must ultimately fall to one side or the other, it would appear that these two ranges must be separated by a sharp boundary such that, if an arbitrarily small adjustment is made to the chute transferring it across the boundary from a position belonging to the one range to a position belonging to the other, then a large change in the effect produced on a ball passing through the apparatus will result. The existence of such a boundary would violate Leibniz's principle. If, however, such an apparatus is set up in a laboratory with the object of establishing the existence of this boundary experimentally, it is invariably found that the results obtained do not support the hypothesis that such a sharp boundary exists. Instead it is found that for a certain range of settings of the chute, it cannot be predicted with certainty which side of the knife edge a ball will fall, although for a given setting in this range the proportions of balls falling on the two sides will be determinate; in other words, the probabilities for the two possible outcomes will be determinate, but the outcomes themselves will not. Thus, the effect of moving the chute over a range of positions including the critical range is to cause a continuous variation of the probabilities that a ball will fall to one side or the other of the knife edge. In the two extreme positions of the chute, the probabilities will be 0 and 1 and the outcome of an experiment will be predictable with certainty; as the chute is moved continuously between these positions, the probabilities will also vary continuously between the values 0 and 1. Our experiment accordingly supports the hypothesis that the Leibniz principle is valid, provided the final effect is regarded as being measured on a probability basis.

Nonetheless, according to the ideas of classical mechanics, the motion of each ball is, in principle, fully determinate and an explanation of the phenomenon we have described is given within this system of ideas by consideration of the fact that the position of the chute is not the only factor affecting the outcome of each experiment; slight variations in the initial conditions of entry of the balls into the chute, vibration of the apparatus, motion of the atmosphere and so on, are all factors which must be taken into account according to classical theory if a successful prediction is to be made when the position of the chute lies within the critical range. It is the neglect of these factors which, according to the classical view, accounts for the randomness of effect experienced in the critical zone; the Leibniz principle is therefore rejected as being invalid in circumstances such as we have been considering.

However, it is evident that it will never be possible to devise an actual experiment which will demonstrate that the Leibniz principle is not universally valid. In every practical case, even though the apparatus is manufactured with precision, a residuum of factors which cannot be exactly allowed for will always remain and a zone, which is small but finite, will be demonstrable within which the outcome of an experiment can only be predicted on a probability basis. The assumption of classical mechanics that it is possible to approach a limiting situation in which the principle is found to be invalid and the result of any experiment can always be predicted with exactitude, can never be established by experiment. Since this assumption has been shown to be incompatible with the Leibniz principle and since this principle is inherently plausible on the grounds that all our experience suggests that nature is in reality a continuous entity which is divided into segments by arbitrary boundaries for the convenience of study by man, it is entirely reasonable to accept the verdict of experiment, to reject the assumption of classical mechanics as representing no more than an approximation to the true principles governing natural phenomena and to accept the Leibniz principle in its amended form as being universally valid. This is the basis upon which the theory of quantum mechanics is constructed.

Thus, in quantum mechanics, we adopt the realistic position that even when the maximum information is available in respect of the initial and other conditions determining the motion of a mechanical system, a prediction of its motion can only be given on a probability basis. However, the possibility that in certain circumstances the probability for a particular outcome may prove to be unity is not excluded; in such a case the outcome would be predictable with absolute certainty. In the next section, we shall cite further considerations of a physical nature which tend to support our expectation that natural phenomena are governed by statistical laws rather than by exact laws of the type with which we become familiar in the classical physical theories.

## 1.2. Heisenberg's Uncertainty Principle

A clearer understanding of why it is, in principle, impossible to have as complete a knowledge of the factors affecting the motion of a physical system as would be required, even according to the classical view, to predict its motion with certainty, can be attained by consideration of the role played by the observing instrument in physical experimentation. All our knowledge of the behaviour of the physical world is obtained by studying the manner in which it interacts with such instruments. These may themselves be complicated physical systems which have been designed on the basis of natural principles already understood, in order to gain information relating to types of behaviour not yet elucidated; or, in the cases when the observations are made directly by the scientific investigator employing no intermediary, they will be part of the sensory equipment of a human being. In either case, the interaction of the system being observed with the observing instrument will inevitably disturb the system and, unless this disturbance can be reduced to negligible proportions or can be allowed for with precision, it will be impossible to predict accurately the motion of the system subsequent to an initial observation of its state. For the type of physical system to which the classical theories were applied, it was reasonable to assume that the reaction of the observing instruments upon a system could be reduced to a completely negligible level by the use of delicate instruments and appropriate amplifying attachments. In principle, therefore, the initial state of a system could be known with precision and its future behaviour calculated exactly. In particular, for a mechanical system of interacting particles, the position and velocity of each particle could be taken to be given with negligible error at some time $t_0$ and its future motion could then be predicted from Newton's laws. However, the disturbance of the system caused by observations being made upon it cannot be reduced indefinitely, since at least one fundamental particle, such as a photon or an electron, employed as a probe, must interact with the system, if information relating to its state is to be obtained and, if the system is of atomic dimensions, such a probing particle possesses a momentum which is comparable with the momenta of the system's elements. Nonetheless, if it were possible to make an exact allowance for the disturbance caused in the system by the probing particle or particles, the subsequent behaviour of the system would still be calculable, in principle, with precision. That such an exact correction can never be made was discovered by Heisenberg and this fact is embodied in quantitative form in his *Uncertainty Principle* (see section 1.11).

In particular, Heisenberg demonstrated that it will never be possible to know with absolute precision both the position and the momentum of a particle at the same instant. For suppose that the momentum of a particle has been determined in some way and that it is then proposed to find its position at some instant. A microscope is set up and the particle illuminated by a parallel beam of light having wavelength $\lambda$. At the instant the particle is observed in the microscope and its position becomes known, let $2\alpha$ be the angle subtended by the diameter of the instrument's aperture at the particle. Then it follows from the theory of the instrument that its resolving power will be insufficient to determine the particle's position to an accuracy greater than about $0{\cdot}6\lambda/\sin\alpha$. Thus, if $\Delta x$ represents the inaccuracy in the determination of the $x$-coordinate of the particle (the $x$-axis being perpendicular to the axis of the microscope), then

$$\Delta x = 0{\cdot}6\lambda/\sin\alpha. \qquad (1.2.1)$$

To minimise the disturbance of the particle's motion which will be caused by the illumination, we shall suppose that the intensity of the beam is diminished to as low a level as possible. However, one photon at least must be scattered into the microscope before a position determination can be made and, if this photon has wavelength $\lambda$, its momentum is known to have magnitude $p$, where

$$p = h/\lambda, \qquad (1.2.2)$$

$h$ being *Planck's constant* ($h = 6{\cdot}6 \times 10^{-27}$ c.g.s. units). After collision with the particle being observed, the photon's path is only known to lie inside a cone having semi-vertical angle $\alpha$ and thus its component of momentum in the direction of the $x$-axis is uncertain by amount

$$\Delta p = \frac{h}{\lambda} \sin \alpha. \tag{1.2.3}$$

Assuming that momentum is conserved during the collision, equation (1.2.3) must also express the uncertainty which, as a result of the observation, has entered into our knowledge of the $x$-component of the particle's momentum. Denoting the momentum uncertainty by $\Delta p_x$, we see from equations (1.2.1), (1.2.3) that this is related to the corresponding uncertainty in position by the equation

$$\Delta x \Delta p_x = 0{\cdot}6h, \tag{1.2.4}$$

i.e. the product of the uncertainties is of the order of Planck's constant. This is a particular case of the general Heisenberg Uncertainty Principle.

A further analysis shows that the process of measuring the $x$-component of the momentum of a particle to an accuracy specified by $\Delta p_x$, itself disturbs the $x$-coordinate of the particle by an unknown amount which cannot be reduced below $\Delta x$. $\Delta p_x$ and $\Delta x$ are again related by equation (1.2.4).

If $m$ is the mass of the particle being considered and $v_x$ is the $x$-component of the particle's velocity, then equation (1.2.4) is equivalent to

$$\Delta x \Delta v_x = 0{\cdot}6h/m. \tag{1.2.5}$$

Since $h$ is a very small constant, it is only when $m$ is also small that this equation places any effective restriction upon the accuracy to which measurements of $x$, $v_x$ can be made. Thus if $m = 1$ g and $\Delta x = 0{\cdot}1$ mm, then $\Delta v_x$ is about $4 \times 10^{-25}$ cm/sec and both $x$ and $v_x$ can be very accurate by ordinary standards. If, however, $m = 9 \times 10^{-28}$ g (the mass of the electron), the same uncertainty in position of $\Delta x = 0{\cdot}1$ mm corresponds to an uncertainty in $v_x$ of about 400 cm/sec, which is clearly highly significant.

The uncertainty principle therefore restricts the possible scope of our information regarding the initial positions and velocities of the particles constituting a mechanical system having atomic dimensions and this upper limit is insufficient to permit a precise prediction to be made of its subsequent behaviour. If a precise knowledge of the positions of the particles is assumed at any instant then, according to this principle, their velocities are completely unknowable; the converse statement is also true. An inaccurate knowledge of their positions can, however, be coupled with an inaccurate knowledge of their velocities and then the subsequent behaviour of the system can be predicted upon a probability basis. This is exactly the conclusion we arrived at in the previous section.

We shall accordingly accept it as a basic property of the physical world that, even when the maximum information is available regarding the state of a certain system at some initial instant, the future behaviour of this system can only be described in terms of probability functions. Although the behaviour of the system will not, therefore, be fully determinate, the manner in which the appropriate probability functions vary with the time will be and the equations expressing this determinacy will constitute the laws governing the physical world.

## 1.3. Pure and mixed states

Suppose that at some instant $t$, a physical system is allowed to interact with certain observing instruments. The information so obtained is said to refer to the *state* of the system at the instant $t$. Such interactions with instruments usually result in values being assigned to certain physical quantities associated with the system, which the instruments are said to measure. Thus, standard instrumental procedures for measuring the linear momentum, angular momentum, energy and position of a particle have been devised by physicists and, if these procedures are followed through in respect of a particular particle, the resulting numbers are, by definition, the values taken by these quantities for the particle in its state at the instant of observation. It is to be emphasised that the meaning to be assigned to a physical quantity is entirely contained in the procedure commonly employed to determine its value in any physical situation. It sometimes happens that two, or more, such procedures for determining the value of a single quantity have become generally accepted by virtue of the fact that it could be shown to be a consequence of classical theory that every such procedure must lead to the same result; the same conclusion may not be derivable from the principles of the new quantum mechanics and then the different procedures must be regarded as defining distinct quantities, which must then be represented by separate symbols in the theory. More will be said on this point later (section 1.9).

In classical mechanics, it is assumed that both the position and the velocity of a particle can be measured with precision at the same instant. Such information relating to the particle's state at the instant is complete in the sense that the value of any other physical quantity associated with the particle (e.g. its energy) can then be derived without it being necessary to perform any further observations. When such complete information relating to the particle's state at an initial instant is available, assuming that it moves through a known force field, the state of the particle at some future instant can be predicted with the same degree of completeness by making use of the principles of classical mechanics. When the initial state of a classical mechanical system is completely specified in this sense, we say that the system is in a *pure state*. When our information relating to the state of the system is deficient, we say that the state is *mixed*. In classical statistical mechanics, the behaviour of a gas is explained on the basis that it comprises a large number of molecules, each of which moves according to Newtonian principles. However, the state of a gas is not defined by specifying the positions and velocities of its molecules exactly at some instant, but only the statistical distributions of these quantities over the population of particles as a whole; each molecule is therefore in a mixed state.

In quantum mechanics, since account is taken of the fact that any observation which is made upon a system disturbs it, the definition of a pure state presents greater difficulties. In particular, an observation of the $x$-coordinate of a particle inevitably alters the $x$-component of the momentum of the particle and this disturbance is, in principle, impossible to determine. We say that observations of $x$ and $p_x$ are *incompatible*. Thus, if we wish to gain as complete information as possible regarding a system's state at a certain instant, we must clearly be careful to carry out only such observations upon the system as are compatible one with another, for otherwise the disturbance caused by one observation will invalidate the result of a preceding observation. Suppose that, for a particular system, a set of mutually compatible measuring procedures has been found, possessing the properties that (a) the results of none of the procedures can be inferred from the results of the others and (b) there exists no other procedure, compatible with every member of the set, whose outcome is not derivable from the results of observations belonging to the set. For a particle moving in a central force field, it will be shown in section 5.4 that observations of the energy, of the square of the angular momentum and of a component of the angular momentum of the particle, constitute such a set. If such a set of observations is made upon the system at the instant $t = t_0$ (the order in which the observations are made is immaterial, since they are compatible), then the information obtained regarding the state of the system at this instant is clearly as complete as the uncertainty principle permits. In quantum mechanics, a system will accordingly be said to be in a pure state for $t = t_0$, when the results of such a maximal set of compatible observations made at $t = t_0$ are available. Since the system will be disturbed in a characteristic way at the instant $t_0$ by the making of this particular series of observations, we shall say that this set of measuring procedures *prepares* the system at $t_0$ in the pure state specified by the results of the observations.

Supposing that a system has been prepared in a pure state at $t = t_0$, the principles of quantum mechanics (like those of Newtonian mechanics) then enable us to predict the state of the system to the same degree of completeness at any later time $t$, it being assumed, of course, that the system is not disturbed by some agent external to it between the instants $t_0$ and $t$. This latter proviso implies that the system must remain unobserved over the whole of the period $(t_0, t)$. Immediately the system is subjected to scrutiny, it is disturbed out of its previous state (pure or otherwise) and must be regarded as having been prepared in a fresh state. Thus the principles of quantum mechanics enable us to relate the results of one set of compatible observations made upon a system, to the results of the *next* set of compatible observations; however, immediately the second set of observations has been made, the system enters a new state and our previous calculations cease to be relevant. The tacit assumption made in the classical theories that the system is maintained under continuous observation as it evolves, cannot be accepted in the new theory.

## 1.4. Transitions between pure states

In this section, we will consider the case when the time which elapses between the making of the two sets of observations referred to at the end of the last section, approaches zero. We shall assume that both the initial and terminal sets are maximal, so that each has the effect of preparing the system under consideration in a pure state. If the terminal set is not equivalent to the initial set, then the two pure states will be distinct and we shall say that the system has undergone a *transition* from the earlier state to the later. Thus, for a system comprising a single particle moving in a given force field, the momentum of the particle might be established with precision at one instant and the position of the particle immediately afterwards; the particle would then be said to have undergone a transition from a state in which its momentum was known, to a state in which its position was known.

Since, by the principle of uncertainty, the information relating to the initial pure state will be insufficient to specify this state to the degree to which we are accustomed in classical theories, we cannot expect that the results of the subsequent set of observations will be precisely predictable. Instead, it is reasonable to assume that only the probabilities of obtaining different sets of results will be calculable. The fundamental problem of quantum mechanics is to provide a mathematical procedure whereby, when the results of the first set of observations are known, these probabilities can be calculated. We proceed to a study of this problem.

Suppose that the observations which prepare the system in its initial pure state are the precise measurements of certain quantities we shall denote by $a$, $b$, .... These measurements must be compatible. At this stage, for simplicity, it will be assumed that the result of measuring $a$ can only be one of a finite set of distinct values $a_1, a_2, \ldots$, the result of measuring $b$ can only be one of a finite set of distinct values $b_1, b_2, \ldots$ and so on. Quantities such as $a$, $b$, etc, whose measurement prepares a system in a pure state, will be termed *observables* of the system and the set of possible results of the measurement of an observable will be referred to as its *spectrum of eigenvalues*. If, as is being assumed, the eigenvalues can be arranged in a sequence, the spectrum is said to be *discrete*; otherwise, if the range of eigenvalues is continuous, the spectrum is also said to be *continuous*. To each of the distinct sets of eigenvalues $(a_i, b_j, \ldots)$, there corresponds a distinct pure state of the system. Such a state is called an *eigenstate* of the system with respect to the compatible observables $a, b, \ldots$ and, in such a pure state, we say that the observables $a$, $b$, ... are *sharp* with values $a_i$, $b_j$, ... respectively. In the case under consideration, since the number of distinct sets of eigenvalues $(a_i, b_j, \ldots)$ is finite, so is the number of distinct eigenstates. We shall denote these states by $\psi^1, \psi^2, \ldots, \psi^m$ and refer to them as constituting a *complete set of eigenstates.*

The second set of observations which prepares the system in its terminal state can be treated similarly and we will suppose that, relative to this set, we can identify $n$ distinct eigenstates $\phi^1, \phi^2, \ldots, \phi^n$ in each of which certain observables assume sharp values. Our problem then is to compute the probability that, if the system is observed in the initial state $\psi^i$, it will immediately afterwards be observed in the state $\phi^j$. This probability will be denoted by $P(\psi^i \to \phi^j)$ and, if every possible transition is to be covered, it will be necessary to know these probabilities for every pair of values of $i$ and $j$, where $i = 1, 2, \ldots, m$, and $j = 1, 2, \ldots, n$.

Writing $P_{ij} = P(\psi^i \to \phi^j)$, it is convenient to arrange these probabilities into the form of an $m \times n$ matrix, $(P_{ij})$. Then the sum of the elements in the $i$th row is the sum of the probabilities that a transition will be observed from the state $\psi^i$ to one at least of the states $\phi^j$. But one such transition must certainly be observed. It follows that the sum of the elements in any row must be unity, i.e.

$$\sum_{j=1}^{n} P_{ij} = 1. \tag{1.4.1}$$

In classical mechanics, if a possible behaviour of a system consists in its passing through a series of states $A, B, C, \ldots$, in that order in time then, by changing the signs of the particle velocities at some instant, the system can be made to 'run in reverse' so that it passes through the same states in the order $\ldots, C, B, A$. This is the *Principle of Reversibility*. In quantum mechanics, we shall take the corresponding principle to be

$$P(\psi^i \to \phi^j) = P(\phi^j \to \psi^i), \tag{1.4.2}$$

and shall adopt this as a fundamental postulate.

It follows from this principle that

$$\sum_{i=1}^{m} P_{ij} = \sum_{i=1}^{m} P(\psi^i \to \phi^j) = \sum_{i=1}^{m} P(\phi^j \to \psi^i) = 1, \tag{1.4.3}$$

i.e. that the sum of the elements in any column is also unity.

Summing all the elements of the matrix by rows, the result is clearly $m$. Similarly, summing by columns, the result is $n$. It follows that $m=n$ and that the matrix is square.

In the special case when we choose the terminal set of measurements to be identical with the initial set, so that $\phi^i \equiv \psi^i$, we shall assume that

$$P(\psi^i \to \psi^j) = \delta_{ij}, \tag{1.4.4}$$

where $\delta_{ij}$ is the Kronecker delta symbol, i.e. we suppose that, if the system is first prepared in the state $\psi^i$ by performing a series of measurements, an immediate repetition of these measurements will yield unchanged results, so that the system will be found to be still in the state $\psi^i$ with 100% probability. In this case, the matrix $(P_{ij})$ is the $n \times n$ unit matrix $\mathbf{I}_n$.

We shall next show how to represent a set of eigenstates $\psi^1, \psi^2, \ldots, \psi^n$ by an abstract mathematical model having the following, clearly necessary, properties. Firstly, if $\psi^i$, $\phi^j$ are two different eigenstates, a universal rule must be stated by which we can calculate $P(\psi^i \to \phi^j)$ when the representatives of these states within the model are known; this rule must be consistent with the principles (1.4.1)–(1.4.4). Secondly, if $\psi^i$, $\phi^i$, $\chi^i$ $(i=1,2,\ldots,n)$ are any three sets of eigenstates, some mathematical relationship will exist between the representatives of $\psi^i$ and $\phi^i$; a relationship of the same type will exist between the representatives of $\phi^i$

and $\chi^i$; it must then be a consequence of these relationships that the representatives of $\psi^i$ and $\chi^i$ are related in the same way. A. Landé† has demonstrated that, by taking account of properties of a general nature such as these, which must be possessed by the mathematical model we are seeking, the form taken by this model is uniquely determined. We shall not repeat his arguments here, but will content ourselves with setting up the model and verifying that it possesses the required properties.

## 1.5. Complex finite-dimensional vector spaces

The geometrical structure we are about to describe is the natural generalisation of the elementary theory of vectors in a three-dimensional Euclidean space to a situation in which the number of dimensions is an arbitrary positive integer $N$ and the vector components are permitted to assume complex values. As in the elementary theory, it is customary to restrict our attention to a particular species of reference frames, viz. rectangular frames, whose axes are mutually orthogonal (the meaning to be assigned to this term in the present context is explained below); relative to such frames, the distinction between covariant and contravariant vectors disappears and the theory is correspondingly simplified. It will appear in the sequel that the imposition of this restriction upon our abstract structure, far from limiting its usefulness as a model for quantum mechanics, is actually the measure which converts it into a completely natural instrument for the expression of the fundamental principles involved.

The $N$-dimensional vector space we shall be describing will be denoted by $\mathscr{S}_N$. A *vector* in this space is defined to be an ordered set of $N$ complex numbers $(\alpha_1, \alpha_2, \ldots, \alpha_N)$. The elements $\alpha_i$ of the set are called the *components* of the vector. The complete set or vector will be represented by $\alpha$.

Two vectors $\alpha$, $\beta$ are said to be *equal* if, and only if, their corresponding components are equal, i.e. if $\alpha_i = \beta_i$ $(i = 1, 2, \ldots, N)$. We then write

$$\alpha = \beta. \tag{1.5.1}$$

The product of a vector $\alpha$ and a complex number $a$ is defined to be the vector $\beta$ whose components are given by $\beta_i = a\alpha_i$ $(i = 1, 2, \ldots, N)$. We write

$$\beta = a\alpha. \tag{1.5.2}$$

The vector $\beta$ is said to be in the same *direction* as the vector $\alpha$.

The *sum* of two vectors $\alpha$, $\beta$ is defined to be the vector $\gamma$ whose components are given by $\gamma_i = \alpha_i + \beta_i$ $(i = 1, 2, \ldots, N)$. We write

$$\gamma = \alpha + \beta. \tag{1.5.3}$$

† *New Foundations of Quantum Mechanics* by A. Landé, C.U.P., 1965.

It is clear that the operations of multiplication and addition so defined, obey the commutative, distributive and associative laws of elementary algebra.

The *difference* of two vectors $\alpha$, $\beta$ is written $\alpha-\beta$ and is defined by the equation

$$\alpha-\beta = \alpha+(-1)\beta. \tag{1.5.4}$$

The components of the difference are evidently $\alpha_i-\beta_i$ $(i=1,2,\ldots,N)$.

The vector whose components are all zero is called the *null vector* and it will be denoted by 0. A vector whose components all vanish except for one which has unit value, is called a *base vector*. There are $N$ distinct base vectors. The base vector whose $i$th component is nonvanishing will be denoted by $\epsilon^i$; it is clear that

$$\epsilon^i_j = \delta_{ij}, \tag{1.5.5}$$

where the $\delta_{ij}$ are Kronecker symbols. Any vector $\alpha$ can be expressed linearly in terms of the base vectors thus,

$$\alpha = \sum_{i=1}^{N} \alpha_i \epsilon^i, \tag{1.5.6}$$

for the vector represented by the right hand member of this equation has $i$th component $\alpha_i$.

When a summation is to be carried out over the integers 1, 2, ..., $N$ with respect to an index which occurs twice in the general term as in equation (1.5.6), the sign of summation will often be omitted. Thus, we may write equation (1.5.6) in the abbreviated form

$$\alpha = \alpha_i \epsilon^i. \tag{1.5.7}$$

This is the *summation convention*. The convention will only apply when the repeated index is a lower case roman letter; thus $\alpha_1 \epsilon^1$, $\alpha_N \epsilon^N$ are not to be summed and represent the first and last terms respectively in the expansion (1.5.6).

If $\alpha^1, \alpha^2, \ldots, \alpha^M$† is a set of $M$ vectors and $M$ complex numbers $a_i$, not all zero, can be found such that

$$\sum_{i=1}^{M} a_i \alpha^i = 0, \tag{1.5.8}$$

† Superscripts will often be employed, as here, to distinguish between members of a set of vectors; subscripts will always be employed to distinguish components of the *same* vector.

then the vectors $\alpha^i$ are said to be *linearly dependent*. Otherwise, the vectors $\alpha^i$ are said to be *linearly independent*. Written in terms of the vectors' components, equation (1.5.8) reads,

$$\sum_{i=1}^{M} a_i \alpha_j^i = 0, \quad j = 1, 2, \ldots, N. \tag{1.5.9}$$

If $M > N$, these equations always possess a non-zero solution in the $a_i$ for any set of vectors $\alpha^i$ and hence it follows that any set of more than $N$ vectors is linearly dependent. However, sets of $N$ linearly independent vectors exist, for the $\epsilon^i$ obviously form such a set.

If a vector $\beta$ can be expressed linearly in terms of a set of vectors $\alpha^1, \alpha^2, \ldots, \alpha^M$, thus,

$$\beta = \sum_{i=1}^{M} \beta_i \alpha^i, \tag{1.5.10}$$

$\beta$ is said to be *linearly dependent on* the $\alpha^i$. It follows that, if $\alpha^1, \alpha^2, \ldots, \alpha^M$ are linearly dependent, then at least one of these vectors is linearly dependent on the remainder. Equation (1.5.7) indicates that every vector is linearly dependent on the base vectors.

The *scalar product* of two vectors $\alpha$, $\beta$ *in that order*, will be denoted by $\langle\alpha|\beta\rangle$ and is the complex number defined by the equation

$$\langle\alpha|\beta\rangle = \alpha_i^* \beta_i.\dagger \tag{1.5.11}$$

The reader will easily verify that this product is *anti-linear* in the factor $\alpha$ and linear in the factor $\beta$, i.e.

$$\langle a_1 \alpha^1 + a_2 \alpha^2|\beta\rangle = a_1^*\langle\alpha^1|\beta\rangle + a_2^*\langle\alpha^2|\beta\rangle, \tag{1.5.12}$$

$$\langle\alpha|b_1 \beta^1 + b_2 \beta^2\rangle = b_1\langle\alpha|\beta^1\rangle + b_2\langle\alpha|\beta^2\rangle. \tag{1.5.13}$$

It is also easy to see that

$$\langle\beta|\alpha\rangle = \langle\alpha|\beta\rangle^*. \tag{1.5.14}$$

The scalar product of a vector $\alpha$ with itself is clearly a positive real number; its positive square root is called the *norm* of the vector and is denoted by $\|\alpha\|$. Hence

$$\|\alpha\|^2 = \langle\alpha|\alpha\rangle. \tag{1.5.15}$$

If $\|\alpha\| = 0$, then every component of $\alpha$ must vanish, i.e. $\alpha = 0$.

The norm of a vector is the natural generalisation to $\mathscr{S}_N$ of the concept of the length of a displacement vector in three-dimensional Euclidean space, $\mathscr{E}_3$. In $\mathscr{E}_3$, the well-known 'triangle inequality' leads to the result

† The conjugate complex of a number will be indicated by an asterisk.

that the magnitude of the resultant of two vectors is always less than or equal to the sum of the magnitudes of the individual vectors. In $\mathscr{S}_N$, we also have that

$$\|\alpha+\beta\| \leqslant \|\alpha\|+\|\beta\|. \tag{1.5.16}$$

This inequality may be proved thus: if $x$ is an arbitrary real number, it follows from equations (1.5.12), (1.5.13), that

$$\langle\alpha+x\beta|\alpha+x\beta\rangle = \langle\alpha|\alpha\rangle+x[\langle\beta|\alpha\rangle+\langle\alpha|\beta\rangle]+x^2\langle\beta|\beta\rangle. \tag{1.5.17}$$

Since this expression must be positive for all $x$, we deduce that

$$[\langle\beta|\alpha\rangle+\langle\alpha|\beta\rangle]^2 \leqslant 4\langle\alpha|\alpha\rangle\langle\beta|\beta\rangle. \tag{1.5.18}$$

Putting $x=1$ in equation (1.5.17) and making use of the last inequality, it follows that

$$\|\alpha+\beta\|^2 \leqslant \|\alpha\|^2+2\|\alpha\|.\|\beta\|+\|\beta\|^2. \tag{1.5.19}$$

This is equivalent to the inequality (1.5.16). It is left as an exercise for the reader to deduce that

$$\|\alpha+\beta\| \geqslant \Big|\|\alpha\|-\|\beta\|\Big|. \tag{1.5.20}$$

The vector whose components are $\alpha_i^*$, i.e. the conjugate complexes of the components of $\alpha$, is denoted by $\alpha^*$. It is easy to see that

$$\|\alpha^*\| = \|\alpha\|. \tag{1.5.21}$$

The inequality

$$(\alpha_i\beta_j-\beta_i\alpha_j)(\alpha_i^*\beta_j^*-\beta_i^*\alpha_j^*) \geqslant 0, \tag{1.5.22}$$

where the left-hand member is to be summed with respect to both $i$ and $j$, is obvious. Multiplying out the left-hand member, we obtain

$$2\|\alpha\|^2\|\beta\|^2-2\langle\alpha|\beta\rangle\langle\beta|\alpha\rangle \geqslant 0, \tag{1.5.23}$$

which is equivalent to

$$\|\alpha\|.\|\beta\| \geqslant |\langle\alpha|\beta\rangle|. \tag{1.5.24}$$

This is *Schwarz's Inequality*. It is clear from (1.5.22) that equality can occur only if

$$\alpha_i\beta_j-\beta_i\alpha_j = 0 \tag{1.5.25}$$

for all $i, j$. This condition is equivalent to the condition

$$\alpha = x\beta, \tag{1.5.26}$$

where $x$ is a complex number.

If the vector $\alpha$ is multiplied by the number $\|\alpha\|^{-1}$ (assumed finite), the resulting vector has unit norm. This process is called *normalising* the vector $\alpha$. If $\alpha$, $\beta$ have been normalised, then it follows from the Schwarz inequality that

$$|\langle\alpha|\beta\rangle| \leqslant 1. \tag{1.5.27}$$

It is clear that, if a normalised vector is multiplied by any complex number having unit modulus (i.e. a number of the form $\exp(\iota\theta)$), it remains normalised (i.e. of unit norm).

In $\mathscr{E}_3$, the scalar product of a pair of perpendicular vectors is zero. In $\mathscr{S}_N$, we shall define two vectors $\alpha$, $\beta$ to be perpendicular or *orthogonal* if

$$\langle\alpha|\beta\rangle = 0. \tag{1.5.28}$$

If $\alpha$ is a vector with unit norm, $\langle\alpha|\beta\rangle$ is called the *orthogonal projection* or simply the *projection* of $\beta$ on the direction of $\alpha$.

The reader will have no difficulty in verifying that the base vectors $\epsilon^i$ are mutually orthogonal and that each has unit norm. Thus,

$$\langle\epsilon^i|\epsilon^j\rangle = \delta_{ij}. \tag{1.5.29}$$

These vectors accordingly play the same role as the unit vectors in the directions of rectangular cartesian axes in $\mathscr{E}_3$. In particular, it follows from the definition of the scalar product that

$$\langle\epsilon^i|\alpha\rangle = \alpha_i, \tag{1.5.30}$$

i.e. the projection of $\alpha$ on the direction of $\epsilon^i$ is the $i$th component of $\alpha$.

## 1.6. Unitary transformations

Consider a vector $\alpha$ with components $\alpha_i$. Let us define another set of $N$ quantities $\alpha_i'$ by the equations

$$\alpha_i' = u_{ij}\,\alpha_j, \tag{1.6.1}$$

the coefficients $u_{ij}$ being complex numbers such that the determinant $|u_{ij}|$ is non-zero. Given the set $(\alpha_i)$, the equations (1.6.1) define the set $(\alpha_i')$ uniquely and, conversely, given the set $(\alpha_i')$, the $\alpha_i$ can be solved for and the solution is unique. The vector $\alpha$ can accordingly be specified by stating the values of the $\alpha_i'$ as well as by giving the values of the $\alpha_i$. The equations (1.6.1) are accordingly said to define a *transformation* in the $\mathscr{S}_N$ from components $\alpha_i$ to new components $\alpha_i'$.

However, in general, the scalar product of two vectors $\alpha$, $\beta$ will not be expressed so simply in terms of the 'dashed' components as it is (equation (1.5.11)) in terms of the original components. We shall next investigate the conditions to be satisfied by the coefficients $u_{ij}$ of the transformation, if the two expressions for the scalar product are to be identical. With this end in view, it will be helpful first to express the transformation in matrix form.

To do this, the components of a vector are first displayed as column matrices, thus,

$$\boldsymbol{\alpha} = \begin{pmatrix} \alpha_1 \\ \alpha_2 \\ \vdots \\ \alpha_N \end{pmatrix}, \quad \boldsymbol{\alpha}' = \begin{pmatrix} \alpha_1' \\ \alpha_2' \\ \vdots \\ \alpha_N' \end{pmatrix}. \tag{1.6.2}$$

As indicated above, in the subsequent development all matrices will be denoted by symbols in bold type. The transformation equation (1.6.1) may then be expressed in matrix form

$$\boldsymbol{\alpha}' = \mathbf{U}\boldsymbol{\alpha}, \tag{1.6.3}$$

$\mathbf{U}$ being the $N \times N$ matrix with the element $u_{ij}$ in its $i$th row and $j$th column.

The scalar product of two vectors $\alpha$, $\beta$ is given in terms of the corresponding column matrices by the equation

$$\langle \alpha | \beta \rangle = \boldsymbol{\alpha}^\dagger \boldsymbol{\beta}, \tag{1.6.4}$$

where $\boldsymbol{\alpha}^\dagger$ is the conjugate transpose of $\boldsymbol{\alpha}$ (i.e. the row matrix $(\alpha_1^*, \alpha_2^*, \ldots, \alpha_N^*)$). Substituting from equation (1.6.3), we find that

$$\boldsymbol{\alpha}'^\dagger \boldsymbol{\beta}' = \boldsymbol{\alpha}^\dagger \mathbf{U}^\dagger \mathbf{U} \boldsymbol{\beta} = \boldsymbol{\alpha}^\dagger \boldsymbol{\beta} \tag{1.6.5}$$

if, and only if, $\mathbf{U}$ satisfies the condition

$$\mathbf{U}^\dagger \mathbf{U} = \mathbf{I}_N, \tag{1.6.6}$$

where $\mathbf{I}_N$ is the $N \times N$ unit matrix. If $\mathbf{U}$ satisfies this condition, the expression for the scalar product $\langle \alpha | \beta \rangle$ takes the same form whether expressed in terms of the 'dashed' or the 'undashed' components. Also, because the transformation is linear, a linear combination of vectors which is represented by the matrix $a\boldsymbol{\alpha} + b\boldsymbol{\beta}$ ($a$, $b$, complex numbers) when the 'undashed' components are employed, will be represented by the matrix $a\boldsymbol{\alpha}' + b\boldsymbol{\beta}'$ when the 'dashed' components are employed.

A transformation satisfying the condition (1.6.6) is said to be *unitary*. Provided transformations of this type alone are permitted, an argument will take the same mathematical form whether expressed in terms of the 'dashed' or the 'undashed' components and hence the 'dashed' components may be thought of as being equally as fundamental as the original components. They will be referred to as the components of the vector referred to a new *reference frame* and a unitary transformation will be regarded as the counterpart for our complex vector space of the familiar *orthogonal transformation* between rectangular cartesian frames in $\mathscr{E}_3$.

The condition (1.6.6) is equivalent to the statement that the inverse of **U** is identical with its conjugate transpose, i.e.

$$\mathbf{U}^{-1} = \mathbf{U}^{\dagger}. \tag{1.6.7}$$

Multiplying both sides of the equation on the left by **U**, we then obtain

$$\mathbf{U}\mathbf{U}^{\dagger} = \mathbf{I}_N. \tag{1.6.8}$$

Conditions (1.6.6)–(1.6.8) are all equivalent. In terms of the elements $u_{ij}$, the conditions (1.6.6), (1.6.8) may be written

$$u_{ji}^* u_{jk} = \delta_{ik}, \tag{1.6.9}$$

$$u_{ij} u_{kj}^* = \delta_{ik}, \tag{1.6.10}$$

respectively.

It is left as an exercise for the reader to show that if **U** is unitary, then so is $\mathbf{U}^{\dagger}$.

Employing the original frame of reference, let $\psi^1, \psi^2, \ldots, \psi^N$ be a set of mutually orthogonal vectors, each having unit norm. The $\psi^i$ are said to constitute a *complete orthonormal set*. Then a unitary transformation to a new frame can always be found relative to which frame these vectors are a set of base vectors. For, if we take

$$u_{ij} = \psi_j^{i*}, \tag{1.6.11}$$

then

$$u_{ij} u_{kj}^* = \psi_j^{i*} \psi_j^k = \langle \psi^i | \psi^k \rangle = \delta_{ik}, \tag{1.6.12}$$

since the $\psi^i$ are mutually orthogonal and normalised. Hence $\mathbf{U} = (u_{ij})$ is a unitary matrix. Also, employing this matrix, the $\psi_j^i$ transform into $\psi_j^{i\prime}$ where

$$\psi_j^{i\prime} = u_{jk} \psi_k^i = \psi_k^{j*} \psi_k^i = \langle \psi^j | \psi^i \rangle = \delta_{ij}. \tag{1.6.13}$$

This proves that the $\psi^i$ are base vectors for the new frame and hence that the new frame has its axes aligned with the vectors $\psi^i$.

Since $\psi_j^i$ is the projection of $\psi^i$ on the direction of the $j$th base vector of the original frame, denoting this base vector by $\phi^j$ it follows from equation (1.6.11) that

$$u_{ij} = \langle \phi^j | \psi^i \rangle^* = \langle \psi^i | \phi^j \rangle. \tag{1.6.14}$$

This is a convenient formula for the coefficients of the unitary transformation which transforms from the frame determined by the orthonormal set $\phi^i$ to the frame determined by the set $\psi^i$.

Let $\alpha'_i$ be the components of a vector $\alpha$ in the new frame. Referring to equation (1.5.6), we note that the vector can be expressed linearly in terms of the base vectors thus,

$$\alpha = \alpha'_i \psi^i. \tag{1.6.15}$$

But this equation is valid in any frame and hence we have proved that an arbitrary vector can be expressed linearly in terms of any complete orthonormal set. Since

$$\langle \psi^j | \psi^i \rangle = \delta_{ij}, \tag{1.6.16}$$

it follows from equation (1.6.15) that

$$\alpha'_j = \langle \psi^j | \alpha \rangle, \tag{1.6.17}$$

i.e. the coefficients $\alpha'_i$ in the expansion (1.6.15) are the projections of $\alpha$ upon the members of the orthonormal set. Equation (1.6.15) can now be written

$$\alpha = \langle \psi^i | \alpha \rangle \psi^i. \tag{1.6.18}$$

This is the generalisation to $\mathscr{S}_N$ of the familiar formula from elementary vector algebra in $\mathscr{E}_3$ expressing a vector in terms of its components in the directions of the mutually perpendicular unit vectors **i**, **j**, **k**.

Taking the scalar product of both members of equation (1.6.18) with a vector $\beta$, we obtain

$$\langle \beta | \alpha \rangle = \langle \beta | \psi^i \rangle \langle \psi^i | \alpha \rangle. \tag{1.6.19}$$

Since $\langle \psi^i | \alpha \rangle$ is the projection of $\alpha$ on the direction of $\psi^i$ and is identical with $\alpha'_i$ equation (1.6.19) may be written

$$\langle \beta | \alpha \rangle = \beta'^*_i \alpha'_i. \tag{1.6.20}$$

It is now seen to be the usual expression for the scalar product of $\alpha$ and $\beta$ in the 'dashed' frame.

Putting $\beta = \alpha$ in equation (1.6.19), we obtain the result

$$\|\alpha\|^2 = \sum_{i=1}^{N} |\langle \psi^i | \alpha \rangle|^2, \tag{1.6.21}$$

which is the form taken by Pythagoras Theorem in $\mathscr{S}_N$. In particular, if $\alpha$ has unit norm, then

$$\sum_{i=1}^{N} |\langle \psi^i | \alpha \rangle|^2 = 1, \tag{1.6.22}$$

corresponding to the Euclidean result that the sum of the squares of the direction cosines of a vector is unity.

The members of any orthonormal set ($\psi^i$) are linearly independent, for if not, numbers $a_i$ must be available such that

$$a_i \psi^i = 0. \tag{1.6.23}$$

Taking the scalar product of both members of this equation with $\psi^j$, we obtain

$$a_j = 0, \tag{1.6.24}$$

i.e. all the numbers $a_i$ vanish.

## 1.7. The vector representation of states

Let $\alpha$, $\beta$, $\gamma$, ... represent pure states in which a certain physical system may be prepared at some instant. We shall assume, as a fundamental postulate, that it is possible to associate with every such state, a vector belonging to a complex vector space, in such a way that the probability of observing a transition from any one of these states $\alpha$ to any other $\beta$ is given by

$$P(\alpha \to \beta) = |\langle\beta|\alpha\rangle|^2, \tag{1.7.1}$$

where the associated vectors are also being denoted by $\alpha$, $\beta$, $\gamma$, .... It is easy to verify that this assumption ensures that the probability matrices have all the properties elucidated in section 1.4.

First, since $\langle\alpha|\beta\rangle = \langle\beta|\alpha\rangle^*$, it follows that

$$P(\alpha \to \beta) = P(\beta \to \alpha), \tag{1.7.2}$$

in accordance with equation (1.4.2).

Also, if $\psi^1$, $\psi^2$, ..., $\psi^N$ represents a complete set of eigenstates with respect to a certain maximal set of compatible observables, in order to satisfy the condition (1.4.4), it is necessary that the corresponding vectors should form an orthonormal set. Further, since no other eigenstate can be found which is distinct from every member of the set, we shall assume that there is no other vector orthogonal to every member of the orthonormal set and hence that the set is complete. Thus $N$ is the dimension of our spatial model.

Let $\alpha$ represent any other state of the system. Then the probability of observing a transition from the state $\alpha$ into one of the eigenstates $\psi^i$ is

$$P(\alpha \to \psi^i) = |\langle\psi^i|\alpha\rangle|^2. \tag{1.7.3}$$

But one of these transitions must be observed and hence

$$\sum_{i=1}^{N} |\langle\psi^i|\alpha\rangle|^2 = 1. \tag{1.7.4}$$

It now follows from equation (1.6.21) that

$$\|\alpha\| = 1. \tag{1.7.5}$$

Thus all vectors representing pure states have unit norms. Also, by replacing $\alpha$ in equation (1.7.4) by $\phi^j$, we obtain equation (1.4.3). Equation (1.4.1) then follows as a consequence of equation (1.7.2). Our mathematical model accordingly possesses all the properties which we have seen are necessarily possessed by any abstract structure which is to act as a suitable vehicle for the expression of quantum mechanical principles.

It is to be noted that if the vector representing a state is multiplied by a number having unit modulus, the modulus of its scalar product with any other vector is unaltered. This implies that the results of all possible observations on the system will have unchanged probabilities. We conclude that a state vector is always arbitrary to the extent of a factor having unit modulus.

As has been explained in the previous section, a vector $\alpha$ can be specified in terms of its components in many ways. Any complete orthonormal set of vectors can be treated as a set of base vectors and the projections of $\alpha$ upon these vectors are then to be taken as the components of $\alpha$ in the frame whose axes are in the directions of the base vectors. Thus, if $\psi^1, \psi^2, \ldots, \psi^N$ are the vectors representing a complete set of eigenstates relative to compatible observables $a, b, \ldots$, these vectors are known to form a complete orthonormal set and may accordingly be employed as base vectors to define a frame $F$. If $\alpha$ is a vector representing an arbitrary state of our physical system, the projections of $\alpha$ upon the vectors $\psi^i$ are the components of $\alpha$ in the frame $F$. Denoting these components by $\alpha_i$, we have

$$\alpha_i = \langle \psi^i | \alpha \rangle. \tag{1.7.6}$$

The components $\alpha_i$ will normally be displayed in the form of a column matrix $\boldsymbol{\alpha}$; we shall say that the matrix specifies the state of the system in the $a, b, \ldots$-representation. Clearly,

$$|\alpha_i|^2 = P(\alpha \to \psi^i), \tag{1.7.7}$$

so that the squares of the moduli of the components of $\alpha$ can be interpreted physically as the probabilities of observing transitions from the state $\alpha$ into the various eigenstates belonging to the representation.

Equation (1.6.18) can now be interpreted as the expansion of the vector associated with the state $\alpha$ in terms of the vectors associated with a complete set of eigenstates for the observables $a$, $b$, .... The coefficients in the expansion are the components of the vector in the directions of the eigenstate vectors and hence determine the probabilities of transitions into these states.

Thus, any complete set of compatible observables leads to a possible representation of the states of a physical system. If $\boldsymbol{\alpha}$, $\boldsymbol{\alpha}'$ are matrices specifying the same state in two different representations then, as explained in the last section, they will be related by a unitary transformation thus:

$$\boldsymbol{\alpha}' = \mathbf{U}\boldsymbol{\alpha}. \tag{1.7.8}$$

## 1.8. Matrices representing observables

Let $a$ be any observable for a certain system. Suppose we associate further observables with $a$ to form a maximal set of compatible observables $a, b, \ldots$. Let $\phi^1, \phi^2, \ldots, \phi^N$ be a complete set of eigenstates for this set and let $a, b, \ldots$ take the sharp values $a_i, b_i, \ldots$ respectively in the state $\phi^i$. The $a_i$ are not, necessarily, all different, but each set $(a_i, b_i, \ldots)$ differs from every other set. Let the matrix $\boldsymbol{\beta}$ specify the state of the system in the $a, b, \ldots$-representation, so that $|\beta_i|^2$ is the probability that, upon measurement, the observables will be found to have values $a_i, b_i, \ldots$. Then, the *expected value* $\bar{a}$ of the observable $a$, when the system is in the state $\boldsymbol{\beta}$, is given by

$$\bar{a} = \sum_{i=1}^{N} a_i|\beta_i|^2 = \sum_{i=1}^{N} \beta_i^* a_i \beta_i = \boldsymbol{\beta}^\dagger \mathbf{A}\boldsymbol{\beta}, \tag{1.8.1}$$

where $\mathbf{A}$ is the diagonal matrix defined by

$$\mathbf{A} = \begin{pmatrix} a_1 & & & \mathbf{0} \\ & a_2 & & \\ & & \ddots & \\ \mathbf{0} & & & a_N \end{pmatrix}. \tag{1.8.2}$$

Now suppose that some other $x, y, \ldots$-representation, based upon the complete orthonormal set $\psi^1, \psi^2, \ldots, \psi^N$, is employed. Let $\boldsymbol{\alpha}$ be the matrix specifying the system's state in this representation. Then, from equation (1.7.8), it follows that

$$\boldsymbol{\alpha} = \mathbf{U}\boldsymbol{\beta}, \quad \boldsymbol{\beta} = \mathbf{U}^\dagger \boldsymbol{\alpha}, \tag{1.8.3}$$

where the elements $u_{ij}$ of the transformation matrix $\mathbf{U}$ are given by equation (1.6.14). Substituting for $\boldsymbol{\beta}$ from the second of equations (1.8.3) into equation (1.8.1), we obtain

$$\bar{a} = \boldsymbol{\alpha}^\dagger \mathbf{U}\mathbf{A}\mathbf{U}^\dagger \boldsymbol{\alpha} = \boldsymbol{\alpha}^\dagger \mathbf{a}\boldsymbol{\alpha}, \tag{1.8.4}$$

where

$$\mathbf{a} = \mathbf{U}\mathbf{A}\mathbf{U}^{\dagger}. \qquad (1.8.5)$$

It is clear from equation (1.8.5) that the $N \times N$ matrix $\mathbf{a}$ depends only upon the observable $a$ and upon the representation being employed; $\mathbf{a}$ will be termed the matrix associated with or representing $a$ in the $x, y, \ldots$-representation. Knowing this matrix, the expected value of the observable $a$ in any state $\alpha$ of the system can be computed by using equation (1.8.4).

In particular, suppose that the system is in one of the eigenstates of the representation, e.g. suppose $\alpha_1 = 1$, $\alpha_2 = \alpha_3 = \ldots = \alpha_N = 0$. Then, from equation (1.8.4), we deduce that $\bar{a} = a_{11}$. Thus, the elements in the principal diagonal of the matrix $\mathbf{a}$ are the expected values of the observable in the eigenstates of the representation. These elements must accordingly be real.

It is clear from equation (1.8.1) that, in a representation based upon a set of observables which includes $a$, the observable $a$ itself is represented by a diagonal matrix $\mathbf{A}$ with the eigenvalues of $a$ arranged along its principal diagonal.

Further, if $q$ is any observable which is compatible with the set $a, b, \ldots$ (and hence, since the set is maximal, must be a function of the quantities $a, b, \ldots$), then $q$ will take a sharp value $q_i$ in each of the eigenstates $\phi^i$. It now follows, as for $a$, that $q$ will be represented by a diagonal matrix with its eigenvalues $q_i$ arranged along a principal diagonal.

An observable which can assume only one value is termed a *constant*. It is evident that a constant takes the same sharp value in every state and is compatible with every observable. Thus, in any representation, a constant $c$ is represented by a diagonal matrix $\mathbf{c}$ having all its principal diagonal elements equal to $c$; thus

$$\mathbf{c} = c\mathbf{I}_N, \qquad (1.8.6)$$

where $\mathbf{I}_N$ is the unit $N \times N$ matrix.

From equation (1.8.5), it follows that

$$\begin{aligned} \mathbf{a}^{\dagger} &= (\mathbf{U}\mathbf{A}\mathbf{U}^{\dagger})^{\dagger} = \mathbf{U}\mathbf{A}^{\dagger}\,\mathbf{U}^{\dagger} = \mathbf{U}\mathbf{A}\mathbf{U}^{\dagger}, \\ &= \mathbf{a}, \end{aligned} \qquad (1.8.7)$$

since the elements of $\mathbf{A}$ are real. A matrix satisfying the condition (1.8.7) is said to be *Hermitian*. All real observables are represented by Hermitian matrices.

Consider the matrix $\mathbf{a}\boldsymbol{\alpha}$. This is a column matrix which we shall denote by $\boldsymbol{\gamma}$. Hence

$$\boldsymbol{\gamma} = \mathbf{a}\boldsymbol{\alpha}. \tag{1.8.8}$$

$\boldsymbol{\gamma}$ represents some vector $\gamma$ in $\mathscr{S}_N$. Being given the matrix $\mathbf{a}$, to each vector $\alpha$ in $\mathscr{S}_N$, there corresponds a unique vector $\gamma$. It will be helpful to think of the transformation from $\alpha$ to $\gamma$ as the performing of a certain operation upon $\alpha$, this operation being determined by the matrix $\mathbf{a}$. Thus we shall write

$$\gamma = \hat{a}\alpha, \tag{1.8.9}$$

where $\hat{a}$ denotes the operator which, when applied to the vector $\alpha$, yields the vector $\gamma$. Equation (1.8.9) is then the vector equivalent of the matrix equation (1.8.8); however, it possesses the advantage that it makes no reference to a specific frame. $\hat{a}$ is said to be the *operator* representing the observable $a$.

Equation (1.8.4) can now be written

$$\bar{a} = \boldsymbol{\alpha}^\dagger \boldsymbol{\gamma}. \tag{1.8.10}$$

The vector form of this equation is

$$\bar{a} = \langle \alpha | \gamma \rangle = \langle \alpha | \hat{a}\alpha \rangle. \tag{1.8.11}$$

If $\alpha$, $\beta$ now denote any vectors, then

$$\begin{aligned} \langle \hat{a}\alpha | \beta \rangle &= (\mathbf{a}\boldsymbol{\alpha})^\dagger \boldsymbol{\beta} = \boldsymbol{\alpha}^\dagger \mathbf{a}^\dagger \boldsymbol{\beta} = \boldsymbol{\alpha}^\dagger \mathbf{a}\boldsymbol{\beta} \\ &= \langle \alpha | \hat{a}\beta \rangle, \end{aligned} \tag{1.8.12}$$

$\mathbf{a}$ being Hermitian. This identity is valid for every Hermitian operator $\hat{a}$.

In particular

$$\bar{a} = \langle \alpha | \hat{a}\alpha \rangle = \langle \hat{a}\alpha | \alpha \rangle. \tag{1.8.13}$$

Consider the equation (1.8.9). In a certain representation, let the matrix form of this equation be $\boldsymbol{\gamma} = \mathbf{a}\boldsymbol{\alpha}$ and, in some other representation, let it be $\boldsymbol{\gamma}' = \mathbf{a}'\boldsymbol{\alpha}'$. Then, if $\mathbf{U}$ is the transformation matrix relating these two representations, we must have

$$\boldsymbol{\alpha}' = \mathbf{U}\boldsymbol{\alpha}, \quad \boldsymbol{\gamma}' = \mathbf{U}\boldsymbol{\gamma}, \quad \boldsymbol{\alpha} = \mathbf{U}^\dagger \boldsymbol{\alpha}', \quad \boldsymbol{\gamma} = \mathbf{U}^\dagger \boldsymbol{\gamma}'. \tag{1.8.14}$$

Substituting for $\boldsymbol{\alpha}$ and $\boldsymbol{\gamma}$ in $\boldsymbol{\gamma} = \mathbf{a}\boldsymbol{\alpha}$, we find that

$$\mathbf{U}^\dagger \boldsymbol{\gamma}' = \mathbf{a}\mathbf{U}^\dagger \boldsymbol{\alpha}' \quad \text{or} \quad \boldsymbol{\gamma}' = \mathbf{U}\mathbf{a}\mathbf{U}^\dagger \boldsymbol{\alpha}', \tag{1.8.15}$$

since $\mathbf{U}^\dagger = \mathbf{U}^{-1}$. It follows that

$$\mathbf{a}' = \mathbf{U}\mathbf{a}\mathbf{U}^\dagger. \tag{1.8.16}$$

This is the transformation equation for any matrix representing an observable. It ensures that the form of any equation relating matrices representing states or observables is preserved upon transformation to a new representation and hence is independent of the frame. The inverse transformation is clearly

$$\mathbf{a} = \mathbf{U}^\dagger \mathbf{a}' \mathbf{U}. \tag{1.8.17}$$

As an example, consider the matrix $c\mathbf{I}_N$ representing a constant $c$. Upon transformation, this matrix becomes

$$\mathbf{U}c\mathbf{I}_N \mathbf{U}^\dagger = c\mathbf{U}\mathbf{U}^\dagger = c\mathbf{I}_N, \tag{1.8.18}$$

since $\mathbf{U}$ is unitary. Thus, as previously noted, $c$ is represented by $c\mathbf{I}_N$ in every representation.

Let $\psi^i$ $(i=1,2,\ldots,N)$ constitute a complete set of eigenstates. Employing the set as a basis for a matrix representation, in this representation we have

$$\boldsymbol{\psi}^1 = \begin{pmatrix} 1 \\ 0 \\ \vdots \\ 0 \end{pmatrix}, \text{ etc.} \tag{1.8.19}$$

In this representation, let $a$ be represented by the matrix $(a_{ij})$. Then $\hat{a}\psi^j$ is represented by the column matrix

$$\begin{pmatrix} a_{1j} \\ a_{2j} \\ \vdots \\ a_{Nj} \end{pmatrix} \tag{1.8.20}$$

and hence

$$\langle \psi^i | \hat{a}\psi^j \rangle = a_{ij}. \tag{1.8.21}$$

But, this scalar product is invariant with respect to a change of reference frame. Hence, employing an arbitrary representation, the elements of the matrix for $a$ in a representation based upon the $\psi^i$ are given by

$$a_{ij} = \langle \psi^i | \hat{a}\psi^j \rangle = \langle \hat{a}\psi^i | \psi^j \rangle. \tag{1.8.22}$$

## 1.9. Representation of functions of observables

Let $a$, $b$ be a pair of compatible observables. Associate with these further observables to form a maximal compatible set $a, b, \ldots$. Let $\phi^1, \phi^2, \ldots, \phi^N$ be a complete set of eigenstates for this maximal set and let $a_i$, $b_i$, ... be the eigenvalues for the $i$th eigenstate. Then, in the $a, b, \ldots$-representation, the observables $a$, $b$ will be represented by diagonal matrices

$$\mathbf{A} = \begin{pmatrix} a_1 & & \mathbf{0} \\ & a_2 & \\ \mathbf{0} & & \ddots & \\ & & & a_N \end{pmatrix}, \quad \mathbf{B} = \begin{pmatrix} b_1 & & \mathbf{0} \\ & b_2 & \\ \mathbf{0} & & \ddots & \\ & & & b_N \end{pmatrix}. \tag{1.9.1}$$

Consider the observable defined by the equation $c=\lambda a+\mu b$, where $\lambda$, $\mu$ are real constants. This is compatible with the set $a, b, \ldots$ and takes the value $\lambda a_i+\mu b_i$ in the state $\phi^i$. Hence it is represented by the diagonal matrix

$$\mathbf{C} = \begin{pmatrix} \lambda a_1+\mu b_1 & & \mathbf{0} \\ & \lambda a_2+\mu b_2 & \\ \mathbf{0} & & \ddots & \\ & & & \lambda a_N+\mu b_N \end{pmatrix}, \tag{1.9.2}$$

$$= \lambda\mathbf{A}+\mu\mathbf{B}.$$

Let us now transform to an arbitrary $x, y, \ldots$-representation by a transformation matrix $\mathbf{U}$. Multiplying equation (1.9.2) on the left by $\mathbf{U}$ and on the right by $\mathbf{U}^\dagger$, we obtain

$$\mathbf{UCU}^\dagger = \lambda\mathbf{UAU}^\dagger+\mu\mathbf{UBU}^\dagger.$$

Hence, by equation (1.8.16), it follows that

$$\mathbf{c} = \lambda\mathbf{a}+\mu\mathbf{b}, \tag{1.9.3}$$

where $\mathbf{a}$, $\mathbf{b}$, $\mathbf{c}$ are the matrices representing $a$, $b$, $c$ respectively in the new representation.

Again, consider the observable defined by the equation $d=ab$. In the $a, b, \ldots$-representation, it is represented by the diagonal matrix

$$\mathbf{D} = \begin{pmatrix} a_1 b_1 & & \mathbf{0} \\ & a_2 b_2 & \\ \mathbf{0} & & \ddots & \\ & & & a_N b_N \end{pmatrix} = \mathbf{AB}. \tag{1.9.4}$$

Multiplying on the left by $\mathbf{U}$ and on the right by $\mathbf{U}^\dagger$, we find that

$$\mathbf{UDU}^\dagger = \mathbf{UABU}^\dagger = (\mathbf{UAU}^\dagger)(\mathbf{UBU}^\dagger),$$

since $\mathbf{U}^\dagger\mathbf{U}=\mathbf{I}_N$. Thus

$$\mathbf{d} = \mathbf{ab}, \tag{1.9.5}$$

where $\mathbf{a}$, $\mathbf{b}$, $\mathbf{d}$ are the matrices representing $a$, $b$, $d$ respectively in the new representation.

Similarly, it may be proved that

$$\mathbf{d} = \mathbf{ba}, \tag{1.9.6}$$

demonstrating that the matrices representing compatible observables must commute.

From the two results (1.9.3), (1.9.5), it now follows that the matrix representing any polynomial function of compatible observables is obtained by replacing the observables in the polynomial by their associated matrices. Since all the matrices commute, the order of the multiplications in a term is of no significance.

Certain combinations of incompatible observables play important roles in classical mechanics and observables which are the counterparts of these quantities may accordingly be expected to occur in the new mechanics. For example, a particle whose coordinates in an inertial frame are $(x, y, z)$ and whose corresponding momentum components are $(p_x, p_y, p_z)$, is taken in classical mechanics to possess an *angular momentum* about the origin of the frame whose $x$-component is given by

$$L_x = yp_z - zp_y. \tag{1.9.7}$$

As explained in section 1.2, the observables $y$, $p_y$ are incompatible, as are the observables $z$, $p_z$. If a counterpart for the observable $L_x$ is to be introduced into quantum mechanics, therefore, this cannot be done, as in the classical theory, by using equation (1.9.7) to define $L_x$, since this equation does not suggest a procedure by which $L_x$ can be measured in any individual case. Thus, it is not possible to measure $L_x$ by measuring $y, z, p_y, p_z$ separately and then substituting in equation (1.9.7). Instead, an alternative prescription for the measurement of $L_x$ not involving the simultaneous measurement of incompatible observables must be agreed upon and this can then be accepted as the definition of $L_x$. This procedure may, of course, be suggested by the fact that, according to the classical theory it will yield the same result as the measurement of $y$, $z$, $p_y$, $p_z$ and substitution in the formula (1.9.7). Nonetheless, according to the new theory, the two procedures are not equivalent definitions of $L_x$. As an example of an acceptable procedure, if the particle possesses an electric charge and is orbiting about a centre of attraction, its angular momentum about this centre will, according to classical ideas, generate a proportional magnetic dipole moment; the measurement of the $x$-component of this moment could therefore be accepted as a procedure for arriving at the value of $L_x$.

Thus, although certain observables are treated in classical mechanics as subsidiary quantities, being derivable from the primary quantities the coordinates and momenta, in quantum mechanics these observables may all enjoy equal status. However, the fact that in the classical theory the different observables are related, suggests that the matrices representing these quantities in the new theory will also be related. As for the case of compatible observables, the relationship between the matrices is (with one amendment) identical with the classical relationship. The reason why this is so will now be explained.

Let $a$, $b$ be incompatible observables and $\lambda$, $\mu$ real constants; let $c$ be an observable which, according to the classical theory, is defined by the equation

$$c = \lambda a + \mu b. \tag{1.9.8}$$

For a particular state of the system under consideration, although $a$, $b$ and $c$ are not simultaneously measurable, we shall take the view that, nonetheless, in any particular instance, these quantities have definite values and that these will be related by equation (1.9.8).

On the grounds that this last statement is not directly verifiable, many physicists will assert that it is of a metaphysical character and therefore has no significance for the physical theory being expounded; some will take the view that the statement is strictly meaningless. For such purists, the fact that the position and momentum of a particle cannot be measured with precision simultaneously, implies that the physical world does not contain such entities as particles possessing exact positions and momenta. We shall take the view that the physical world is an abstract creation of the mind of man, modelling for him the pattern of his sense perceptions and so assisting him to understand and predict the course taken by this stream of events; he is therefore free to build into this model any features which render the model effective for its purpose, requiring only that the resulting structure shall be internally consistent and that those of its elements which possess an interpretation in terms of sense perceptions shall be in accord with experience. It is certainly not necessary that every element should possess a correlate in the flux of sense perception. Some of the elements will be introduced with the sole intention of simplifying the logical structure of the model and need not be directly observable; such are the vectors associated with the states of a physical system, since it is only the squares of the moduli of their scalar products with one another that are observable. Such, also, we take to be the concept of a system, all of whose associated observables (compatible or not) are thought of as taking precise values at some instant. Provided this concept leads us to conclusions which are not contradicted by experiment, we shall regard it as an acceptable component of the theory.

Returning to the system possessing observables $a$, $b$, $c$ related by equation (1.9.8), suppose that it is prepared in a pure state $\alpha$ a very large number of times $n$ and that, on a third of these occasions $a$ is measured, on another third $b$ is measured and on the remaining third $c$ is measured. Assuming that, on the $r$th occasion, $a$, $b$, $c$ took precise values $a_r$, $b_r$, $c_r$ (only one of which was actually measured), we shall also suppose that

$$c_r = \lambda a_r + \mu b_r \tag{1.9.9}$$

and hence that

$$\frac{1}{n}\sum_{r=1}^{n} c_r = \lambda \frac{1}{n}\sum_{r=1}^{n} a_r + \mu \frac{1}{n}\sum_{r=1}^{n} b_r. \tag{1.9.10}$$

Now, if $n$ is very large, we shall have

$$\frac{1}{n}\sum_{r=1}^{n} a_r = \bar{a}, \text{ etc.} \tag{1.9.11}$$

and it follows that

$$\bar{c} = \lambda \bar{a} + \mu \bar{b}. \tag{1.9.12}$$

But $\bar{a}, \bar{b}, \bar{c}$ can be derived from the sets of observations made on the system and hence equation (1.9.12), which is a consequence of the non-verifiable assumptions we have made, can be subjected to an observational check. In fact, the check is invariably confirmatory and we shall regard this as evidence in support of our hypothesis.

The experimentally verifiable fact that a relationship such as (1.9.8) existing between observables in the classical theory can be reinterpreted in quantum mechanics as a relationship between expected values, is referred to as the *Correspondence Principle*. Many instances of the validity of this principle will be met in the later sections.

Now let $\mathbf{a}$, $\mathbf{b}$, $\mathbf{c}$ denote the matrices representing the three observables of the system considered above. If $\alpha$ is the column matrix specifying the system's state, equation (1.9.12) is equivalent to

$$\begin{aligned} \boldsymbol{\alpha}^{\dagger}\,\mathbf{c}\boldsymbol{\alpha} &= \lambda\boldsymbol{\alpha}^{\dagger}\,\mathbf{a}\boldsymbol{\alpha} + \mu\boldsymbol{\alpha}^{\dagger}\,\mathbf{b}\boldsymbol{\alpha}, \\ &= \boldsymbol{\alpha}^{\dagger}(\lambda\mathbf{a} + \mu\mathbf{b})\,\boldsymbol{\alpha}. \end{aligned} \tag{1.9.13}$$

Since $\mathbf{c}$ and $\lambda\mathbf{a}+\mu\mathbf{b}$ are known to be Hermitian and $\boldsymbol{\alpha}$ is arbitrary, this equation implies that

$$\mathbf{c} = \lambda\mathbf{a} + \mu\mathbf{b}. \tag{1.9.14}$$

Thus the correspondence principle implies that a classical equation such as (1.9.8) can be reinterpreted in quantum mechanics as an equation relating the corresponding matrices.

Consider, next, the matrix representing the observable $c^2$. From a previous result, we know that this is $\mathbf{c}^2$, where

$$\mathbf{c}^2 = \lambda^2 \mathbf{a}^2 + \mu^2 \mathbf{b}^2 + \lambda\mu(\mathbf{ab} + \mathbf{ba}), \tag{1.9.15}$$

remembering that $\mathbf{a}$, $\mathbf{b}$ do not necessarily commute. But, classically, $c^2$ is given in terms of the observables $a$, $b$ by the equation

$$c^2 = \lambda^2 a^2 + \mu^2 b^2 + 2\lambda\mu ab \tag{1.9.16}$$

and hence, by the result just proved, $\mathbf{c}^2$ must be given by

$$\mathbf{c}^2 = \lambda^2 \mathbf{a}^2 + \mu^2 \mathbf{b}^2 + 2\lambda\mu \mathbf{d}, \tag{1.9.17}$$

where $\mathbf{d}$ is the matrix representing the observable $ab$. Comparing equations (1.9.15), (1.9.17), we see that

$$\mathbf{d} = \tfrac{1}{2}(\mathbf{ab} + \mathbf{ba}). \tag{1.9.18}$$

The right-hand member of this equation is called the *symmetrised product* of the matrices $\mathbf{a}$, $\mathbf{b}$. We have proved, therefore, that the matrix representing the product of two incompatible observables is not the product of their matrices, but the symmetrised product.

The observables $ab$, $ba$ are identical but, unless the matrices $\mathbf{a}$, $\mathbf{b}$ commute, the products $\mathbf{ab}$, $\mathbf{ba}$ are distinct. It follows immediately that the observable $ab$ cannot, in general, be represented by the simple matrix product $\mathbf{ab}$. If $\mathbf{a}$, $\mathbf{b}$ do commute, then equation (1.9.18) shows that $\mathbf{d} = \mathbf{ab}$ and the simple product is then the proper representative of the observable $ab$.

Since $ab$ is a real observable, its matrix representative must be Hermitian. This also follows from equation (1.9.18), for

$$\begin{aligned} \mathbf{d}^\dagger &= \tfrac{1}{2}(\mathbf{b}^\dagger \mathbf{a}^\dagger + \mathbf{a}^\dagger \mathbf{b}^\dagger) = \tfrac{1}{2}(\mathbf{ba} + \mathbf{ab}), \\ &= \mathbf{d}. \end{aligned} \tag{1.9.19}$$

The matrix $\mathbf{ab}$ alone is not, in general, Hermitian.

It is now possible to write down the matrix representing any polynomial expression involving compatible or incompatible observables. The resulting matrix expression is identical with the original polynomial, except that any products of incompatible observables must be symmetrised. This can sometimes be done in more than one way. For example, the observable $a^2b^2$ can be symmetrised in either of the forms $(a^2)(b^2)$ or $(ab)(ab)$, yielding distinct matrix expressions

$$\tfrac{1}{2}(\mathbf{a}^2\mathbf{b}^2 + \mathbf{b}^2\mathbf{a}^2), \quad \tfrac{1}{4}(\mathbf{ab} + \mathbf{ba})^2, \tag{1.9.20}$$

respectively. In these circumstances, which only occur rarely, the appropriate matrix can only be decided by comparison of the physical consequences of accepting each expression with experiment. Alternatively, it may be that either of two measurement procedures (classically equivalent) is acceptable as a specification for the observable $a^2b^2$ and then the first matrix may be appropriate for one procedure and the second matrix appropriate for the other.

## 1.10. Characteristic equations

Let $a$ be an observable of the system under consideration and suppose the system is prepared in a pure state with $a$ taking the sharp value $a_1$. Let $b, c, \ldots$ be the remaining observables whose values have been measured to generate the pure state and let their sharp values be $b_1, c_1, \ldots$ respectively. Then we can regard the pure state as being the first eigenstate $\phi^1$ for the complete set of compatible observables $a, b, \ldots$ and, employing the $a, b, \ldots$-representation, $\phi^1$ is represented by the column matrix

$$\boldsymbol{\alpha}' = \begin{pmatrix} 1 \\ 0 \\ \vdots \\ 0 \end{pmatrix}, \tag{1.10.1}$$

all elements of which are zero, except that in the first row, which is unity. In this representation, the matrix $\mathbf{a}'$ representing the observable $a$ is diagonal, thus

$$\mathbf{a}' = \begin{pmatrix} a_1 & & & \mathbf{0} \\ & a_2 & & \\ & & \ddots & \\ \mathbf{0} & & & a_N \end{pmatrix}. \tag{1.10.2}$$

It follows that

$$\mathbf{a}'\boldsymbol{\alpha}' = \begin{pmatrix} a_1 \\ 0 \\ \vdots \\ 0 \end{pmatrix} = a_1\boldsymbol{\alpha}'. \tag{1.10.3}$$

Transforming to any other $x, y, \ldots$-representation, let $\boldsymbol{\alpha}$ be the matrix representing the state $\phi^1$ and let $\mathbf{a}$ also denote the matrix representing the observable $a$. Then, by equations (1.8.14), (1.8.16), we have

$$\boldsymbol{\alpha}' = \mathbf{U}\boldsymbol{\alpha}, \quad \mathbf{a}' = \mathbf{U}\mathbf{a}\mathbf{U}^\dagger. \tag{1.10.4}$$

Substituting for $\boldsymbol{\alpha}'$ and $\mathbf{a}'$ in equation (1.10.3), we find that

$$\mathbf{U}\mathbf{a}\mathbf{U}^\dagger\,\mathbf{U}\boldsymbol{\alpha} = a_1\mathbf{U}\boldsymbol{\alpha}. \tag{1.10.5}$$

Multiplying on the left by $\mathbf{U}^\dagger$, we get

$$\mathbf{a}\boldsymbol{\alpha} = a_1 \boldsymbol{\alpha}. \tag{1.10.6}$$

In general, if in the state $\boldsymbol{\alpha}$ an observable $a$ is known to take a sharp value $\lambda$, then

$$\mathbf{a}\boldsymbol{\alpha} = \lambda\boldsymbol{\alpha}. \tag{1.10.7}$$

This will be called the *characteristic equation* for the observable $a$ and a non-zero column matrix $\boldsymbol{\alpha}$ satisfying the equation will be called an *eigenvector* for $a$ corresponding to the eigenvalue $\lambda$. Writing the equation in the form

$$(\mathbf{a}-\lambda\mathbf{I}_N)\,\boldsymbol{\alpha} = 0, \tag{1.10.8}$$

it is well known that this equation possesses a non-zero solution for $\boldsymbol{\alpha}$ only if

$$|\mathbf{a}-\lambda\mathbf{I}_N| = 0. \tag{1.10.9}$$

This is an equation of the $N$th degree in $\lambda$ and its roots are the only possible eigenvalues for the observable $a$.

Conversely, if $\boldsymbol{\alpha}$ is a normalised eigenvector satisfying the equation (1.10.7), it represents a state of the system in which $a$ is sharp with the value $\lambda$. For, in the state $\boldsymbol{\alpha}$,

$$\bar{a} = \boldsymbol{\alpha}^\dagger\,\mathbf{a}\boldsymbol{\alpha} = \boldsymbol{\alpha}^\dagger\,\lambda\boldsymbol{\alpha} = \lambda, \tag{1.10.10}$$

$\boldsymbol{\alpha}$ being normalised. Now, the variance of the observable $a$ is the expected value of $(a-\bar{a})^2=(a-\lambda)^2$; the matrix representing $(a-\lambda)^2$ is $(\mathbf{a}-\lambda\mathbf{I}_N)^2$ and hence, in the state $\boldsymbol{\alpha}$,

$$\operatorname{var} a = \boldsymbol{\alpha}^\dagger(\mathbf{a}-\lambda\mathbf{I}_N)^2\,\boldsymbol{\alpha} = 0, \tag{1.10.11}$$

by equation (1.10.8). But $\operatorname{var} a$ can only vanish if $a=\bar{a}=\lambda$ with certainty, i.e. if $a$ is sharp.

In particular, if $\mathbf{a}$ is diagonal as in equation (1.10.2), the characteristic equation (1.10.7) is equivalent to the $N$ equations

$$a_1\,\alpha_1 = \lambda\alpha_1, \quad a_2\,\alpha_2 = \lambda\alpha_2, \ldots, \quad a_N\,\alpha_N = \lambda\alpha_N. \tag{1.10.12}$$

It is clear that the base vectors of the representation, viz. $(1,0,\ldots,0)$, $(0,1,\ldots,0),\ldots,(0,0,\ldots,1)$, satisfy these equations with $\lambda=a_1, a_2, \ldots, a_N$ respectively. It follows that the base states of the representation are eigenstates of $a$, in which this observable takes the eigenvalues $a_1, a_2, \ldots, a_N$. This is the converse of the result stated at equation (1.8.2).

The characteristic equation (1.10.7) can also be expressed in an alternative vector form thus: Let $\alpha$ be the vector represented by the matrix $\boldsymbol{\alpha}$. Then the vector represented by the matrix $\mathbf{a}\boldsymbol{\alpha}$ is $\hat{a}\alpha$ (equation (1.8.9)) and the vector represented by the matrix $\lambda\boldsymbol{\alpha}$ is $\lambda\alpha$. Hence

$$\hat{a}\alpha = \lambda\alpha. \tag{1.10.13}$$

Since $\hat{a}$ is a linear operator, it follows that if $\alpha_1$, $\alpha_2$ are vectors which both satisfy the characteristic equation (1.10.13) with the same value of $\lambda$ and so represent states of the system in which $a = \lambda$, then $A_1\alpha_1 + A_2\alpha_2$ ($A_1, A_2$ any complex numbers) also satisfies this equation with the same value of $\lambda$. This implies that the vector $A_1\alpha_1 + A_2\alpha_2$ represents a state of the system in which $a$ takes the sharp value $\lambda$ for all values of $A_1$ and $A_2$ giving a vector with unit norm. This is the *principle of superposition* for state vectors.

We can also deduce that if $\alpha_1$, $\alpha_2$ are any two states in which the observable $a$ takes *different* sharp values $a_1$, $a_2$, then the vectors representing these states are orthogonal. For we have

$$\hat{a}\alpha_1 = a_1\alpha_1 \tag{1.10.14}$$

and hence

$$\langle\alpha_2|\hat{a}\alpha_1\rangle = a_1\langle\alpha_2|\alpha_1\rangle. \tag{1.10.15}$$

Similarly, we have

$$\hat{a}\alpha_2 = a_2\alpha_2, \tag{1.10.16}$$

from which it follows that

$$\langle\hat{a}\alpha_2|\alpha_1\rangle = a_2^*\langle\alpha_2|\alpha_1\rangle. \tag{1.10.17}$$

But the left-hand members of equations (1.10.15), (1.10.17) are identical since $\hat{a}$ is Hermitian. Thus

$$(a_1 - a_2^*)\langle\alpha_2|\alpha_1\rangle = 0. \tag{1.10.18}$$

Since $a_1$, $a_2$ are different, this equation implies that $\alpha_1$, $\alpha_2$ are orthogonal.

## 1.11. The Uncertainty Principle

Let $a$ be any observable of a system. When the system is in a state $\alpha$, let the variance of $a$ be $\sigma_a^2$. Then the standard deviation $\sigma_a$ measures the 'spread' to be expected in a series of measurements of $a$ made upon the system when it is in the state $\alpha$, i.e. it measures the uncertainty in our knowledge of the precise value of $a$ in this state.

Now

$$\begin{aligned}\sigma_a^2 &= \boldsymbol{\alpha}^\dagger(\mathbf{a} - \bar{a}\mathbf{I}_N)^2\boldsymbol{\alpha},\\ &= \boldsymbol{\alpha}^\dagger(\mathbf{a}^\dagger - \bar{a}\mathbf{I}_N^\dagger)(\mathbf{a} - \bar{a}\mathbf{I}_N)\boldsymbol{\alpha},\end{aligned} \tag{1.11.1}$$

since $\mathbf{a}$, $\mathbf{I}_N$ are both Hermitian. Putting

$$\boldsymbol{\psi} = (\mathbf{a}-\bar{a}\mathbf{I}_N)\,\boldsymbol{\alpha}, \tag{1.11.2}$$

we have

$$\sigma_a^2 = \boldsymbol{\psi}^\dagger \boldsymbol{\psi} = \|\psi\|^2. \tag{1.11.3}$$

Similarly, if $b$ is another observable of the system and $\sigma_b$ is its standard deviation in the state $\alpha$,

$$\sigma_b^2 = \|\phi\|^2, \tag{1.11.4}$$

where

$$\boldsymbol{\phi} = (\mathbf{b}-\bar{b}\mathbf{I}_N)\,\boldsymbol{\alpha}. \tag{1.11.5}$$

Employing the Schwarz inequality (1.5.24), we deduce that

$$\sigma_a\,\sigma_b = \|\psi\|.\|\phi\| \geqslant |\langle\psi|\phi\rangle|. \tag{1.11.6}$$

We now define matrices $\mathbf{p}$, $\mathbf{q}$ by the equations

$$\mathbf{p} = \mathbf{a}-\bar{a}\mathbf{I}_N, \quad \mathbf{q} = \mathbf{b}-\bar{b}\mathbf{I}_N. \tag{1.11.7}$$

Then, $\mathbf{p}$ and $\mathbf{q}$ are the Hermitian matrices representing the observables $a-\bar{a}$, $b-\bar{b}$ respectively and, using equations (1.11.2), (1.11.5),

$$\langle\psi|\phi\rangle = \boldsymbol{\psi}^\dagger\boldsymbol{\phi} = \boldsymbol{\alpha}^\dagger\,\mathbf{p}^\dagger\,\mathbf{q}\boldsymbol{\alpha} = \boldsymbol{\alpha}^\dagger\,\mathbf{pq}\boldsymbol{\alpha}. \tag{1.11.8}$$

$\mathbf{pq}$ is not, in general, Hermitian, but it can be analysed into real and imaginary parts which are, thus:

$$\mathbf{pq} = \mathbf{P}+\iota\mathbf{Q}, \tag{1.11.9}$$

where

$$\mathbf{P} = \tfrac{1}{2}(\mathbf{pq}+\mathbf{qp}), \quad \mathbf{Q} = \frac{1}{2\iota}(\mathbf{pq}-\mathbf{qp}). \tag{1.11.10}$$

(See equation (1.9.19) and Ex. 1, Chap. 1.) $\mathbf{P}$ and $\mathbf{Q}$ are accordingly matrices representing real observables $P$, $Q$ respectively and

$$\begin{aligned}\langle\psi|\phi\rangle &= \boldsymbol{\alpha}^\dagger\,\mathbf{P}\boldsymbol{\alpha}+\iota\boldsymbol{\alpha}^\dagger\,\mathbf{Q}\boldsymbol{\alpha},\\ &= \bar{P}+\iota\bar{Q}.\end{aligned} \tag{1.11.11}$$

The inequality (1.11.6) can now be expressed in the form

$$\sigma_a\,\sigma_b \geqslant (\bar{P}^2+\bar{Q}^2)^{\frac{1}{2}}. \tag{1.11.12}$$

We note that $\mathbf{P}$ is the symmetrised product of $\mathbf{p}$ and $\mathbf{q}$ and hence is the matrix representing the observable

$$pq = (a-\bar{a})\,(b-\bar{b}). \tag{1.11.13}$$

The expected value of this observable is termed the *covariance* of the observables $a$, $b$ (in the state $\alpha$). If $a$, $b$ are statistically independent, it is proved in works devoted to the mathematical theory of statistics that the covariance is zero (see e.g. *An Introduction to Mathematical Statistics* by C. E. Weatherburn, C.U.P., 1949, p. 32). In our case, the variates $a$, $b$ may not be statistically independent, but a value for $\bar{P}$ in the general case is not easy to obtain. However, in all cases, the inequality (1.11.12) certainly allows us to assert that

$$\sigma_a \sigma_b \geqslant |\bar{Q}|. \tag{1.11.14}$$

Now

$$\mathbf{Q} = \frac{1}{2\iota}[(\mathbf{a}-\bar{a}\mathbf{I}_N)(\mathbf{b}-\bar{b}\mathbf{I}_N)-(\mathbf{b}-\bar{b}\mathbf{I}_N)(\mathbf{a}-\bar{a}\mathbf{I}_N)],$$

$$= \frac{1}{2\iota}(\mathbf{ab}-\mathbf{ba}). \tag{1.11.15}$$

The *commutator* of two matrices $\mathbf{a}$, $\mathbf{b}$ will be denoted by $[\mathbf{a}, \mathbf{b}]$ and will be defined by the equation

$$[\mathbf{a}, \mathbf{b}] = \mathbf{ab}-\mathbf{ba}. \tag{1.11.16}$$

Clearly, if $\mathbf{a}$, $\mathbf{b}$ commute, their commutator vanishes. In general, the commutator is not Hermitian, but $\iota[\mathbf{a}, \mathbf{b}]$ is Hermitian (see Ex. 1, Chap. 1) and can accordingly represent a real observable; if this observable is denoted by $c$, we shall write

$$c \sim \iota[\mathbf{a}, \mathbf{b}]. \tag{1.11.17}$$

Then $Q = -\frac{1}{2}c$ and hence, finally,

$$\sigma_a \sigma_b \geqslant \tfrac{1}{2}|\bar{c}|. \tag{1.11.18}$$

This is the *General Heisenberg Uncertainty Principle.*

If $\mathbf{a}$, $\mathbf{b}$ commute, then $c=0$ and the inequality (1.11.18) is trivial. However, if $\mathbf{a}$, $\mathbf{b}$ do not commute, then the possibility arises that the right-hand member of this inequality may be non-zero for every state. In such a case, it will be impossible to know the values of both $a$ and $b$ at the same instant with absolute precision. The implication would then be that, if $a$ is measured at some instant and $b$ is then measured immediately afterwards, the act of measuring $b$ must always disturb the value of $a$ by an unknown amount, so that its value becomes uncertain to the degree required by the Heisenberg inequality. The observables $a$, $b$ are then said to be *incompatible* as previously explained. The problem of the compatibility of observables will be examined more closely in the next section.

## 1.12. Compatibility of observables

The uncertainty principle established in the previous section suggests that, if the matrices representing two observables do not commute, then it is possible that no state exists in which the observables are sharp together, i.e. there is no state which is simultaneously an eigenstate for both observables. However, that this is not generally true is easily seen, for consider the observables $a$, $b$ represented by the Hermitian matrices

$$\mathbf{a} = \begin{pmatrix} 1 & 0 & 0 \\ 0 & 0 & 1 \\ 0 & 1 & 0 \end{pmatrix}, \quad \mathbf{b} = \begin{pmatrix} 1 & 0 & 0 \\ 0 & 0 & \iota \\ 0 & -\iota & 0 \end{pmatrix}. \tag{1.12.1}$$

It is easily verified that these matrices do not commute. Nevertheless, it may also be verified that the state represented by the vector

$$\boldsymbol{\alpha} = \begin{pmatrix} 1 \\ 0 \\ 0 \end{pmatrix} \tag{1.12.2}$$

is an eigenstate for both observables, in which $a$, $b$ both take the eigenvalue 1.

On the other hand, it is possible to prove that, if $\mathbf{a}$ and $\mathbf{b}$ commute, then simultaneous eigenstates for the observables $a$ and $b$ certainly do exist. Further a complete set of such eigenstates can be found.

To prove this, associate with the observable $a$ further observables $a'$, $a''$, etc, so as to form a maximal set of compatible observables. Let $\psi^1, \psi^2, \ldots, \psi^N$ be a complete orthonormal set of vectors representing the eigenstates of this set and suppose $a$ takes the eigenvalues $a_1, a_2, \ldots, a_N$ respectively in these states. Let $\phi$ be any eigenvector of $b$. Then $\phi$ can be expanded in terms of the $\psi^i$ thus,

$$\phi = c_i \psi^i. \tag{1.12.3}$$

If the $a_i$ are not all distinct, we shall suppose that terms in this expansion corresponding to the same eigenvalue are combined to form a single term; when this has been done, any two terms in the expansion will correspond to distinct eigenvalues of $a$. Now

$$(\hat{b}-b)\,\phi = 0 \tag{1.12.4}$$

and hence

$$c_i(\hat{b}-b)\,\psi^i = 0. \tag{1.12.5}$$

But $\hat{b}\psi^i$ is an eigenvector of $a$ corresponding to the eigenvalue $a_i$, for

$$\hat{a}(\hat{b}\psi^i) = \hat{b}(\hat{a}\psi^i) = a_i\hat{b}\psi^i \text{ (not summed)}, \tag{1.12.6}$$

since it is being assumed that $\hat{a}$, $\hat{b}$ commute. It follows that $(\hat{b}-b)\psi^i$ is also an eigenvector of $a$ for the eigenvalue $a_i$. Since we can assume the eigenvalues $a_i$ to be all different, the eigenvectors $(\hat{b}-b)\psi^i$ must be orthogonal. Hence, if we take the scalar product of both members of equation (1.12.5) with the vector $(\hat{b}-b)\psi^1$, we obtain

$$c_1\|(\hat{b}-b)\psi^1\|^2 = 0. \tag{1.12.7}$$

Thus, either

$$c_1 = 0 \quad \text{or} \quad (\hat{b}-b)\psi^1 = 0. \tag{1.12.8}$$

Similar results follow for $i=2, 3, \ldots, N$. This result implies that the non-vanishing terms in the expansion (1.12.3) are eigenvectors of $b$ for the eigenvalue $b$.

We have proved, therefore, that any eigenvector of $b$ can be expanded in terms of simultaneous eigenvectors of $a$ and $b$. But an arbitrary state vector can certainly be expanded in terms of eigenvectors of $b$ (associate with $b$ compatible observables $b'$, $b''$, etc. to form a maximal set and employ the corresponding complete orthonormal set of eigenvectors). It follows that an arbitrary state vector can also be expanded in terms of simultaneous eigenvectors of $a$ and $b$ and hence that a complete set of such eigenvectors can be found.

Conversely, if an arbitrary vector can be expanded in terms of a set of simultaneous eigenvectors of $a$ and $b$, then the matrices **a**, **b** commute. For, let $\chi^1, \chi^2, \ldots, \chi^N$ be such a complete set and let $\psi$ be an arbitrary vector whose expansion is

$$\psi = c_i\chi^i. \tag{1.12.9}$$

Now

$$(\hat{a}\hat{b}-\hat{b}\hat{a})\chi^1 = (a_1b_1-b_1a_1)\chi^1 = 0, \tag{1.12.10}$$

$a_1$, $b_1$ being the eigenvalues of $a$, $b$ respectively in the state $\chi^1$. A similar result can be obtained in respect of each of the $\chi^i$. Hence

$$(\hat{a}\hat{b}-\hat{b}\hat{a})\psi = 0. \tag{1.12.11}$$

But $\psi$ is arbitrary and hence

$$\hat{a}\hat{b} = \hat{b}\hat{a}, \tag{1.12.12}$$

with a corresponding result for matrices.

More generally, if **a**, **b** and **c** commute in pairs, there exists a set of simultaneous eigenvectors for the observables $a$, $b$, $c$ and it is complete.

The argument proving this result is similar to that given above except that we commence with a complete set of simultaneous eigenvectors of $a$ and $b$ and then prove that any eigenvector of $c$ can be expanded in terms of simultaneous eigenvectors of $a$, $b$, $c$. The argument can clearly be extended to any number of observables whose matrices commute.

The ideas we have been studying are directly related to the concept of *degeneracy*. If $a, b, \ldots$ constitutes a maximal set of independent compatible observables, a complete set of simultaneous eigenvectors $\psi^1, \psi^2, \ldots, \psi^N$ can be found and these will correspond to the eigenstates in which $a, b, \ldots$ all have sharp values. Since any pair of these vectors are eigenvectors corresponding to different eigenvalues of at least one of the observables, the vectors form an orthonormal set. Consider a pure state in which $a$ is sharp with the value $a_1$, but $b$ and the remaining observables of the maximal set are not necessarily sharp. If $\phi$ is the vector representing this state, then $\phi$ is expansible thus,

$$\phi = x_i \psi^i. \tag{1.12.13}$$

But $\phi$ must satisfy the characteristic equation for $a$, viz.

$$\hat{a}\phi = a_1 \phi. \tag{1.12.14}$$

Since the $\psi^i$ are eigenvectors of $a$,

$$\hat{a}\psi^1 = a_1 \psi^1, \quad \hat{a}\psi^2 = a_2 \psi^2, \text{ etc.} \tag{1.12.15}$$

Hence, substituting from equation (1.12.13) into equation (1.12.14), we obtain

$$\sum_{i=1}^{N} (a_i - a_1) x_i \psi^i = 0. \tag{1.12.16}$$

Since the $\psi^i$ form an orthonormal set, they are linearly independent and it follows that

$$\text{either } a_i - a_1 = 0 \quad \text{or} \quad x_i = 0. \tag{1.12.17}$$

This means that $\phi$ is expansible in terms of the subset of eigenvectors corresponding to the eigenvalue $a = a_1$. However, the coefficients in this expansion are not determined by the measurement $a = a_1$. We say that the system is *degenerate* with respect to the observable $a$, meaning thereby that a precise measurement of this observable alone is insufficient to specify the state vector uniquely.

More generally, if $a$ and $b$ are both measured to take the values $a_1$, $b_1$ precisely, $\phi$ will be expressible linearly in terms of the smaller subset of eigenvectors corresponding to these eigenvalues $a_1$, $b_1$. The number of unknown coefficients will accordingly be less than before and the degree of the degeneracy will have been reduced. If every one of the set of observables $a$, $b$, ... is measured precisely, the form of $\phi$ is no longer in doubt and must be identical with one of the $\psi^i$; in this case, the degeneracy has been completely eliminated.

## 1.13. Combination of systems

Consider two systems $A$ and $B$ which do not interact.† Let $a, a', \ldots$ be a maximal set of observables for $A$ and $b, b', \ldots$ a maximal set for $B$. Also let $\psi^i$ $(i=1,2,\ldots,M)$ represent a complete set of eigenstates for $A$ in which $a, a', \ldots$ are sharp with values $a_i, a'_i, \ldots$ $(i=1,2,\ldots,M)$ respectively. Similarly, let $\phi^j$ $(j=1,2,\ldots,N)$ represent a complete set of eigenstates for $B$ in which $b, b', \ldots$ take the eigenvalues $b_j, b'_j, \ldots$ $(j=1,2,\ldots,N)$ respectively. Then the two systems may be thought of as comprising two parts of a single combined system, which we shall denote by $A+B$. For this system, the observables $a, a', \ldots, b, b', \ldots$ form a maximal set and the $MN$ states in which these observables take the eigenvalues $a_i, a'_i, \ldots, b_j, b'_j, \ldots$ $(i=1,2,\ldots,M;\ j=1,2,\ldots,N)$ constitutes a complete set of eigenstates for this system. We will represent these eigenstates by vectors $\chi^{ij}$ $(i=1,2,\ldots,M; j=1,2,\ldots,N)$ in an $MN$-dimensional space, the state $\chi^{ij}$ being that in which the observables $a, a', \ldots$ take values $a_i, a'_i, \ldots$ and the observables $b, b', \ldots$ take values $b_j, b'_j, \ldots$.

Consider a state of the combined system in which the state of the part $A$ is specified by a vector $\alpha$ and the state of the part $B$ is represented by a vector $\beta$. Suppose $\alpha$ and $\beta$ have the expansions

$$\alpha = \alpha_i \psi^i, \quad \beta = \beta_j \phi^j, \tag{1.13.1}$$

summations with respect to $i$ and $j$ being assumed over the appropriate integer ranges. Then the state of the combined system can be represented by a vector $\gamma$ whose expansion in terms of the eigenvectors $\chi^{ij}$ is

$$\gamma = \gamma_{ij}\chi^{ij} = \alpha_i \beta_j \chi^{ij}. \tag{1.13.2}$$

For, consider the probability of a transition being observed to an arbitrary state in which the system $A$ is characterised by a vector $\alpha'$ and the system $B$ by a vector $\beta'$. In this state the system $A+B$ would be characterised by a vector $\gamma'$, where

$$\gamma' = \alpha'_i \beta'_j \chi^{ij}, \tag{1.13.3}$$

† We also assume that the systems are distinguishable, for otherwise the considerations raised in section 2.6 are relevant.

and the probability would be given by the modulus squared of the vector product

$$\langle\gamma|\gamma'\rangle = \alpha_i^* \beta_j^* \alpha_i' \beta_j' = \langle\alpha|\alpha'\rangle\langle\beta|\beta'\rangle. \qquad (1.13.4)$$

Hence

$$|\langle\gamma|\gamma'\rangle|^2 = |\langle\alpha|\alpha'\rangle|^2 \, |\langle\beta|\beta'\rangle|^2, \qquad (1.13.5)$$

implying that

$$P(\gamma \to \gamma') = P(\alpha \to \alpha')P(\beta \to \beta'), \qquad (1.13.6)$$

which is the product law for probabilities (the systems being independent). This proves that the vector $\gamma$ yields the correct probability for the occurrence of an arbitrary transition and is accordingly a suitable vector to represent the state of $A+B$.

Employing the complete set of eigenvectors $\chi^{ij}$ as a basis, the state $\gamma$ of the system $A+B$ is now seen to be specified by a column matrix $\boldsymbol{\gamma}$ having $MN$ elements $\alpha_i \beta_j$. The order in which these elements are arranged is open to choice and is determined by the order of arrangement of the $\chi^{ij}$. The order $\chi^{11}, \chi^{12}, \ldots, \chi^{1N}, \chi^{21}, \chi^{22}, \ldots, \chi^{2N}, \ldots, \chi^{M1}, \chi^{M2}, \ldots, \chi^{MN}$, will be termed the *natural* order and then

$$\boldsymbol{\gamma} = \begin{pmatrix} \alpha_1 \beta_1 \\ \alpha_1 \beta_2 \\ \vdots \\ \alpha_M \beta_N \end{pmatrix}. \qquad (1.13.7)$$

Next, we will calculate the form taken by the $MN \times MN$ matrix representing an observable $x$ of the system $A$, when it is considered as an observable of the system $A+B$ and the eigenvectors $\chi^{ij}$ are employed as the basis. Let $x_{ik}$ $(i,k=1,2,\ldots,M)$ be the elements of the matrix representing $x$, when the observable is considered in relation to the system $A$ alone, the $\psi^i$ being employed as a basis. Let $x_{ijkl}$ $(i,k=1,2,\ldots,M;$ $j,l=1,2,\ldots,N)$ be the elements of the matrix representing $x$ considered as an observable of $A+B$ and employing the $\chi^{ij}$ as a basis. This element is to be placed in the $ij$-row and the $kl$-column; the precise positions of this row and column will depend upon the ordering of the $\chi^{ij}$ and will be considered further later in this section. Suppose $A$ is in the state $\alpha$ and $B$ is in the state $\beta$. Then, by equation (1.8.4),

$$\bar{x} = x_{ik} \alpha_i^* \alpha_k. \qquad (1.13.8)$$

But $A+B$ is in the state $\gamma$ and hence

$$\bar{x} = x_{ijkl} \gamma_{ij}^* \gamma_{kl} = x_{ijkl} \alpha_i^* \beta_j^* \alpha_k \beta_l. \qquad (1.13.9)$$

Now, the coefficients $x_{ijkl}$ are Hermitian, i.e.

$$x_{klij} = x^*_{ijkl}. \tag{1.13.10}$$

Putting

$$x_{ijkl}\,\alpha_i^*\,\alpha_k = a_{jl}, \tag{1.13.11}$$

it is easy to prove that the coefficients $a_{jl}$ are Hermitian. Equation (1.13.9) can then be written

$$a_{jl}\,\beta_j^*\,\beta_l = \bar{x}. \tag{1.13.12}$$

It is evident from equations (1.13.8), (1.13.11), that $\bar{x}$ and $a_{jl}$ are independent of the vector $\beta$. Hence, since equation (1.13.12) is valid for arbitrary normalised vectors $\beta$, it follows that

$$a_{jl} = \bar{x}\delta_{jl}, \tag{1.13.13}$$

$\delta_{jl}$ being the Kronecker symbol (see Ex. 6, Chap. 1). Thus, equations (1.13.8), (1.13.11) now yield the identity

$$x_{ijkl}\,\alpha_i^*\,\alpha_k = x_{ik}\,\alpha_i^*\,\alpha_k\,\delta_{jl}, \tag{1.13.14}$$

valid for arbitrary normalised vectors $\alpha$ and hence, also, for quite arbitrary $\alpha$. It follows that

$$x_{ijkl} = x_{ik}\,\delta_{jl}. \tag{1.13.15}$$

This equation relates the matrices representing $x$ when the eigenvectors $\chi^{ij}$ are employed as a basis and when the eigenvectors $\psi^i$ are employed as a basis.

Similarly, if $y$ is an observable of the system $B$, the corresponding equation is

$$y_{ijkl} = y_{jl}\,\delta_{ik}. \tag{1.13.16}$$

Matrices representing any polynomial expression in observables of $A$ and $B$ can now be written down.

Consider the structure of the matrix $(x_{ijkl})$ in the case when the basic eigenvectors $\chi^{ij}$ are arranged in their natural order (see above). Then the $ij$-row (column) is the $[(i-1)N+j]$th row (column). If the matrix is partitioned into $M \times M$ matrices by lines drawn between the $N$th and $(N+1)$th rows, between the $2N$th and $(2N+1)$th rows and so on and similarly with regard to the columns, the array will take the form

$$(x_{ijkl}) = \begin{pmatrix} \mathbf{x}_{11} & \mathbf{x}_{12} & \cdots & \mathbf{x}_{1M} \\ \mathbf{x}_{21} & \mathbf{x}_{22} & \cdots & \mathbf{x}_{2M} \\ \cdots & \cdots & \cdots & \cdots \\ \mathbf{x}_{M1} & \mathbf{x}_{M2} & \cdots & \mathbf{x}_{MM} \end{pmatrix}. \tag{1.13.17}$$

The indices $i$, $k$ now determine the 'block' $\mathbf{x}_{ik}$ in which the element $x_{ijkl}$ will be found and the indices $j$, $l$ determine its position within this block. Equation (1.13.15) shows that each block $\mathbf{x}_{ik}$ is a diagonal matrix with all its diagonal elements equal to $x_{ik}$.

The structure of the matrix $(y_{ijkl})$ in the case when the base vectors $\chi^{ij}$ are arranged in their natural order may be found similarly. Thus

$$(y_{ijkl}) = \begin{pmatrix} \mathbf{y}_{11} & \mathbf{y}_{12} & \cdots & \mathbf{y}_{1M} \\ \mathbf{y}_{21} & \mathbf{y}_{22} & \cdots & \mathbf{y}_{2M} \\ \cdots & \cdots & \cdots & \cdots \\ \mathbf{y}_{M1} & \mathbf{y}_{M2} & \cdots & \mathbf{y}_{MM} \end{pmatrix}. \tag{1.13.18}$$

Equation (1.13.16) indicates that $\mathbf{y}_{ik} = \mathbf{0}$ if $i \neq k$ and that $\mathbf{y}_{11}, \mathbf{y}_{22}, \ldots, \mathbf{y}_{MM}$ are all identical with the matrix $(y_{jl}) = \mathbf{y}$ representing $y$ as an observable of the system $B$ employing the vectors $\phi^i$ as a basis. Thus,

$$(y_{ijkl}) = \begin{pmatrix} \mathbf{y} & & \mathbf{0} \\ & \mathbf{y} & \\ \mathbf{0} & \ddots & \\ & & \mathbf{y} \end{pmatrix}. \tag{1.13.19}$$

CHAPTER 2

# Spin

## 2.1. The Stern-Gerlach experiment

A magnetic dipole placed in a uniform magnetic field experiences a couple tending to align its axis with the direction of the field. However, there is no resultant force tending to cause a translation of the dipole, unless the field is non-uniform. In 1921, Stern and Gerlach designed a pair of pole pieces which, when magnetised, generated a field whose departure from uniformity was very pronounced. The apparatus was evacuated and a parallel beam of silver atoms was then directed between the pole pieces. Each silver atom has the properties of a magnetic dipole and, as it passed between the pole pieces, was subjected to a force whose direction was perpendicular to the axis of the beam. Taking a $z$-axis in the direction of this force, calculations indicated that the net impulse applied to an atom would be proportional to the $z$-component of the atom's dipole moment, $\mu_z$. The deflection caused in the atom's direction of motion by the magnetic field would accordingly also be proportional to $\mu_z$ and a measurement of this deflection would therefore provide a means of arriving at the value of $\mu_z$. According to classical ideas, since no action was taken to align the axes of the dipoles forming the beam, the directions of these axes could be expected to be distributed randomly and thus, if $\mu$ is the magnitude of the magnetic moment of a silver atom, values of $\mu_z$ throughout the range $-\mu \leqslant \mu_z \leqslant \mu$ could be expected to be observed. It was to be expected, therefore, that after passage between the pole pieces the beam would fan out and illuminate a continuous band upon a sensitive screen. Instead, it was found that the effect of the field was to split the beam evenly into two components, one corresponding to a deflection in the positive $z$-direction and one to an equal deflection in the negative $z$-direction; only two spots were illuminated on the screen.

The inference to be made from this experiment is that the observable $\mu_z$ possesses but two eigenvalues $+\mu_0$ and $-\mu_0$ and hence that any experiment carried through with the intention of measuring $\mu_z$ can only yield one of these values as its result. No objection is raised against the idea that, before an atom enters the deflecting field and is thereby forced to reveal its magnetic state, the $z$-component of its moment may be taking any value within a range $(-\mu, \mu)$. However, the disturbing effect upon the atom of any apparatus capable of measuring $\mu_z$ is taken to be such that one of the values $\pm\mu_0$ is invariably observed.

To explain this result in terms of the known structure of the silver atom and for other reasons, the hypothesis was proposed by Uhlenbeck and Goudsmit in 1926 that the electron possesses an intrinsic angular momentum or *spin* $\mathfrak{s}$ and that this spin is responsible for its behaving like a magnetic dipole; in crude terms, the electron was to be thought of as a small, negatively charged, spinning sphere which, according to Maxwellian theory, would generate the magnetic field of a dipole with its axis aligned with the axis of spin. The component of $\mathfrak{s}$ in an arbitrary $z$-direction, viz. $s_z$, is taken to have two eigenvalues $\pm h/4\pi$ (a reason why the eigenvalues should be related to Planck's constant in this way will be given in section 5.1). The $z$-component of the associated magnetic moment also, therefore, possesses two eigenvalues and these are denoted by $\pm\mu_0$. It is found that, in e.m.u.,

$$\mu_0 = \frac{eh}{4\pi mc}, \tag{2.1.1}$$

where $h$ is Planck's constant, $m$ is the mass of the electron and $c$ is the velocity of light, all in c.g.s. units, and $e$ is the magnitude of the charge on the electron in e.s.u. $\mu_0$ is called the *Bohr magneton*. This relationship between $s_z$ and $\mu_z$ will be derived in section 7.3. The spin of one of the electrons orbiting the nucleus of the silver atom is held to be responsible for the magnetic property of the atom and for the observed fact that any component of its dipole moment possesses the two eigenvalues $\pm\mu_0$.

The observables $s_x, s_y, s_z$ prove to be incompatible with one another and hence only one can be observed to possess a sharp value at a given instant. Thus, if $s_z$ has been measured to take the value $+h/4\pi$, the values of $s_x, s_y$ will be unknown, although the probabilities that either will be found upon subsequent measurement to yield the values $\pm h/4\pi$ can be calculated.

Since each of the observables $s_x, s_y, s_z$ possesses but two eigenvalues, the theory of spin provides the simplest possible illustration of the ideas presented in Chapter 1. In the following sections, we shall therefore apply the results obtained in the previous chapter to elucidate the properties of these observables.

## 2.2. Spinors. Pauli matrices

For an electron having spin $\mathfrak{s}$, it will be convenient to define a vector $\mathfrak{z}$ termed the *spin axis* by means of the equation

$$\mathfrak{s} = \frac{h}{4\pi}\mathfrak{z}. \tag{2.2.1}$$

The components of $\mathfrak{z}$ will be denoted by $\sigma_x$, $\sigma_y$, $\sigma_z$ and will clearly each possess eigenvalues $\pm 1$.

If we agree, for the moment, to neglect all aspects of the particle's state (e.g. its position and momentum) apart from its spin, since the observables $\sigma_x$, $\sigma_y$, $\sigma_z$ are mutually incompatible, each alone constitutes a maximal set of compatible observables. In particular $(\sigma_z)$ is such a set and its spectrum of eigenvalues is simply $(+1, -1)$. Let $\psi^+$, $\psi^-$ be the vectors representing the two eigenstates of the particle in which $\sigma_z$ takes the sharp values $+1$, $-1$ respectively. Taking these as base vectors, an arbitrary spin state represented by the vector $\alpha$ is specified in the $\sigma_z$-representation by the column matrix $\boldsymbol{\alpha}$, where

$$\boldsymbol{\alpha} = \begin{pmatrix} \alpha_+ \\ \alpha_- \end{pmatrix}. \tag{2.2.2}$$

As explained in section (1.7), $|\alpha_+|^2$ and $|\alpha_-|^2$ will give the probabilities of observing transitions from the state $\alpha$ into the states $\psi^+$, $\psi^-$ respectively, i.e. the probabilities of measuring $\sigma_z = +1$ and $-1$ respectively. $\boldsymbol{\alpha}$ is arbitrary to the extent of a multiplier $\exp(\iota\theta)$ and is called a *spinor*. Since, by equation (1.7.5), $\|\alpha\| = 1$, we have

$$\boldsymbol{\alpha}^\dagger \boldsymbol{\alpha} = \alpha_+^* \alpha_+ + \alpha_-^* \alpha_- = 1, \tag{2.2.3}$$

and the sum of these two probabilities is unity, as is clearly necessary.

We will next calculate the $2 \times 2$ matrices representing the observables $\sigma_x$, $\sigma_y$, $\sigma_z$. Since we are employing a $\sigma_z$-representation, the matrix representing $\sigma_z$ will be diagonal with its eigenvalues forming the principal diagonal. Thus

$$\boldsymbol{\sigma}_z = \begin{pmatrix} 1 & 0 \\ 0 & -1 \end{pmatrix}. \tag{2.2.4}$$

The matrix representing $\sigma_z^2$ is then

$$\boldsymbol{\sigma}_z^2 = \begin{pmatrix} 1 & 0 \\ 0 & 1 \end{pmatrix} = \mathbf{I}_2. \tag{2.2.5}$$

This is obviously correct, since $\sigma_z^2$ can take only one value, viz. $+1$, i.e. it is a constant (equation (1.8.6)).

But $\sigma_x^2$, $\sigma_y^2$ are also both unity in every state and hence

$$\boldsymbol{\sigma}_x^2 = \boldsymbol{\sigma}_y^2 = \mathbf{I}_2. \tag{2.2.6}$$

Now let $\sigma_0$ be the component of $\mathfrak{z}$ in a direction having direction cosines $(l, m, n)$ relative to the axes $Oxyz$. Then, classically,

$$\sigma_0 = l\sigma_x + m\sigma_y + n\sigma_z. \tag{2.2.7}$$

Hence, if $\boldsymbol{\sigma}_0$ is the matrix representing $\sigma_0$, then

$$\boldsymbol{\sigma}_0 = l\boldsymbol{\sigma}_x + m\boldsymbol{\sigma}_y + n\boldsymbol{\sigma}_z. \tag{2.2.8}$$

Since $\sigma_0^2 = 1$ in every state,

$$\boldsymbol{\sigma}_0^2 = \mathbf{I}_2 \tag{2.2.9}$$

and hence, substituting from equation (2.2.8), we obtain the equation

$$l^2\boldsymbol{\sigma}_x^2 + m^2\boldsymbol{\sigma}_y^2 + n^2\boldsymbol{\sigma}_z^2 + mn(\boldsymbol{\sigma}_y\boldsymbol{\sigma}_z + \boldsymbol{\sigma}_z\boldsymbol{\sigma}_y) + nl(\boldsymbol{\sigma}_z\boldsymbol{\sigma}_x + \boldsymbol{\sigma}_x\boldsymbol{\sigma}_z) + lm(\boldsymbol{\sigma}_x\boldsymbol{\sigma}_y + \boldsymbol{\sigma}_y\boldsymbol{\sigma}_x) = \mathbf{I}_2. \tag{2.2.10}$$

Substituting $\mathbf{I}_2$ for $\boldsymbol{\sigma}_x^2, \boldsymbol{\sigma}_y^2, \boldsymbol{\sigma}_z^2$, this equation reduces to

$$mn(\boldsymbol{\sigma}_y\boldsymbol{\sigma}_z + \boldsymbol{\sigma}_z\boldsymbol{\sigma}_y) + nl(\boldsymbol{\sigma}_z\boldsymbol{\sigma}_x + \boldsymbol{\sigma}_x\boldsymbol{\sigma}_z) + lm(\boldsymbol{\sigma}_x\boldsymbol{\sigma}_y + \boldsymbol{\sigma}_y\boldsymbol{\sigma}_x) = \mathbf{0}. \tag{2.2.11}$$

Taking $l = 0$, $m \neq 0$, $n \neq 0$, this gives the result

$$\boldsymbol{\sigma}_y\boldsymbol{\sigma}_z + \boldsymbol{\sigma}_z\boldsymbol{\sigma}_y = \mathbf{0}. \tag{2.2.12}$$

Similarly, we obtain

$$\boldsymbol{\sigma}_z\boldsymbol{\sigma}_x + \boldsymbol{\sigma}_x\boldsymbol{\sigma}_z = \mathbf{0}, \tag{2.2.13}$$

$$\boldsymbol{\sigma}_x\boldsymbol{\sigma}_y + \boldsymbol{\sigma}_y\boldsymbol{\sigma}_x = \mathbf{0}. \tag{2.2.14}$$

It is clear that no pair of the matrices $\boldsymbol{\sigma}_x$, $\boldsymbol{\sigma}_y$, $\boldsymbol{\sigma}_z$ commute. It follows (Ex. 2, Chap. 1) that only one of the observables $\sigma_x$, $\sigma_y$, $\sigma_z$ can be sharp in any spin state. However, $\boldsymbol{\sigma}_x\boldsymbol{\sigma}_y = -\boldsymbol{\sigma}_y\boldsymbol{\sigma}_x$ and hence we say that $\boldsymbol{\sigma}_x$, $\boldsymbol{\sigma}_y$ *anti-commute*.

Taking

$$\boldsymbol{\sigma}_x = \begin{pmatrix} a & b \\ b^* & c \end{pmatrix}, \tag{2.2.15}$$

noting that $\boldsymbol{\sigma}_z$ is given by equation (2.2.4) and substituting in equation (2.2.13), we find that

$$a = c = 0. \tag{2.2.16}$$

A similar conclusion is arrived at in respect of the matrix $\boldsymbol{\sigma}_y$. Thus, taking

$$\boldsymbol{\sigma}_y = \begin{pmatrix} 0 & d \\ d^* & 0 \end{pmatrix} \tag{2.2.17}$$

and substituting in equation (2.2.14), we obtain

$$bd^*+b^*d = 0. \tag{2.2.18}$$

Also, from the equation (2.2.6), it follows that

$$bb^* = dd^* = 1. \tag{2.2.19}$$

Thus, we can write

$$b = \exp(\iota\alpha), \quad d = \exp(\iota\beta) \tag{2.2.20}$$

and then equation (2.2.18) requires that

$$\exp(2\iota\beta) = -\exp(2\iota\alpha), \tag{2.2.21}$$

i.e.

$$\exp(\iota\beta) = -\iota\exp(\iota\alpha). \tag{2.2.22}$$

(A sign ambiguity arises in passing from equation (2.2.21) to (2.2.22); we take the negative sign to conform with the usual convention.) The matrices representing $\sigma_x$, $\sigma_y$ have now been reduced to the form

$$\boldsymbol{\sigma}_x = \begin{pmatrix} 0 & \exp(\iota\alpha) \\ \exp(-\iota\alpha) & 0 \end{pmatrix}, \quad \boldsymbol{\sigma}_y = \begin{pmatrix} 0 & -\iota\exp(\iota\alpha) \\ \iota\exp(-\iota\alpha) & 0 \end{pmatrix}. \tag{2.2.23}$$

A further reduction is possible by rotating the axes $Ox$, $Oy$ in their plane through an angle $-\alpha$. Let $Ox'$, $Oy'$ denote the new axes. Then

$$\left.\begin{aligned} \sigma_{x'} &= \sigma_x\cos\alpha+\sigma_y\sin\alpha, \\ \sigma_{y'} &= -\sigma_x\sin\alpha+\sigma_y\cos\alpha. \end{aligned}\right\} \tag{2.2.24}$$

The same equations relate the matrices representing these components of the spin axis and it follows that

$$\boldsymbol{\sigma}_{x'} = \begin{pmatrix} 0 & 1 \\ 1 & 0 \end{pmatrix}, \quad \boldsymbol{\sigma}_{y'} = \begin{pmatrix} 0 & -\iota \\ \iota & 0 \end{pmatrix}. \tag{2.2.25}$$

Omitting the primes, the final form of the matrices representing the three components of the spin axis are:

$$\boldsymbol{\sigma}_x = \begin{pmatrix} 0 & 1 \\ 1 & 0 \end{pmatrix}, \quad \boldsymbol{\sigma}_y = \begin{pmatrix} 0 & -\iota \\ \iota & 0 \end{pmatrix}, \quad \boldsymbol{\sigma}_z = \begin{pmatrix} 1 & 0 \\ 0 & -1 \end{pmatrix}. \tag{2.2.26}$$

These are the *Pauli Spin Matrices.*

It is now easy to verify the identities

$$\boldsymbol{\sigma}_x\boldsymbol{\sigma}_y = \iota\boldsymbol{\sigma}_z, \text{ etc.,} \tag{2.2.27}$$

$$[\boldsymbol{\sigma}_x, \boldsymbol{\sigma}_y] = 2\iota\boldsymbol{\sigma}_z, \text{ etc.} \tag{2.2.28}$$

From equation (2.2.1), we obtain

$$s_x = \tfrac{1}{2}\hbar\sigma_x, \text{ etc.,} \tag{2.2.29}$$

where we have adopted the common notation

$$\hbar = h/2\pi. \tag{2.2.30}$$

It now follows that the counterparts of the identities (2.2.27), (2.2.28) for the spin observables $s_x$, $s_y$, $s_z$ are

$$\mathbf{s}_x\,\mathbf{s}_y = \tfrac{1}{2}\iota\hbar\mathbf{s}_z, \text{ etc.,} \tag{2.2.31}$$

$$[\mathbf{s}_x, \mathbf{s}_y] = \iota\hbar\mathbf{s}_z, \text{ etc.} \tag{2.2.32}$$

Also,

$$\mathbf{s}_x\,\mathbf{s}_y + \mathbf{s}_y\,\mathbf{s}_x = \mathbf{0}, \text{ etc.} \tag{2.2.33}$$

The equations (2.2.32) are called the *commutation relations* for spin.

Denoting the magnitude of the spin by $s$, we have

$$\mathbf{s}^2 = \mathbf{s}_x^2 + \mathbf{s}_y^2 + \mathbf{s}_z^2 = \tfrac{3}{4}\hbar^2\mathbf{I}_2, \tag{2.2.34}$$

indicating that $s^2 = \frac{3}{4}\hbar^2$ in every state.

## 2.3. Illustrations of the theory

The general theory developed in Chapter 1 receives its simplest illustration in terms of spin states. In this case, the unitary vector space which is employed to map the states has only two dimensions ($N=2$). Thus, employing the $\sigma_z$-representation, if the particle's spin state is given by the spinor $\{\alpha_+, \alpha_-\}$ (curly brackets will, in future, be used to indicate that the elements between the brackets are to be arranged to form a column matrix), the probability of observing a transition to the state $\{\alpha'_+, \alpha'_-\}$ is

$$|\langle\alpha|\alpha'\rangle|^2 = |\alpha_+^*\alpha'_+ + \alpha_-^*\alpha'_-|^2. \tag{2.3.1}$$

In particular, if the particle is in the eigenstate $\sigma_z = +1$ so that $\boldsymbol{\alpha} = \{1, 0\}$, the probability of a transition to the other eigenstate $\sigma_z = -1$ for which $\boldsymbol{\alpha}' = \{0, 1\}$ is zero; this verifies that these eigenstates are orthogonal.

Consider next the observable $\sigma_0$ defined by equation (2.2.7); this is the component of spin axis in the direction $(l, m, n)$. Its matrix representation is given by equation (2.2.8). Substituting the Pauli matrices in this equation, we find that

$$\boldsymbol{\sigma}_0 = \begin{pmatrix} n & l-\iota m \\ l+\iota m & -n \end{pmatrix}. \tag{2.3.2}$$

Let $\theta$ be the angle between $Oz$ and the direction $(l, m, n)$ and let $\phi$ be the angle made by the plane containing $Oz$ parallel to this direction and the $Oxz$ plane. Then

$$l = \sin\theta\cos\phi, \quad m = \sin\theta\sin\phi, \quad n = \cos\theta. \tag{2.3.3}$$

Substituting in equation (2.3.2), we obtain

$$\boldsymbol{\sigma}_0 = \begin{pmatrix} \cos\theta & \exp(-\iota\phi)\sin\theta \\ \exp(\iota\phi)\sin\theta & -\cos\theta \end{pmatrix}. \tag{2.3.4}$$

The characteristic equation for $\sigma_0$ is

$$\boldsymbol{\sigma}_0\boldsymbol{\alpha} = \lambda\boldsymbol{\alpha} \tag{2.3.5}$$

and this is now seen to be equivalent to the equations

$$\alpha_+\cos\theta + \alpha_-\exp(-\iota\phi)\sin\theta = \lambda\alpha_+, \tag{2.3.6}$$

$$\alpha_+\exp(\iota\phi)\sin\theta - \alpha_-\cos\theta = \lambda\alpha_-. \tag{2.3.7}$$

The condition that these equations possess a non-zero solution in the unknowns $\alpha_+$, $\alpha_-$ is that

$$\begin{vmatrix} \cos\theta-\lambda & \exp(-\iota\phi)\sin\theta \\ \exp(\iota\phi)\sin\theta & -\cos\theta-\lambda \end{vmatrix} = 0. \tag{2.3.8}$$

This reduces to

$$\lambda^2 = 1 \tag{2.3.9}$$

and hence the eigenvalues of $\sigma_0$ are $\pm 1$ (as we already know to be the case).

If $\sigma_0 = \lambda = +1$ then, from either of the equations (2.3.6), (2.3.7), it follows that

$$\alpha_+ : \alpha_- = \exp(-\iota\phi)\cos\tfrac{1}{2}\theta : \sin\tfrac{1}{2}\theta. \tag{2.3.10}$$

But $\boldsymbol{\alpha}$ is to be normalised and hence

$$\alpha_+\alpha_+^* + \alpha_-\alpha_-^* = 1. \tag{2.3.11}$$

Thus the eigenvector for this eigenstate is given by

$$\boldsymbol{\alpha} = \exp(\iota\gamma)\begin{pmatrix} \exp(-\iota\phi)\cos\tfrac{1}{2}\theta \\ \sin\tfrac{1}{2}\theta \end{pmatrix}, \tag{2.3.12}$$

where $\gamma$ is arbitrary. Since vectors representing states are invariably arbitrary to the extent of a factor of modulus unity, there is no loss of generality if we take $\gamma = 0$.

Similarly, a normalised eigenvector for the eigenstate in which $\sigma_0 = -1$ is found to be

$$\boldsymbol{\alpha} = \exp(\iota\delta)\begin{pmatrix} -\exp(-\iota\phi)\sin\frac{1}{2}\theta \\ \cos\frac{1}{2}\theta \end{pmatrix}. \tag{2.3.13}$$

In the state described by equation (2.3.12), the probabilities of observing transitions into the two eigenstates of $\sigma_z$ are seen to be $\cos^2\frac{1}{2}\theta$ and $\sin^2\frac{1}{2}\theta$. This implies that if $\sigma_0$ is measured to be $+1$, the probabilities that an immediate subsequent measurement of $\sigma_z$ will yield the values $+1$, $-1$ are $\cos^2\frac{1}{2}\theta$, $\sin^2\frac{1}{2}\theta$ respectively. In the state (2.3.13), the corresponding probabilities are $\sin^2\frac{1}{2}\theta$, $\cos^2\frac{1}{2}\theta$ respectively.

Putting $\theta = \frac{1}{2}\pi$, $\phi = 0$ in equations (2.3.12), (2.3.13), we calculate that eigenvectors for the two $\sigma_x$-eigenstates are

$$\sqrt{(\tfrac{1}{2})}\begin{pmatrix} 1 \\ 1 \end{pmatrix}, \quad \sqrt{(\tfrac{1}{2})}\begin{pmatrix} -1 \\ 1 \end{pmatrix}. \tag{2.3.14}$$

The probability of a transition from the eigenstate $\sigma_z = +1$ to the eigenstate $\sigma_x = +1$ is accordingly the square of the modulus of

$$\sqrt{(\tfrac{1}{2})}\,(1, 1)\begin{pmatrix} 1 \\ 0 \end{pmatrix} = \sqrt{(\tfrac{1}{2})}, \tag{2.3.15}$$

i.e. is $\frac{1}{2}$.

Next, let us calculate the expected value of $\sigma_y$ in the eigenstate $\sigma_0 = +1$. From equation (1.8.4), this is

$$\begin{aligned} \bar{\sigma}_y &= (\exp(\iota\phi)\cos\tfrac{1}{2}\theta, \sin\tfrac{1}{2}\theta)\begin{pmatrix} 0 & -\iota \\ \iota & 0 \end{pmatrix}\begin{pmatrix} \exp(-\iota\phi)\cos\frac{1}{2}\theta \\ \sin\frac{1}{2}\theta \end{pmatrix} \\ &= \sin\theta\sin\phi. \end{aligned} \tag{2.3.16}$$

The reader may verify that, in the same state, $\bar{\sigma}_x = \sin\theta\cos\phi$. In particular, in the eigenstate $\sigma_z = +1$, $\theta = 0$ and hence $\bar{\sigma}_x = \bar{\sigma}_y = 0$. Thus, in this state,

$$\operatorname{var}\sigma_x = (\overline{\sigma_x^2}) - (\bar{\sigma}_x)^2 = 1, \tag{2.3.17}$$

since $\sigma_x^2 = 1$ in every state. Similarly $\operatorname{var}\sigma_y = 1$.

Finally, in this section, consider the form taken by the uncertainty principle (1.11.18) in the case when $a = \sigma_x$, $b = \sigma_y$. Then, by equation (2.2.28),

$$\iota[\mathbf{a}, \mathbf{b}] = -2\boldsymbol{\sigma}_z \tag{2.3.18}$$

and hence

$$c = -2\sigma_z. \tag{2.3.19}$$

The uncertainty principle then states that

$$\text{var}\,\sigma_x\,.\,\text{var}\,\sigma_y \geqslant \bar{\sigma}_z^2. \tag{2.3.20}$$

In particular, in the eigenstate $\sigma_z = +1$, this becomes

$$\text{var}\,\sigma_x\,.\,\text{var}\,\sigma_y \geqslant 1. \tag{2.3.21}$$

In fact, we have already proved that, in this state $\text{var}\,\sigma_x = \text{var}\,\sigma_y = 1$.

## 2.4. Unitary transformation of spinors

Suppose that the spin state of a particle is characterised by the spinor $\boldsymbol{\alpha} = \{\alpha_+, \alpha_-\}$ in the $\sigma_z$-representation. Let $Ox'y'z'$ be another set of rectangular cartesian axes with $O$ as origin and let $(\theta, \phi)$ be the spherical polar angles determining the direction of $Oz'$ relative to the frame $Oxyz$. If $\sigma_{z'}$ is the component of spin axis in the direction $Oz'$, we may employ its eigenstates as a basis for a representation. In the $\sigma_{z'}$-representation, let the particle's state be determined by the spinor $\boldsymbol{\alpha}' = \{\alpha'_+, \alpha'_-\}$. Then $\boldsymbol{\alpha}$, $\boldsymbol{\alpha}'$ must be related by a unitary transformation.

The coefficients of this transformation may be calculated from equation (1.6.14). $\phi^1$, $\phi^2$ are the vectors representing the eigenstates of $\sigma_z$ and $\psi^1$, $\psi^2$ are the vectors representing the eigenstates of $\sigma_{z'}$. Thus, in the $\sigma_z$-representation,

$$\boldsymbol{\phi}^1 = \begin{pmatrix}1\\0\end{pmatrix}, \quad \boldsymbol{\phi}^2 = \begin{pmatrix}0\\1\end{pmatrix} \tag{2.4.1}$$

and, by equations (2.3.12), (2.3.13),

$$\boldsymbol{\psi}^1 = \exp(\iota\gamma)\begin{pmatrix}\exp(-\iota\phi)\cos\frac{1}{2}\theta\\ \sin\frac{1}{2}\theta\end{pmatrix}, \quad \boldsymbol{\psi}^2 = \exp(\iota\delta)\begin{pmatrix}-\exp(-\iota\phi)\sin\frac{1}{2}\theta\\ \cos\frac{1}{2}\theta\end{pmatrix}. \tag{2.4.2}$$

We now find that the matrix of the required transformation is

$$\mathbf{U} = \begin{pmatrix}\exp[\iota(\phi-\gamma)]\cos\frac{1}{2}\theta & \exp(-\iota\gamma)\sin\frac{1}{2}\theta\\ -\exp[\iota(\phi-\delta)]\sin\frac{1}{2}\theta & \exp(-\iota\delta)\cos\frac{1}{2}\theta\end{pmatrix}. \tag{2.4.3}$$

It is left as an exercise for the reader to verify that $\mathbf{U}$ is unitary. The relationship between $\boldsymbol{\alpha}$, $\boldsymbol{\alpha}'$ is

$$\boldsymbol{\alpha}' = \mathbf{U}\boldsymbol{\alpha}. \tag{2.4.4}$$

The quantities $\gamma$, $\delta$ are at present unknown. However, specific values may be found for them thus:

Consider the observable $\sigma_z$. Since the spin axis is a vector, this observable is related to the observables $\sigma_{x'}$, $\sigma_{y'}$, $\sigma_{z'}$ by the equation

$$\sigma_z = l_{zx'}\,\sigma_{x'} + l_{zy'}\,\sigma_{y'} + l_{zz'}\,\sigma_{z'}, \tag{2.4.5}$$

where $l_{zx'}$ is the cosine of the angle between $Oz$ and $Ox'$. Employing the $\sigma_{z'}$-representation, equation (2.4.5) leads to a corresponding matrix equation

$$\boldsymbol{\sigma}'_z = l_{zx'}\boldsymbol{\sigma}'_{x'} + l_{zy'}\boldsymbol{\sigma}'_{y'} + l_{zz'}\boldsymbol{\sigma}'_{z'}. \tag{2.4.6}$$

Now, if the $\sigma_z$-representation is employed, $\sigma_z$ is represented by the Pauli matrix given at (2.2.26). However, if the $\sigma_{z'}$ representation is used, it is the observables $\sigma_{x'}$, $\sigma_{y'}$, $\sigma_{z'}$ which are represented by Pauli matrices. It follows from equation (2.4.6) that, in this representation, $\sigma_z$ has the matrix

$$\boldsymbol{\sigma}'_z = \begin{pmatrix} l_{zz'} & l_{zx'} - \iota l_{zy'} \\ l_{zx'} + \iota l_{zy'} & -l_{zz'} \end{pmatrix}. \tag{2.4.7}$$

But, by equation (1.8.16),

$$\boldsymbol{\sigma}'_z = \mathbf{U}\boldsymbol{\sigma}_z\mathbf{U}^\dagger. \tag{2.4.8}$$

Substituting for $\boldsymbol{\sigma}_z$, $\boldsymbol{\sigma}'_z$ and $\mathbf{U}$, this leads to

$$\begin{pmatrix} l_{zz'} & l_{zx'} - \iota l_{zy'} \\ l_{zx'} + \iota l_{zy'} & -l_{zz'} \end{pmatrix} = \begin{pmatrix} \cos\theta & -\exp[\iota(\delta-\gamma)]\sin\theta \\ -\exp[\iota(\gamma-\delta)]\sin\theta & -\cos\theta \end{pmatrix}. \tag{2.4.9}$$

It follows that

$$l_{zz'} = \cos\theta, \tag{2.4.10}$$

$$\exp[\iota(\delta-\gamma)](l_{zx'} + \iota l_{zy'}) = -\sin\theta. \tag{2.4.11}$$

Equation (2.4.10) follows from the geometry of the two sets of axes.

Suppose we regard the axes $Ox'y'z'$ as being brought into their final position from one of initial coincidence with the frame $Oxyz$, by the following series of rigid body rotations: First rotate through an angle $\phi$ about $Oz$; secondly, rotate through an angle $\theta$ about $Oy'$; thirdly, rotate through an angle $\psi$ about $Oz'$. Then $(\theta, \phi, \psi)$ are the *Euler Angles* specifying the position of the frame $Ox'y'z'$ relative to the frame $Oxyz$ and it may be shown that

$$l_{zx'} = -\sin\theta\cos\psi, \quad l_{zy'} = \sin\theta\sin\psi. \tag{2.4.12}$$

Equation (2.4.11) is accordingly equivalent to the equation

$$\exp[\iota(\delta-\gamma-\psi)] = 1 \tag{2.4.13}$$

and this implies that

$$\delta-\gamma = \psi+2n\pi, \tag{2.4.14}$$

where $n$ is an integer or zero.

Consideration of the observables $\sigma_x$, $\sigma_y$ in a similar fashion leads to the same result.

Equation (2.4.3) can now be written

$$\mathbf{U} = \exp(-\iota\delta)\begin{pmatrix} \exp[\iota(\phi+\psi)]\cos\frac{1}{2}\theta & \exp(\iota\psi)\sin\frac{1}{2}\theta \\ -\exp(\iota\phi)\sin\frac{1}{2}\theta & \cos\frac{1}{2}\theta \end{pmatrix}. \tag{2.4.15}$$

Since spinors are arbitrary to the extent of a factor of modulus unity, $\mathbf{U}$ is also arbitrary to the same degree. $\delta$ is accordingly open to choice and we shall take

$$\delta = \tfrac{1}{2}(\phi+\psi). \tag{2.4.16}$$

Then

$$\mathbf{U} = \begin{pmatrix} \exp[\frac{1}{2}\iota(\phi+\psi)]\cos\frac{1}{2}\theta & \exp[-\frac{1}{2}\iota(\phi-\psi)]\sin\frac{1}{2}\theta \\ -\exp[\frac{1}{2}\iota(\phi-\psi)]\sin\frac{1}{2}\theta & \exp[-\frac{1}{2}\iota(\phi+\psi)]\cos\frac{1}{2}\theta \end{pmatrix}. \tag{2.4.17}$$

It is easy to verify that the determinant of $\mathbf{U}$ is now unity and we say that $\mathbf{U}$ is the matrix of a unitary *unimodular* transformation. If $\theta=\phi=\psi=0$, then $\mathbf{U}=\mathbf{I}_2$ and the transformation is the identity.

Consider the *rotation group* $\mathscr{R}_3$ of rigid body rotations about the point $O$. To each element of the group there correspond two matrices $\mathbf{U}$ (note that increasing $\theta$, $\phi$ or $\psi$ by $2\pi$ does not lead to a new element of $\mathscr{R}_3$, but it changes the sign of $\mathbf{U}$). Let $R$, $S$ be two rotations of the group and let $SR$ be the product of these rotations, i.e. the result of first performing the rotation $R$ and then the rotation $S$. Then, if $\mathbf{U}$, $\mathbf{V}$ are unitary matrices corresponding to the rotations $R$, $S$ respectively, the spinor transformation generated by the resultant rotation $SR$ will have a matrix $\mathbf{VU}$. Also, $\mathbf{U}^{-1}$ is a matrix corresponding to the inverse rotation $R^{-1}$. It follows that the group $\mathscr{U}_2$ of unitary unimodular $2\times 2$ matrices $\mathbf{U}$ is homomorphic to the group $\mathscr{R}_3$ and provides a representation of this latter group. A geometrical relationship between the set of unitary unimodular transformations $\mathscr{U}_2$ and the set of rotations $\mathscr{R}_3$ is suggested in Ex. 7, Chap. 2.

Transformation properties of vectors having more dimensions than the spinors considered in this section will arise later (section 2.6). These lead, in a similar fashion, to further matrix representations of the rotation group $\mathscr{R}_3$. The theory of spinors and other vectors representing quantum mechanical states may, in fact, be developed from this aspect *ab initio* (see, e.g. *Group Theory, The Application to Quantum Mechanics* by P. H. E. Meijer and E. Bauer, North-Holland, 1962).

## 2.5. A two-particle system

Consider a system comprising two particles, each having eigenvalues for the $z$-component of spin of $\pm\frac{1}{2}\hbar$. We assume that the particles are distinguishable, i.e. the system does not comprise two particles of the same type (such as two electrons or two protons, etc.) for, otherwise, the possible spin states of the system will be limited by additional constraints which we shall study in the next section. The general theory of section 1.13 can be illustrated by applying it to such a system, account being taken of the spin observables alone.

Taking rectangular axes $Oxyz$, let $s_{z1}$, $s_{z2}$ be the $z$-components of spin of the particles $P_1$, $P_2$ respectively. Then a complete set of eigenstates for the combined system is defined by the four pairs of eigenvalues

$$\left.\begin{aligned}
&(1)\ s_{z1} = \tfrac{1}{2}\hbar, s_{z2} = \tfrac{1}{2}\hbar, \text{ eigenstate } \chi^{11},\\
&(2)\ s_{z1} = \tfrac{1}{2}\hbar, s_{z2} = -\tfrac{1}{2}\hbar, \text{ eigenstate } \chi^{12},\\
&(3)\ s_{z1} = -\tfrac{1}{2}\hbar, s_{z2} = \tfrac{1}{2}\hbar, \text{ eigenstate } \chi^{21},\\
&(4)\ s_{z1} = -\tfrac{1}{2}\hbar, s_{z2} = -\tfrac{1}{2}\hbar, \text{ eigenstate } \chi^{22}.
\end{aligned}\right\} \tag{2.5.1}$$

Employing this set as a basis for a matrix representation, a state of the system will be specified by a $4\times 1$ column matrix having elements $\gamma_{ij}$ $(i=1,2; j=1,2)$, where we shall suppose that these elements are taken in their natural order. $|\gamma_{ij}|^2$ is, of course, the probability of observing a transition from the given state of the system to the state $\chi^{ij}$.

Consider the quantity $s_{x1}$, regarded as an observable of the whole system. According to equation (1.13.17), the $4\times 4$ matrix representing this observable is

$$\mathrm{s}_{x1} = \tfrac{1}{2}\hbar \begin{pmatrix} 0 & 0 & 1 & 0\\ 0 & 0 & 0 & 1\\ 1 & 0 & 0 & 0\\ 0 & 1 & 0 & 0 \end{pmatrix}. \tag{2.5.2}$$

Similarly, by equation (1.13.19), the matrix representing $s_{x2}$ is

$$\mathrm{s}_{x2} = \tfrac{1}{2}\hbar \begin{pmatrix} 0 & 1 & 0 & 0\\ 1 & 0 & 0 & 0\\ 0 & 0 & 0 & 1\\ 0 & 0 & 1 & 0 \end{pmatrix}. \tag{2.5.3}$$

Now suppose that we define the net spin of the system to be the vector sum of the spins of the constituent particles. Then, if $s_x$ is the $x$-component of the net spin,

$$s_x = s_{x1} + s_{x2} \tag{2.5.4}$$

and the matrix representing $s_x$ is

$$\mathbf{s}_x = \mathbf{s}_{x1}+\mathbf{s}_{x2},$$
$$= \tfrac{1}{2}\hbar \begin{pmatrix} 0 & 1 & 1 & 0 \\ 1 & 0 & 0 & 1 \\ 1 & 0 & 0 & 1 \\ 0 & 1 & 1 & 0 \end{pmatrix}. \tag{2.5.5}$$

It is evident from equation (2.5.4) that $s_x$ possesses eigenvalues $\hbar, 0, -\hbar$. It is left as an exercise for the reader to check that these are the eigenvalues calculated from the characteristic equation for the matrix $\mathbf{s}_x$ as given by equation (2.5.5); the eigenvalue 0 will be found to occur twice; this arises from the circumstance that $s_x=0$ in both the eigenstates (i) $s_{x1}=\frac{1}{2}\hbar$, $s_{x2}=-\frac{1}{2}\hbar$, (ii) $s_{x1}=-\frac{1}{2}\hbar$, $s_{x2}=\frac{1}{2}\hbar$.

The matrices representing $s_y$, $s_z$ may be calculated similarly. They prove to be

$$\mathbf{s}_y = \tfrac{1}{2}\hbar \begin{pmatrix} 0 & -\iota & -\iota & 0 \\ \iota & 0 & 0 & -\iota \\ \iota & 0 & 0 & -\iota \\ 0 & \iota & \iota & 0 \end{pmatrix}, \quad \mathbf{s}_z = \hbar \begin{pmatrix} 1 & 0 & 0 & 0 \\ 0 & 0 & 0 & 0 \\ 0 & 0 & 0 & 0 \\ 0 & 0 & 0 & -1 \end{pmatrix}. \tag{2.5.6}$$

The eigenvalues for both these observables are $+\hbar, 0, -\hbar$.

The reader should verify that $\mathbf{s}_x$, $\mathbf{s}_y$, $\mathbf{s}_z$ satisfy the commutation equations (2.2.32), but that they do not anti-commute. He should also verify that if

$$s^2 = s_x^2+s_y^2+s_z^2 \tag{2.5.7}$$

is the square of the magnitude of the resultant spin, then the matrix representing $s^2$ is

$$\mathbf{s}^2 = \hbar^2 \begin{pmatrix} 2 & 0 & 0 & 0 \\ 0 & 1 & 1 & 0 \\ 0 & 1 & 1 & 0 \\ 0 & 0 & 0 & 2 \end{pmatrix}. \tag{2.5.8}$$

$s^2$ will be found to possess two eigenvalues, viz. 0 and $2\hbar^2$.

As a further exercise, simultaneous eigenstates for the observables $s_z$, $s^2$ may be computed. It will be found that, if $s^2=0$, there is only one such eigenstate; in this state $s_z=0$ and the eigenspinor is

$$\sqrt{(\tfrac{1}{2})} \begin{pmatrix} 0 \\ 1 \\ -1 \\ 0 \end{pmatrix}. \tag{2.5.9}$$

This is called a *singlet state*. However, if $s^2 = 2\hbar^2$, it will be found that there are three simultaneous eigenstates corresponding to the $s_z$-values $-\hbar, 0, +\hbar$. The corresponding eigenspinors are

$$\begin{pmatrix}0\\0\\0\\1\end{pmatrix}, \quad \sqrt{(\tfrac{1}{2})}\begin{pmatrix}0\\1\\1\\0\end{pmatrix}, \quad \begin{pmatrix}1\\0\\0\\0\end{pmatrix}, \tag{2.5.10}$$

respectively. These states are said to be members of a *triplet*.

## 2.6. Indistinguishable particles. Exclusion Principle

In this section, we will again study the spin states of a two-particle system, but will suppose that the particles are of the same type (e.g. two electrons) and are in identical states with respect to all but the spin observables. In these circumstances, since the particles are indistinguishable, if the spin state of $P_1$ is changed to that of $P_2$ and vice versa, then the state of the system as a whole must remain unaltered.

When the spin states are switched, let the components of the vector determining the spin state of the whole system change from $\gamma_{ij}$ to $\gamma'_{ij}$. Then, since the relationship between the new state of the system and the eigenstate $\chi^{ij}$ must be the same as the relationship between the original state of the system and the eigenstate $\chi^{ji}$, we can take

$$\gamma'_{ij} = \gamma_{ji}. \tag{2.6.1}$$

But $\gamma'_{ij}$ represents a state which is observationally indistinguishable from the state represented by $\gamma_{ij}$ and hence

$$\gamma'_{ij} = \exp(\iota\theta)\,\gamma_{ij}. \tag{2.6.2}$$

From the last two equations, we deduce that

$$\gamma_{ij} = \exp(\iota\theta)\,\gamma_{ji} = \exp(2\iota\theta)\,\gamma_{ij}. \tag{2.6.3}$$

Thus

$$\exp(2\iota\theta) = 1 \quad \text{and} \quad \exp(\iota\theta) = \pm 1. \tag{2.6.4}$$

The components $\gamma_{ij}$ are accordingly constrained to satisfy one of the conditions

$$\gamma_{ij} = \gamma_{ji} \quad \text{or} \quad \gamma_{ij} = -\gamma_{ji}, \tag{2.6.5}$$

i.e. $\gamma_{ij}$ is either symmetric with respect to its indices or anti-symmetric.

This new constraint arising from the indistinguishability of the two particles, implies that the theory of section 1.13 cannot be applied to the system without some modification. For, if $\alpha_i$ $(i=1,2)$ are the components of the spinor determining the spin state of $P_1$ before the exchange takes place and $\beta_i$ $(i=1,2)$ are the spinor components for $P_2$ in the same circumstances, according to the theory of section 1.13, we would expect that

$$\gamma_{ij} = \alpha_i \beta_j. \tag{2.6.6}$$

However, in general, $\gamma_{ij}$ as given by this equation will be neither symmetric nor anti-symmetric. Suppose the spin states are exchanged, so that $P_1$ has spin $\beta_i$ and $P_2$ has spin $\alpha_i$. Then, by similar reasoning, we would expect that the spin state of the system would now be given by

$$\gamma'_{ij} = \beta_i \alpha_j. \tag{2.6.7}$$

But, since the states (2.6.6), (2.6.7) are to be regarded as identical, it follows from the principle of superposition (section 1.10), that the state

$$\gamma''_{ij} = A\alpha_i \beta_j + B\alpha_j \beta_i \tag{2.6.8}$$

is also identical with the original state for all values of $A$ and $B$. Now, by proper choice of $A$ and $B$, we can ensure that $\gamma''_{ij}$ satisfies the new constraint. Thus, taking $A=B$, we have

$$\gamma''_{ij} = A(\alpha_i \beta_j + \alpha_j \beta_i) \tag{2.6.9}$$

and $\gamma''_{ij}$ is symmetric. Also, taking $A=-B$, we have

$$\gamma''_{ij} = A(\alpha_i \beta_j - \alpha_j \beta_i) \tag{2.6.10}$$

and $\gamma''_{ij}$ is anti-symmetric. $A$ will, of course, now be chosen to normalise $\gamma''_{ij}$.

If a state in which $\gamma_{ij}$ is symmetric is superposed on a state in which $\gamma_{ij}$ is anti-symmetric, the result is a state in which $\gamma_{ij}$ is neither symmetric nor anti-symmetric. Since such a non-symmetric state is impossible we conclude that, for a given system, $\gamma_{ij}$ is either always symmetric or always anti-symmetric, i.e. symmetry or anti-symmetry is a characteristic of every particle. Particles which lead to anti-symmetric $\gamma_{ij}$ are called *Fermi particles* or *fermions* and particles which lead to symmetric $\gamma_{ij}$ are called *Bose particles* or *bosons*. The electron, proton and neutron are all fermions and deuterons in the stable state are bosons (each with three spin eigenstates, however; see below).

Consider a system comprising two fermions of the same type, e.g. two electrons. If the two particles are assumed to be in the same spin state, then $\alpha_i = \beta_i$ and it follows from equation (2.6.10) that $\gamma_{ij}$ vanishes identically. Since no state is represented by a null vector, we conclude that *a pair of identical fermions can never occupy the same state simultaneously.* This is the *Exclusion Principle* first formulated by Pauli. If the particles are not in the same state then, since $\gamma_{ij}$ is anti-symmetric, we must have $\gamma_{11}=\gamma_{22}=0$, $\gamma_{12}=-\gamma_{21}=\gamma$. Thus, any column matrix representing a state of the system will take the form

$$\begin{pmatrix} 0 \\ \gamma \\ -\gamma \\ 0 \end{pmatrix}. \qquad (2.6.11)$$

This implies that, in any state, there is zero probability of a transition being observed into either of the eigenstates $\chi^{11}$, $\chi^{22}$ and, therefore, that these states will never be observed. This, of course, follows immediately from the exclusion principle.

It will further be noted from the matrix (2.6.11) that the probabilities of transitions to the states $\chi^{12}$, $\chi^{21}$ are always equal. This is to be expected, since these two eigenstates are indistinguishable. Thus, a system comprising a pair of fermions of the same type and in the same state except for spin has, essentially, only one spin eigenstate for which $s_z=0$. For such a system, $s_z$ is zero with complete certainty and similarly, of course, $s_x=s_y=0$; i.e. the system behaves like a particle with no spin.

Now consider a system comprising two bosons. In this case, $\gamma_{ij}$ is given by equation (2.6.9) and it is possible for the particles to occupy identical states; thus the exclusion principle is not applicable. Any state of the system is determined by a column matrix having the form

$$\begin{pmatrix} \gamma_1 \\ \gamma_2 \\ \gamma_3 \\ \gamma_4 \end{pmatrix}, \qquad (2.6.12)$$

where $\gamma_2=\gamma_3$. Since the eigenstates $\chi^{12}$, $\chi^{21}$ are identical, we can calculate the net probability of a transition being observed into this eigenstate as

$$\gamma_2\gamma_2^*+\gamma_3\gamma_3^* = 2\gamma_2\gamma_2^*. \qquad (2.6.13)$$

Since the three states $\chi^{11}$, $\chi^{12}$, $\chi^{22}$ form a complete set of eigenstates for our system, the state (2.6.12) can equally well be represented by a column matrix with only three elements, viz.

$$\begin{pmatrix} \gamma_1 \\ \sqrt{2}\gamma_2 \\ \gamma_4 \end{pmatrix}. \qquad (2.6.14)$$

Employing this contracted basis, the question now arises: what is the appropriate matrix representation of the spin observables $s_x$, $s_y$, $s_z$?

Referring to equations (2.5.5), (2.5.6), it will be noted that each of these matrix representations has identical second and third rows and identical second and third columns. It is left as an exercise for the reader to prove that the sums and products of matrices possessing this characteristic, also possess the same characteristic. Hence, if $o$ is any spin observable derived from the observables $s_x$, $s_y$, $s_z$, its $4 \times 4$ matrix representation will be $(o_{ij})$ $(i,j = 1,2,3,4)$, where

$$o_{2j} = o_{3j}, \quad o_{i2} = o_{i3}. \tag{2.6.15}$$

The expected value of $o$ in the state (2.6.12) will therefore be given by

$$\begin{aligned} \bar{o} &= o_{ij}\gamma_i^* \gamma_j, \\ &= o_{11}\gamma_1^*\gamma_1 + \sqrt{2}o_{12}\gamma_1^*(\sqrt{2}\gamma_2) + o_{14}\gamma_1^*\gamma_4 \\ &\quad + \sqrt{2}o_{21}(\sqrt{2}\gamma_2)^*\gamma_1 + 2o_{22}(\sqrt{2}\gamma_2)^*(\sqrt{2}\gamma_2) + \sqrt{2}o_{24}(\sqrt{2}\gamma_2)^*\gamma_4 \\ &\quad + o_{41}\gamma_4^*\gamma_1 + \sqrt{2}o_{42}\gamma_4^*(\sqrt{2}\gamma_2) + o_{44}\gamma_4^*\gamma_4. \end{aligned} \tag{2.6.16}$$

This expression for $\bar{o}$ implies that the matrix which represents $o$ when the contracted basis is used is

$$\mathbf{o} = \begin{pmatrix} o_{11} & \sqrt{2}o_{12} & o_{14} \\ \sqrt{2}o_{21} & 2o_{22} & \sqrt{2}o_{24} \\ o_{41} & \sqrt{2}o_{42} & o_{44} \end{pmatrix}. \tag{2.6.17}$$

Since we are assuming that the particles have spin $\frac{1}{2}\hbar$, it follows from equations (2.5.5), (2.5.6) that

$$\mathbf{s}_x = \sqrt{(\tfrac{1}{2})}\hbar \begin{pmatrix} 0 & 1 & 0 \\ 1 & 0 & 1 \\ 0 & 1 & 0 \end{pmatrix}, \tag{2.6.18}$$

$$\mathbf{s}_y = \sqrt{(\tfrac{1}{2})}\hbar \begin{pmatrix} 0 & -\iota & 0 \\ \iota & 0 & -\iota \\ 0 & \iota & 0 \end{pmatrix}, \tag{2.6.19}$$

$$\mathbf{s}_z = \hbar \begin{pmatrix} 1 & 0 & 0 \\ 0 & 0 & 0 \\ 0 & 0 & -1 \end{pmatrix}. \tag{2.6.20}$$

Now, it is found by observation that all fermions possess spins which are odd multiples of $\frac{1}{2}\hbar$ and all bosons possess no spin or spins which are even multiples of $\frac{1}{2}\hbar$ (i.e. multiples of $\hbar$). Hence, a system of the kind we have postulated in the previous paragraph does not exist in nature. Nevertheless, if it could exist, it would behave like a particle possessing a complete set of three spin eigenstates $s_z = \hbar$, 0, $-\hbar$ and so would be expected to behave like a boson. It is, in fact, found that a boson of spin $\hbar$ (e.g. a deuteron) has three such spin eigenstates and that, employing these as a basis, equations (2.6.18)–(2.6.20) are correct matrix representations of its observables $s_x$, $s_y$ and $s_z$.

The reader may again verify that each of the matrices (2.6.18)–(2.6.20) possesses characteristic roots $\pm\hbar$, 0 and that they obey the commutation rules (2.2.32). It may also be verified that, if $s$ is the magnitude of the spin defined by the equation

$$s^2 = s_x^2 + s_y^2 + s_z^2, \tag{2.6.21}$$

then the matrix representing $s^2$ is $2\hbar^2\mathbf{I}_3$. This implies that $s^2$ is a constant, assuming the value $2\hbar^2$ in every state.

CHAPTER 3

# Observables having continuous spectra

## 3.1. Spectra with an infinity of eigenvalues

In the first chapter, we considered a mode of representation of the states of a physical system in the cases when a specification of such states could be achieved in terms of complete sets of observables, each of which possessed a finite number of eigenvalues only. In this section, we will generalise our mathematical model to include within its scope systems whose states are described in terms of observables some, or all of which, possess an infinity of eigenvalues. This extension is found to raise problems of a pure mathematical nature, many of which are the subject of current research and to which no complete solution can, as yet, be given. In any case, an attempt to present a rigorous justification for this generalisation of the model would require us to give a detailed account of the modern theory of function spaces and such an account would inevitably occupy a space considerably greater than that which it is our intention to allot to the physical theory which is the prime topic of this book. For these reasons, we shall content ourselves with the enunciation of results, which appear to be the obvious generalisations for these more complex systems, of principles which were introduced earlier in their application to relatively simple systems.

The case when the sets of eigenvalues of the observables comprising a maximal set for a system, are infinite but countable (enumerable), so that the spectra are still discrete, is a generalisation of the finite case which is comparatively easy to deal with and we shall dismiss it after only a brief examination. In this case, the complete set of eigenstates corresponding to the maximal set of observables must also be countable and hence the dimension $N$ of the vector space of our model is also countable. In other words, a complete orthonormal set of vectors $\psi^1$, $\psi^2$, ... (infinite bu countable) can be found, which will constitute basis for a matrix representation. Employing this representation, states of the system will be specified by column matrices possessing an infinity of elements and observables will be represented by matrices possessing an infinity of rows and columns. It will be assumed that such matrices can be manipulated algebraically in the same manner as are finite matrices, although the elements of the product of two such matrices will have to be calculated by the summation of infinite series; the convergence of such series will be taken for granted. Equations such as (1.8.4) for the expected value of an observable, will now also involve infinite series, but will still be assumed valid provided the series converge.

Examples of observables possessing this type of spectrum are (i) any component of the angular momentum of a particle about a point (see section 5.2) and (ii) the energy of a particle subject to a simple harmonic restoring force (see section 3.10).

However, many of the most important observables associated with physical systems possess continuous spectra of eigenvalues. For example, the observables employed most frequently in classical mechanics to describe the state of a particle, namely its coordinates and components of momentum, belong to this class. Denoting the cartesian coordinates of a particle by $(x,y,z)$ and the corresponding momentum components by $p_x, p_y, p_z$, each of these quantities can be measured to take any value in the range $(-\infty,\infty)$; its eigenvalues accordingly form a continuous spectrum extending to infinity in both the positive and negative senses. However, by the uncertainty principle (section 1.2), only one observable in each of the pairs $(x,p_x)$, $(y,p_y)$, $(z,p_z)$ can be measured precisely (i.e. can be sharp) at a given instant and it follows that $(x,y,z)$ form a maximal set of compatible observables for a single particle; a particle whose position is known precisely at some instant is accordingly in a pure state at this instant. Such a pure state is an eigenstate with respect to the observables $(x,y,z)$ and the set of all such eigenstates (which will not be countable) may be employed as a basis for a representation of an arbitrary state of the particle and of its associated observables. The momentum observables $(p_x,p_y,p_z)$ may, similarly, be used as a basis for a representation, as may, for example, the mixed set $(x,p_y,z)$. In the next section, we study the character of such representations based upon a non-countable infinity of eigenstates.

## 3.2. Continuous representations. Impulse function

For simplicity consider first a system for which a single quantity $q$, possessing a continuous spectrum, forms a maximal set of compatible observables, e.g. a particle which is constrained to move parallel to the $x$-axis under known forces and for which we can take $q=x$ (or $q=p_x$). Suppose a definite procedure for measuring $q$ has been decided upon,

Then, even though we assume that the instruments being employed are perfectly precise, a reading of the result will only locate the value of $q$ in one of a set of small, but finite, intervals $(a, a+\eta)$, $(a+\eta, a+2\eta)$, ..., $(a+(n-1)\eta, a+n\eta)$, where $a \leqslant q \leqslant b = a+n\eta$ represents the range of possible variation of $q$ ($a, b$ may be infinite) and $\eta$ is the smallest interval between graduations which can be distinguished on the $q$-meter. It follows that, in practice, the $q$-eigenstates will always be finite in number or countable and may be denoted by $\psi^1, \psi^2, \ldots, \psi^n$ ($n$ may be infinite).

Now suppose that the system is in a state represented by a vector $\alpha$. Relative to the frame determined by the $q$-eigenstates, $\alpha$ will possess components which we shall denote by $\alpha_1\eta^{\frac{1}{2}}, \alpha_2\eta^{\frac{1}{2}}, \ldots, \alpha_n\eta^{\frac{1}{2}}$; the factor $\eta^{\frac{1}{2}}$ has been introduced since it is evident that the probability that $q$ will be measured to lie in a particular one of the intervals $(a+(r-1)\eta, a+r\eta)$ $(r = 1, 2, \ldots, n)$ must be proportional to $\eta$ (if $\eta$ is small) and the squares of the moduli of the components are known to equal these probabilities. We now define a function $\alpha_\eta(q)$ by the equations

$$\alpha_\eta(q) = \alpha_r, \quad a+(r-1)\eta \leqslant q < a+r\eta, \quad r = 1, 2, \ldots, n, \tag{3.2.1}$$

and assume that, as $\eta \to 0$,

$$\alpha_\eta(q) \to \alpha(q), \tag{3.2.2}$$

where $\alpha(q)$ is a complex function of $q$ defined over the interval $[a, b]$. $\alpha(q)$ will be termed the *wave function* of the system in the state $\alpha$, employing the $q$-representation. For a representation based upon an observable possessing a continuous spectrum, it is the counterpart of the column matrix specifying a state when observables possessing discrete spectra are being used as a basis.

The probability that, upon measurement, $q$ will be found to lie in the interval $(a+(r-1)\eta, a+r\eta)$ is $|\alpha_r|^2\eta$. Letting $\eta \to 0$, this leads to the statement that the probability that $q$ will be found to lie in an infinitesimal interval $(q, q+dq)$ is

$$|\alpha(q)|^2\,dq. \tag{3.2.3}$$

Thus, the *probability density* for $q$ is given by the square of the modulus of the wave function. The probability that $q$ will be found in a finite interval $(q_1, q_2)$ is therefore

$$\int_{q_1}^{q_2} \alpha^*(q)\,\alpha(q)\,dq. \tag{3.2.4}$$

Consider the probability of a transition between two states $\alpha$ and $\beta$; this is determined by the scalar product $\langle\alpha|\beta\rangle$. Now

$$\langle\alpha|\beta\rangle = \sum_i \alpha_i^* \beta_i \eta \to \int_a^b \alpha^*(q)\,\beta(q)\,dq, \tag{3.2.5}$$

as $\eta \to 0$. This integral accordingly defines the scalar product of two vectors in the $q$-representation.

The probability of a transition from a state $\alpha$ into itself is unity. It follows that

$$\langle\alpha|\alpha\rangle = \int_a^b \alpha^*(q)\,\alpha(q)\,dq = 1. \tag{3.2.6}$$

This result is expressed by saying that the vector $\alpha$ is normalised. Equation (3.2.6) also follows from the fact that the integral represents the probability that a measurement of $q$ will yield *some* value in the interval $[a, b]$.

We shall denote the eigenvector representing the eigenstate in which $q$ takes the sharp value $c$ by $\psi^c$. A difficulty now arises when we attempt to determine the wave function corresponding to $\psi^c$. Clearly, when the system is in this eigenstate, the probability that $q$ will be measured to take any value other than $c$ is zero. It follows that the probability density for $q$ must be zero for all values except $c$; for $q = c$, the density must be infinite. The wave function will possess the same properties. Consider the form taken by the function $\alpha_\eta(q)$ in this case. If $c$ lies in the $r$th interval, the components $\alpha_i \eta^{\frac{1}{2}}$ of the vector $\alpha$ representing the state will all be zero, with the exception of the $r$th, which will be unity. It follows that $\alpha_\eta(q)$ vanishes for all values of $q$ except those in a neighbourhood of the point $q = c$ of length $\eta$, where its value is $1/\eta^{\frac{1}{2}}$. Hence, as $\eta \to 0$,

$$\left.\begin{aligned} \alpha_\eta(q) &\to 0 \text{ for } q \neq c, \\ &\to \infty \text{ for } q = c, \end{aligned}\right\} \tag{3.2.7}$$

and $\alpha(q)$ is not a function in the ordinary sense of that term. Nonetheless if, as will invariably be the case in practice, $\eta$ is small but non-vanishing, the definition of $\alpha_\eta(q)$ will be complete; we shall accordingly accept such a function $\alpha_\eta(q)$ as the wave function for the eigenstate $q = c$, letting $\eta \to 0$ at any point in an associated calculation where this procedure leads to a definite result.

With this understanding, consider the probability density $|\alpha(q)|^2$ for the eigenstate $q = c$. This is zero outside a neighbourhood of length $\eta$ of the point $q=c$, within which it takes the value $1/\eta$. If $c=0$, this function is denoted by $\delta(q)$ and is called *Dirac's Unit Impulse Function*. In the eigenstate $q=c$, therefore,

$$|\alpha(q)|^2 = \delta(q-c) \tag{3.2.8}$$

and, since in this case $\alpha(q)$ has been defined to be real and positive, this implies that

$$\alpha(q) = \delta^{\frac{1}{2}}(q-c). \tag{3.2.9}$$

The function $\delta(q)$ has an 'integral' property which is of great importance for the manipulation of equations in which it appears. Let $f(q)$ be any function which is continuous in some neighbourhood of $q=c$. Then, if $a<c<b$,

$$\int_a^b f(q)\,\delta(q-c)\,dq = \frac{1}{\eta}\int_{c-\eta_1}^{c+\eta_2} f(q)\,dq, \tag{3.2.10}$$

where $\eta_1$, $\eta_2$ are positive quantities such that $\eta_1+\eta_2=\eta$. Since, provided $\eta$ is sufficiently small, $f(q)$ is continuous in $(c-\eta_1, c+\eta_2)$, by the mean value theorem

$$\int_{c-\eta_1}^{c+\eta_2} f(q)\,dq = \eta f(c+\eta'), \tag{3.2.11}$$

where $\eta'\to 0$ as $\eta\to 0$. Thus, in the limit

$$\int_a^b f(q)\,\delta(q-c)\,dq = f(c). \tag{3.2.12}$$

On the other hand, if $c$ lies outside the interval $[a,b]$,

$$\int_a^b f(q)\,\delta(q-c)\,dq = 0. \tag{3.2.13}$$

Equations (3.2.12), (3.2.13) constitute the 'integral' property of $\delta(q)$. In particular, taking $c=0$, $f(q)\equiv 1$, we have

$$\left.\begin{aligned} \int_a^b \delta(q)\,dq &= 1, \text{ if } a<0<b, \\ &= 0, \text{ if } a<b<0, \\ &\quad \text{or } 0<a<b. \end{aligned}\right\} \tag{3.2.14}$$

This can be expressed by saying that the 'area under the $\delta(q)$-graph' is unity. This result also follows from the fact that this area represents the probability that $q$ takes *some* value in its range, when the system is in the state $\psi^c$.

These ideas can now be generalised to cover the case when the maximal set of compatible observables being employed as a basis for a representation includes more than one observable having a continuous spectrum and a number of observables having discrete spectra. Let $p$, $q$, ... be the observables having continuous spectra and let $a$, $b$, ... be the observables having discrete spectra with eigenvalues $a_i$ $(i=1,2,\ldots,m)$, $b_j$ $(j=1,2,\ldots,n)$, ... $(m,n,\ldots$ may be infinite). Then the state of the system will be specified by a set of wave functions

$$\alpha_{ij\ldots}(p,q,\ldots). \tag{3.2.15}$$

Physically, these functions have the following significance:

$$|\alpha_{ij\ldots}(p,q,\ldots)|^2\,dp\,dq\ldots \tag{3.2.16}$$

is the probability that a measurement of the observables $p$, $q$, ... will yield values in the intervals $(p,p+dp)$, $(q,q+dq)$, ... respectively and that a simultaneous measurement of the observables $a$, $b$, ... will yield values $a_i$, $b_j$, ... respectively.

The probability of observing a transition from the state $\alpha$ into the state $\beta$ for such a system using this representation is $|\langle\alpha|\beta\rangle|^2$, where

$$\langle\alpha|\beta\rangle = \sum_{i,j,\ldots}\int \alpha^*_{ij\ldots}(p,q,\ldots)\,\beta_{ij\ldots}(p,q,\ldots)\,dp\,dq\ldots, \tag{3.2.17}$$

the ranges of summation and integration being the complete ranges for the corresponding observables.

The set of wave functions appropriate to the eigenstate $p=p_0$, $q=q_0,\ldots$, $a=a_r$, $b=b_s$, ... is determined by the equation

$$\alpha_{ij\ldots} = \delta^{\frac{1}{2}}(p-p_0)\,\delta^{\frac{1}{2}}(q-q_0)\ldots\delta_{ir}\,\delta_{js}\ldots. \tag{3.2.18}$$

It will often be convenient to arrange the members of the set $\alpha_{ij\ldots}$ as the elements of a column matrix, as was done when all the observables were assumed to possess discrete spectra. This will lead to a matrix representation in which the matrix elements are functions of the observables $p$, $q$, ....

As a final generalisation, we are led to consider the case when the observables of the maximal set possess spectra which are partly discrete and partly continuous. Thus, over a certain range or ranges of values $q$ might have a continuous spectrum, whereas over other ranges its spectrum might be discrete with eigenvalues $q_1$, $q_2$, .... Then, ignoring all other observables of the maximal set, a state of the system would be associated with a wave function $\alpha(q)$ defined over the continuous spectrum and a sequence $\alpha_1, \alpha_2, \ldots$ corresponding to the discrete eigenvalues. $|\alpha(q)|^2 dq$ would then represent the probability of measuring $q$ to take a value in the interval $(q, q+dq)$ and $|\alpha_i|^2$ would represent the probability of measuring $q = q_i$. The probability of a transition occurring between states $\alpha$ and $\beta$ would be determined, as usual, by $\langle\alpha|\beta\rangle$, where

$$\langle\alpha|\beta\rangle = \int \alpha^*(q)\,\beta(q)\,dq + \sum_i \alpha_i^*\,\beta_i, \tag{3.2.19}$$

the integration being carried out over the range of the continuous spectrum.

## 3.3. Transformation of representations

We will study first the case of a system for which a single observable $q$ constitutes a maximal set of compatible observables and will suppose that $q$ has a continuous spectrum of eigenvalues. Let $\alpha(q)$ be the wave function for the system in a certain state. Suppose $q'$ is another observable, also possessing a continuous spectrum, which forms a maximal set and let $\alpha'(q')$ be the corresponding wave function for the system in the same state as previously. We thus have two representations, a $q$-representation and a $q'$-representation, and our object is to determine the relationship existing between them.

As an example, consider the system comprising a particle which is constrained to move parallel to the $x$-axis in a given force field. Then, $q = x$ is a maximal set of compatible observables and $q' = p_x$ is another. The state of the particle at any instant can be described by a wave function $\psi(x)$ or by a wave function $\phi(p_x)$ and it will be proved later (section 3.5) that these functions are related by the equations

$$\phi(p_x) = \frac{1}{h^{\frac{1}{2}}} \int_{-\infty}^{\infty} \psi(x) \exp\left(-\iota x p_x/\hbar\right) dx, \tag{3.3.1}$$

$$\psi(x) = \frac{1}{h^{\frac{1}{2}}} \int_{-\infty}^{\infty} \phi(p_x) \exp\left(\iota x p_x/\hbar\right) dp_x. \tag{3.3.2}$$

In the case of two representations based upon observables having discrete spectra, we know that the relationship between the components of the vector associated with the state, relative to the corresponding pair of frames is given by the unitary transformation (1.6.1). The counterpart of this transformation when the representations are continuous is clearly

$$\alpha'(q') = \int U(q', q)\, \alpha(q)\, dq, \tag{3.3.3}$$

the range of integration being the spectrum of $q$. The function $U(q', q)$ which replaces the unitary matrix $\mathbf{U}$, is called the *kernel* of the transformation. The inverse transformation may be expected to take a form which is the counterpart of the inverse unitary transformation; recalling that $\mathbf{U}$ satisfies the condition (1.6.7), it is evident that this should be

$$\alpha(q) = \int U^*(q', q)\, \alpha(q')\, dq'. \tag{3.3.4}$$

In the special case leading to equation (3.3.1),

$$U(p_x, x) = \exp(-\iota x p_x/\hbar) \tag{3.3.5}$$

and the kernel of the inverse transformation should be

$$U^*(p_x, x) = \exp(\iota x p_x/\hbar). \tag{3.3.6}$$

This is in agreement with equation (3.3.2). It may be shown that, under certain conditions of a general nature, if the relationship between the wave functions $\alpha(q)$, $\alpha'(q')$ is as given by equation (3.3.3) and the inverse transformation takes the form of equation (3.3.4), then the scalar product $\langle\alpha|\beta\rangle$ is invariant with respect to a change of representation (see e.g. *Functional Analysis* by F. Riesz and B. Sz.-Nagy (trans. L. F. Baron), p. 291, Ungar Publishing Co., New York, 1955).

If the observable $q$ possesses a discrete spectrum of eigenvalues $q_1, q_2, \ldots$, whereas the spectrum of $q'$ is continuous, in the $q$-representation the state of the system will be specified by a column matrix $\{\alpha_1, \alpha_2, \ldots\}$ and equations (3.3.3), (3.3.4) must be replaced by

$$\alpha'(q') = \sum_i U_i(q')\, \alpha_i, \tag{3.3.7}$$

$$\alpha_i = \int U_i^*(q')\, \alpha'(q')\, dq'. \tag{3.3.8}$$

Suppose that the system is in the eigenstate $q = q_j$. Then $\alpha_j = 1$, $\alpha_i = 0$ $(i \neq j)$ and hence, substituting in equation (3.3.7),

$$\alpha'(q') = U_j(q'). \tag{3.3.9}$$

$U_j(q')$ can accordingly be identified with the wave function in the $q'$-representation for the system in the $q$-eigenstate, $q = q_j$. We shall refer to the sequence $U_1(q')$, $U_2(q')$, ... as a set of *q-eigenfunctions* in the $q'$-representation; equation (3.3.7) then gives an expansion of any wave function

$\alpha'(q')$ in a series of eigenfunctions of $q$, the coefficients of the expansion being calculable from equation (3.3.8). Since $|\alpha_i|^2$ is the probability of measuring $q=q_i$, the coefficients in such an eigenfunction expansion are seen to possess a straightforward physical interpretation; this also follows from equation (3.3.8), which can be written in the form

$$\alpha_i = \langle U_i|\alpha'\rangle. \tag{3.3.10}$$

If we substitute for $\alpha'(q')$ from equation (3.3.7) into equation (3.3.8), we obtain the result

$$\alpha_i = \sum_j \alpha_j \int U_i^*(q')\, U_j(q')\, dq', \tag{3.3.11}$$

assuming that the order of the summation and integration operations can be reversed. Since equation (3.3.11) is valid for arbitrary $\alpha_i$, it follows that

$$\int U_i^*(q')\, U_j(q')\, dq' = \langle U_i|U_j\rangle = \delta_{ij}. \tag{3.3.12}$$

This implies that the eigenfunctions $U_1$, $U_2$, ... are normalised and orthogonal; this result also follows from the fact that they are the wave functions of different eigenstates.

$U_r(q') = (2\pi)^{-\frac{1}{2}}\exp(\iota r q')$ is a special case which will be familiar to the reader. If $q'$ is restricted to take values in the interval $(-\pi, \pi)$, equation (3.3.7) represents the complex Fourier expansion of the function $\alpha'(q')$ and equation (3.3.8) then gives the usual formula for the coefficients. Equation (3.3.12) expresses the well-known orthogonality property which relates the Fourier harmonics.

The generalisation of equations (3.3.3), (3.3.4), to the case when the one representation is based upon a maximal set of compatible observables $p, q, \ldots$ and the other is based upon a maximal set $p', q', \ldots$ (all observables being assumed to possess continuous spectra), is straightforward. We have

$$\alpha'(p',q',\ldots) = \int U(p',q',\ldots,p,q,\ldots)\,\alpha(p,q,\ldots)\,dp\,dq\,\ldots, \tag{3.3.13}$$

$$\alpha(p,q,\ldots) = \int U^*(p',q',\ldots,p,q,\ldots)\,\alpha'(p',q',\ldots)\,dp'\,dq'\,\ldots. \tag{3.3.14}$$

The further generalisation to the case where some observables in the maximal sets possess discrete spectra, some continuous spectra and some spectra which are partly discrete and partly continuous, also presents no difficulty, but the result will not be written out at length.

## 3.4. Representation of observables

Employing the $q$-representation, where $q$ possesses a continuous spectrum, if $a$ is any observable of the system and $\alpha(q)$ is the wave function defining the state of the system, we shall assume that the appropriate replacement for equation (1.8.4) is

$$\bar{a} = \int \alpha^*(q)\, a(q,q')\, \alpha(q')\, dq\, dq', \tag{3.4.1}$$

where $a(q,q')$ is the function representing $a$ in the $q$-representation. $a(q,q')$ will possess the Hermitian property (1.8.7), i.e.

$$a^*(q,q') = a(q',q). \tag{3.4.2}$$

Writing

$$\int a(q,q')\, \alpha(q')\, dq' = \beta(q), \tag{3.4.3}$$

we can regard $\beta(q)$ as having been obtained from $\alpha(q)$ by acting upon it with an operator $\hat{a}$; thus $\beta(q) = \hat{a}\alpha(q)$. Equation (3.4.1) can then be written in the form

$$\begin{aligned} \bar{a} &= \int \alpha^*(q)\, \hat{a}\alpha(q)\, dq, \\ &= \langle \alpha | \hat{a}\alpha \rangle, \end{aligned} \tag{3.4.4}$$

which is identical with the equation (1.8.11). $\hat{a}$ will be termed the operator representing $a$ in the $q$-representation. Its Hermitian character is expressed by the equation

$$\langle \alpha | \hat{a}\beta \rangle = \int \alpha^*(q)\, \hat{a}\beta(q)\, dq = \int \beta(q)\, [\hat{a}\alpha(q)]^*\, dq = \langle \hat{a}\alpha | \beta \rangle, \tag{3.4.5}$$

where $\alpha(q)$, $\beta(q)$ are arbitrary wave functions.

If the representation is based upon a number of observables having continuous spectra, viz. $p, q, \ldots$, then equation (3.4.1) must be amended to read

$$\bar{a} = \int \alpha^*(p,q,\ldots)\, a(p,q,\ldots,p',q',\ldots)\, \alpha(p',q',\ldots)\, dp\, dq \ldots dp'\, dq' \ldots. \tag{3.4.6}$$

Introducing an operator $\hat{a}$ as before, this equation can also be written

$$\bar{a} = \int \alpha^*(p,q,\ldots)\, \hat{a}\alpha(p,q,\ldots)\, dp\, dq \ldots. \tag{3.4.7}$$

In the special case when $a$ is a function of the observables defining the representation, e.g. $a = f(p,q,\ldots)$, the physical significance of $\alpha\alpha^*$ as a probability density implies that

$$\bar{a} = \int \alpha\alpha^* f\, dp\, dq \ldots. \tag{3.4.8}$$

Comparing this with equation (3.4.7), it is evident that

$$\hat{a} = f(p, q, \ldots), \tag{3.4.9}$$

where $\hat{a}$ is now to be interpreted as a simple multiplier.

If the representation is based upon a set of observables having both discrete and continuous spectra, both summations and integrations will be present in the formula for $\bar{a}$. The generalisation to this case is sufficiently transparent to require no explicit statement.

Consider a system for which there are two representations, a $q$-representation which is discrete and a $q'$-representation which is continuous. Equations (3.3.7), (3.3.8) relate the column matrix $\{\alpha_i\}$ specifying the state in the discrete representation and the wave function $\alpha'(q')$ determining the state in the continuous representation. We shall put $U_i(q') = \psi^i(q')$, the wave functions $\psi^i$ corresponding to the $q$-eigenstates in the $q'$-representation and satisfying the orthonormality conditions (3.3.12). Then, if $a$ is any observable of the system, by equation (3.4.7)

$$\begin{aligned} \bar{a} &= \int \alpha'^*(q')\,\hat{a}\alpha'(q')\,dq', \\ &= \int \{\alpha_i^* \psi^{i*}(q')\}\{\alpha_j \hat{a}\psi^j(q')\}\,dq', \\ &= \alpha_i^* \alpha_j \int \psi^{i*}(q')\,\hat{a}\psi^j(q')\,dq', \\ &= \alpha_i^* a_{ij} \alpha_j, \end{aligned} \tag{3.4.10}$$

where

$$\begin{aligned} a_{ij} &= \int \psi^{i*}(q')\,\hat{a}\psi^j(q')\,dq' \\ &= \langle \psi^i | \hat{a}\psi^j \rangle. \end{aligned} \tag{3.4.11}$$

Equation (3.4.10) indicates that $a_{ij}$ is the $ij$th element of the matrix representing $a$ in the $q$-representation. We have accordingly succeeded in generalising equation (1.8.22) to the case of a continuous representation.

It will be assumed, as was the case for representations based upon observables possessing discrete spectra, that the representation of the sum of two observables is the sum of their representations. Thus $a+b$ is represented by the function $a(q,q')+b(q,q')$ or by the operator $\hat{a}+\hat{b}$. Further, the function representing the product $ab$ will be taken to be the 'symmetrised matrix product'

$$\tfrac{1}{2}\int [a(q,q'')\,b(q'',q') + b(q,q'')\,a(q'',q')]\,dq'' \tag{3.4.12}$$

and the operator will be taken to be

$$\tfrac{1}{2}(\hat{a}\hat{b}+\hat{b}\hat{a}). \tag{3.4.13}$$

The representation of any polynomial function of a set of observables can now be written down.

Finally, the theory of section 1.10 will be assumed extensible to the general case. Thus, if $a$ is represented by the function $a(q,q')$, the characteristic equation for this observable is

$$\int a(q,q')\,\alpha(q')\,dq' = \lambda\alpha(q). \tag{3.4.14}$$

A wave function $\alpha(q)$ satisfying this equation for a particular value of $\lambda$ will be the wave function appropriate to the eigenstate $a=\lambda$. This equation can also be written in the form

$$\hat{a}\alpha(q) = \lambda\alpha(q). \tag{3.4.15}$$

## 3.5. Momentum and coordinate representations

In this section, we shall study the representations of the state of a system comprising a single particle based upon (i) the rectangular cartesian coordinates $(x_1, x_2, x_3)$ of the particle with respect to an inertial frame $S$ and (ii) the components $(p_1, p_2, p_3)$ of the linear momentum of the particle relative to the same frame. It will first be assumed that the particle possesses no spin. Then from what has been said in section 1.2, it follows that the coordinates form a maximal set of compatible observables and the components of momentum form another set; we will calculate the relationship which exists between representations of the same state of the particle based upon these two sets.

Let $\psi(x_i)=\psi(x_1,x_2,x_3)$ be the wave function describing a particular state $\alpha$ of the particle when the $x$-representation is being employed, and let $\phi(p_i)$ be the wave function describing the same state in the $p$-representation. Then, by equations (3.3.3), (3.3.4), we may assume that

$$\phi(p_i) = \int U(p_i, x_i)\,\psi(x_i)\,dx, \tag{3.5.1}$$

$$\psi(x_i) = \int U^*(p_i, x_i)\,\phi(p_i)\,dp, \tag{3.5.2}$$

where $dx=dx_1\,dx_2\,dx_3$, $dp=dp_1\,dp_2\,dp_3$ and the integrations are all over the range $(-\infty, +\infty)$.

In conformity with the special principle of relativity, we shall assume that the form of this relationship is independent of the inertial frame being employed, i.e. the form of the kernel $U$ does not alter if we transform from one inertial frame to another. Consider, therefore, a second inertial frame $S'$ relative to which the particle's coordinates are $x_i'$ and its components of momentum are $p_i'$. Then, we have the relationships

$$x_i' = a_{ij}x_j + b_i, \tag{3.5.3}$$

$$p_i' = a_{ij}p_j + c_i, \tag{3.5.4}$$

where $a_{ij}$ are the coefficients of an orthogonal transformation determined by the inclinations of the axes of $S'$ to the axes of $S$ (summation with respect to repeated indices over the range $i=1, 2, 3$, is understood). In the $x'$-representation, let the state $\alpha$ be determined by a wave function $\psi'(x_i')$ and, in the $p'$-representation, by a wave function $\phi'(p_i')$. Then, the probability of finding the particle in the neighbourhood of a certain point must be the same whether calculated in the $S$-frame or the $S'$-frame and hence

$$|\psi| = |\psi'|. \tag{3.5.5}$$

Similarly,

$$|\phi| = |\phi'|. \tag{3.5.6}$$

Hence we can write

$$\psi' = \psi \exp(\iota\alpha), \quad \phi' = \phi \exp(\iota\beta), \tag{3.5.7}$$

where

$$\alpha = \alpha(x_i, a_{jk}, b_l, c_n), \qquad \beta = \beta(p_i, a_{jk}, b_l, c_n). \tag{3.5.8}$$

According to our initial hypothesis,

$$\phi'(p_i') = \int U(p_i', x_i')\,\psi'(x_i')\,dx'. \tag{3.5.9}$$

Substituting from the transformation equations (3.5.3), (3.5.4) and from (3.5.7), we get

$$\phi \exp(\iota\beta) = \int U(a_{ij}p_j + c_i, a_{ij}x_j + b_i)\,\psi \exp(\iota\alpha)\,dx, \tag{3.5.10}$$

since

$$\frac{\partial(x_1', x_2', x_3')}{\partial(x_1, x_2, x_3)} = |a_{ij}| = \pm 1. \tag{3.5.11}$$

In the case of the negative sign (indirect transformation), we replace $\beta$ by $\beta + \pi$. It now follows from equation (3.5.1) that

$$\int U(p_i, x_i)\,\psi(x_i) \exp(\iota\beta)\,dx = \int U(a_{ij}p_j + c_i, a_{ij}x_j + b_i)\,\psi(x_i) \exp(\iota\alpha)\,dx, \tag{3.5.12}$$

since $\beta$ is not dependent on the $x_i$. This last equation is to be valid for arbitrary wave functions $\psi(x_i)$ which result in convergent integrals. For this to be so, it is necessary that

$$U(p_i, x_i)\exp(\iota\beta) = U(a_{ij}p_j + c_i, a_{ij}x_j + b_i)\exp(\iota\alpha) \qquad (3.5.13)$$

identically in the variables $x_i$, $p_i$ and parameters $a_{ij}$, $c_i$, $b_i$. The form of $U$ can now be deduced as follows:

First, we note that

$$|U(p_i, x_i)| = |U(a_{ij}p_j + c_i, a_{ij}x_j + b_i)|. \qquad (3.5.14)$$

Since $b_i$, $c_i$ are arbitrary, this identity implies that

$$|U(p_i, x_i)| = \text{constant} = U_0. \qquad (3.5.15)$$

Taking arguments of both sides of the identity (3.5.13), we find

$$\theta(p_i, x_i) + \beta = \theta(a_{ij}p_j + c_i, a_{ij}x_j + b_i) + \alpha, \qquad (3.5.16)$$

where

$$\theta = \arg U. \qquad (3.5.17)$$

Differentiating equation (3.5.16) partially with respect to $p_i$ and $x_j$, since $\alpha$ is independent of $p_i$ and $\beta$ is independent of $x_j$, it follows that

$$\frac{\partial^2\theta}{\partial p_i\,\partial x_j} = \frac{\partial^2\theta}{\partial p_r'\,\partial x_s'} a_{ri}\, a_{sj}. \qquad (3.5.18)$$

This shows that the quantities $\partial^2\theta/\partial p_i\,\partial x_j$ transform, between two sets of rectangular axes, like the elements of a cartesian tensor of the second rank. The inverse relationship is

$$\frac{\partial^2\theta}{\partial p_i'\,\partial x_j'} = a_{ir}\, a_{js} \frac{\partial^2\theta}{\partial p_r\,\partial x_s}. \qquad (3.5.19)$$

Putting $x_i = 0$, $p_i = 0$ $(i = 1, 2, 3)$ in this identity, we obtain

$$\frac{\partial^2\theta}{\partial c_i\,\partial b_j} = a_{ir}\, a_{js} \left(\frac{\partial^2\theta}{\partial p_r\,\partial x_s}\right)_0, \qquad (3.5.20)$$

where the subscript zero indicates that the arguments $p_r$, $x_s$ are put equal to zero after differentiation. But the right-hand member is independent of $c_i$, $b_j$ and it follows, therefore, that the left-hand member is also. Thus

$$\frac{\partial^2\theta}{\partial p_i\,\partial x_j} = \chi_{ij}, \qquad (3.5.21)$$

where $\chi_{ij}$ are constants. Equation (3.5.19) can now be written

$$\chi_{ij} = a_{ir}\, a_{js}\, \chi_{rs}, \qquad (3.5.22)$$

implying that $\chi_{ij}$ is a second rank tensor whose elements are the same in every frame. It is shown in Appendix A that any such tensor must be an invariant multiple of the fundamental tensor, i.e.

$$\chi_{ij} = \gamma\delta_{ij}. \tag{3.5.23}$$

We have proved, therefore, that

$$\frac{\partial^2 \theta}{\partial p_i \partial x_j} = \gamma\delta_{ij}. \tag{3.5.24}$$

Integrating, we find that

$$\theta = \gamma p_i x_i + P + X, \tag{3.5.25}$$

where $P$ is a function of the $p_i$ alone and $X$ is a function of the $x_i$ alone. Hence

$$U = U_0 \exp[\iota(\gamma p_i x_i + P + X)]. \tag{3.5.26}$$

The transformation equations (3.5.1), (3.5.2) can now be written as

$$\exp(-\iota P)\phi(p_i) = \int U_0 \exp(\iota\gamma p_i x_i) \exp(\iota X) \psi(x_i)\, dx, \tag{3.5.27}$$

$$\exp(\iota X)\psi(x_i) = \int U_0 \exp(-\iota\gamma p_i x_i) \exp(-\iota P)\phi(p_i)\, dp. \tag{3.5.28}$$

But the only connection between the physical world and the mathematical symbolism we have introduced is that $|\psi|^2$, $|\phi|^2$ are interpretable as probability densities. This connection is unaltered if we now absorb a factor $\exp(\iota X)$ in $\psi$ and a factor $\exp(-\iota P)$ in $\phi$, to yield new wave functions, which will continue to be denoted by $\psi$, $\phi$ respectively. $\gamma$ is a fundamental constant of the physical world and its value can only be decided by comparing the physical implications of equations (3.5.27), (3.5.28), with actual observations. It is found that agreement between theory and experiment can only be obtained if we take $\gamma = -1/\hbar$ (the negative sign is purely conventional), where $\hbar = h/2\pi$ and $h$ is Planck's constant. This we shall take to be the fundamental definition of Planck's constant, equations such as (1.2.2), (1.2.4) being regarded as derived results which will be obtained later. Finally, $U_0$ is determined by the requirement that the equations (3.5.27), (3.5.28), should be consistent; these equations are almost exactly those relating a function and its Fourier transform and it follows from the theory of this relationship that $U_0 = h^{-3/2}$. Thus

$$\phi(p_i) = h^{-3/2} \int \exp(-\iota p_i x_i/\hbar)\, \psi(x_i)\, dx, \tag{3.5.29}$$

$$\psi(x_i) = h^{-3/2} \int \exp(\iota p_i x_i/\hbar)\, \phi(p_i)\, dp. \tag{3.5.30}$$

If, now, we suppose the particle has spin and that $s_z$ possesses two eigenvalues $\pm\frac{1}{2}\hbar$, a maximal set of compatible observables is $(x_1, x_2, x_3, s_z)$. Employing this set as a basis, a pure state of the particle can be specified by two wave functions $\psi_+(x_i)$, $\psi_-(x_i)$ which will be arranged as a column matrix thus:

$$\begin{pmatrix} \psi_+(x_i) \\ \psi_-(x_i) \end{pmatrix}, \tag{3.5.31}$$

i.e. a spinor which is now a function of the coordinates. $|\psi_+|^2 dx_1 dx_2 dx_3$ measures the probability of observing $s_z$ to take the value $+\frac{1}{2}\hbar$ and, at the same time, $x_i$ to lie in the interval $(x_i, x_i + dx_i)$ $(i = 1, 2, 3)$. $\psi_-$ possesses a similar physical interpretation. It is now evident that the net probability that $s_z = +\frac{1}{2}\hbar$ is given by the integral

$$\int |\psi_+|^2\, dx \tag{3.5.32}$$

calculated over the whole of the coordinate space.

$(p_1, p_2, p_3, s_z)$ also constitutes a maximal set of compatible observables, leading to a representation in which a state of the particle is determined by a spinor in the form

$$\begin{pmatrix} \phi_+(p_i) \\ \phi_-(p_i) \end{pmatrix}. \tag{3.5.33}$$

Since the basis for the representation of the spin is the same in these two representations, $\psi_+$, $\phi_+$ will be related by the transformation equations (3.5.29), (3.5.30); $\psi_-$, $\phi_-$ will also be related by these equations.

## 3.6. Momentum and coordinate representation of observables

In this section, we shall be considering a system comprising a single particle without spin. Let $f(x_i)$ be any function of the particle's coordinates $x_1$, $x_2$, $x_3$. Then, working in the coordinate representation, by equation (3.4.9) the operator representing $f$ is given by

$$\hat{f} = f. \tag{3.6.1}$$

In particular,

$$\hat{x}_i = x_i. \tag{3.6.2}$$

Consider, next, the observable $p_1$. If $\phi(p_i)$ is the wave function in the momentum representation, we have

$$\bar{p}_1 = \int \phi\phi^* p_1\, dp. \tag{3.6.3}$$

Substituting for $\phi$ from equation (3.5.29), we now obtain

$$\bar{p}_1 = h^{-3/2} \int \phi^* p_1 \exp(-\iota p_i x_i/\hbar) \psi \, dx \, dp, \tag{3.6.4}$$

there being six integrations over the range $(-\infty, \infty)$. Consider the integration with respect to $x_1$; we have

$$\int_{-\infty}^{\infty} p_1 \exp(-\iota p_i x_i/\hbar) \psi \, dx_1$$

$$= -\int_{-\infty}^{\infty} \frac{\hbar}{\iota} \frac{\partial}{\partial x_1} [\exp(-\iota p_i x_i/\hbar)] \psi \, dx_1$$

$$= \left| -\frac{\hbar}{\iota} \exp(-\iota p_i x_i/\hbar) \psi \right|_{-\infty}^{\infty} + \int_{-\infty}^{\infty} \exp(-\iota p_i x_i/\hbar) \left( \frac{\hbar}{\iota} \frac{\partial \psi}{\partial x_1} \right) dx_1$$

$$= \int_{-\infty}^{\infty} \exp(-\iota p_i x_i/\hbar) \left( \frac{\hbar}{\iota} \frac{\partial \psi}{\partial x_1} \right) dx_1, \tag{3.6.5}$$

assuming that $\psi \to 0$ as $x_1 \to \pm\infty$ (this is necessarily the case if

$$\int |\psi|^2 \, dx = 1, \tag{3.6.6}$$

i.e. if the probability of finding the particle somewhere is to be unity). It now follows that

$$\begin{aligned} \bar{p}_1 &= h^{-3/2} \int \phi^* \exp(-\iota p_i x_i/\hbar) \left( \frac{\hbar}{\iota} \frac{\partial \psi}{\partial x_1} \right) dx \, dp, \\ &= \int \psi^* \left( \frac{\hbar}{\iota} \frac{\partial \psi}{\partial x_1} \right) dx, \end{aligned} \tag{3.6.7}$$

having made use of the conjugate of equation (3.5.30).

Equation (3.6.7) can be written in the form

$$\bar{p}_1 = \int \psi^* \hat{p}_1 \psi \, dx, \tag{3.6.8}$$

where

$$\hat{p}_1 = \frac{\hbar}{\iota} \frac{\partial}{\partial x_1}. \tag{3.6.9}$$

Comparing equations (3.4.7), (3.6.8), we deduce that $\hat{p}_1$ is the operator representing $p_1$ in the $x$-representation. The operators representing $p_2$, $p_3$ can now be written down immediately and then the operators associated with any polynomial function of the $x_i, p_i$ follow by application of our earlier rules. For example, the operator representing the $x_1$-component of the angular momentum of the particle about the origin of coordinates is

$$\hat{x}_2\hat{p}_3 - \hat{x}_3\hat{p}_2 = \frac{\hbar}{\iota}\left(x_2\frac{\partial}{\partial x_3} - x_3\frac{\partial}{\partial x_2}\right). \tag{3.6.10}$$

(N.B. Since $\hat{x}_2$, $\hat{p}_3$ commute, their product need not be replaced by a symmetrised product.)

Similarly, employing the $p$-representation, it can be shown that

$$\hat{x}_i = -\frac{\hbar}{\iota}\frac{\partial}{\partial p_i}, \quad \hat{p}_i = p_i. \tag{3.6.11}$$

It was remarked above that the operators $\hat{x}_2$, $\hat{p}_3$ commute. However, the operators $\hat{x}_1$, $\hat{p}_1$ do not commute and, working with the coordinate representation, their commutator is calculated thus: If $\psi(x_i)$ is an arbitrary wave function,

$$\begin{aligned}[\hat{p}_1, \hat{x}_1]\psi &= \frac{\hbar}{\iota}\left[\frac{\partial}{\partial x_1}(x_1\psi) - x_1\frac{\partial\psi}{\partial x_1}\right], \\ &= \frac{\hbar}{\iota}\psi. \end{aligned} \tag{3.6.12}$$

$\psi$ being arbitrary, this implies that

$$[\hat{p}_1, \hat{x}_1] = \frac{\hbar}{\iota}. \tag{3.6.13}$$

This result can also be obtained by using the $p$-representation; the calculation is left as an exercise for the reader. The commutators $[\hat{p}_2, \hat{x}_2]$, $[\hat{p}_3, \hat{x}_3]$ are each calculated similarly to be equal to $\hbar/\iota$. To summarise, we can write

$$[\hat{p}_i, \hat{x}_j] = \frac{\hbar}{\iota}\delta_{ij}. \tag{3.6.14}$$

If $p_i$, $x_j$ can be represented by matrices $\mathbf{p}_i$, $\mathbf{x}_j$ then, in this representation, the constant $\hbar/\iota$ will be represented by the matrix $\hbar\mathbf{I}/\iota$ and equation (3.6.14) will be equivalent to the matrix equation

$$[\mathbf{p}_i, \mathbf{x}_j] = \frac{\hbar}{\iota}\mathbf{I}\delta_{ij}. \tag{3.6.15}$$

As an application of this last result, we may calculate the form taken by the uncertainty principle inequality (1.11.18) for the two observables $a = p_1$, $b = x_1$. We find that $c = \hbar$ and hence

$$\sigma_{p_1} \sigma_{x_1} \geqslant \tfrac{1}{2}\hbar. \tag{3.6.16}$$

This is the precise statement of the approximate result (1.2.4).

## 3.7. Eigenstates for momentum and position of a particle

Consider a system comprising a single particle devoid of spin. If the components of the momentum of the particle are measured at some instant to be $(p_1', p_2', p_3')$ precisely, then the particle is put into a momentum eigenstate for which the wave function in the momentum representation is given by

$$\phi(p_1, p_2, p_3) = \delta^{\frac{1}{2}}(p_1 - p_1')\,\delta^{\frac{1}{2}}(p_2 - p_2')\,\delta^{\frac{1}{2}}(p_3 - p_3'). \tag{3.7.1}$$

Suppose, however, we wish to calculate the wave function in the coordinate representation. If this is denoted by $\psi(x_1, x_2, x_3)$, then $\psi$ must satisfy the characteristic equations

$$\hat{p}_1 \psi = p_1' \psi, \text{ etc.,} \tag{3.7.2}$$

where $\hat{p}_1$ is given by equation (3.6.9). Omitting the primes, it follows that $\psi$ satisfies three partial differential equations, viz.

$$\frac{\partial \psi}{\partial x_1} = \frac{\iota p_1}{\hbar} \psi, \text{ etc.} \tag{3.7.3}$$

or

$$\frac{\partial}{\partial x_1} (\log \psi) = \frac{\iota p_1}{\hbar}, \text{ etc.} \tag{3.7.4}$$

Hence

$$\psi = A \exp(\iota p_i x_i / \hbar), \tag{3.7.5}$$

where $A$ is an unknown constant. This is the eigenfunction in the $x$-representation corresponding to the eigenstate $(p_1, p_2, p_3)$. The probability density for the position of the particle in this state is clearly

$$|\psi|^2 = |A|^2, \tag{3.7.6}$$

i.e. is constant. This is expected, for if the momentum is sharp, i.e. $\sigma_{p_1} = \sigma_{p_2} = \sigma_{p_3} = 0$, by the uncertainty principle (3.6.16), $\sigma_{x_1}, \sigma_{x_2}, \sigma_{x_3}$ must all be infinite and hence we can have no knowledge whatsoever regarding the particle's position; this implies that all sets of coordinates are equally probable and that the probability distribution over the $x$-space is uniform.

It should be noted that it is possible to find an eigenfunction corresponding to any set of $p$-eigenvalues $(p_1, p_2, p_3)$. Thus, the spectrum of $p$-eigenvalues is continuous; we say that the $p_i$ are not *quantised*.

It is evident that we have been studying a somewhat ideal case, which will never be encountered in reality. In practice, the particle will be known to be confined within some neighbourhood of our observing apparatus and hence its position cannot be completely unknown. This implies that its momentum will never be absolutely precise. The fact that we are dealing with an unreal situation creates difficulties when we attempt to normalise $\psi$; this requires that

$$\int |\psi|^2 \, dx = 1 \tag{3.7.7}$$

and since this volume integral is to be taken over the whole of $x$-space and the integrand is constant, it is clear that the integral diverges. $\psi$ cannot, therefore, be normalised and $A$ must remain arbitrary. Nevertheless, $A$ can be given a physical interpretation in certain situations, as will be demonstrated in section 4.7.

The problem of a particle whose position is sharp may be analysed similarly. In the $x$-representation, its wave function is

$$\psi(x_1, x_2, x_3) = \delta^{\frac{1}{2}}(x_1 - x_1')\, \delta^{\frac{1}{2}}(x_2 - x_2')\, \delta^{\frac{1}{2}}(x_3 - x_3') \tag{3.7.8}$$

and in the $p$-presentation, its wave function will be found to be given by

$$\phi(p_1, p_2, p_3) = B \exp(-\iota p_i x_i/\hbar). \tag{3.7.9}$$

Again, the constant $B$ cannot be determined by a normalisation requirement. The spectrum of $x$-eigenvalues is also continuous.

Equations (3.5.29), (3.5.30) can now be given the following interpretation: The first of these equations represents the expansion of any wave function in terms of $x$-eigenfunctions employing the $p$-representation and the second gives the expansion of any wave function in terms of $p$-eigenfunctions using the $x$-representation.

## 3.8. Energy eigenstates for a particle

Consider a particle of mass $m$ having no spin, free to move under the action of a conservative force. Let $V(x_1, x_2, x_3)$ be the potential energy (P.E.) of the particle with respect to the force field when its coordinates are $(x_1, x_2, x_3)$. Then, if $(p_1, p_2, p_3)$ are the components of momentum of the particle, its total energy $H$ is given by the equation

$$H = \frac{1}{2m}(p_1^2 + p_2^2 + p_3^2) + V. \tag{3.8.1}$$

This expression for the energy of the particle in terms of its coordinates and momenta is referred to as the *Hamiltonian* in classical analytical mechanics. Employing the $x$-representation, the operator representing the observable $H$ is given by

$$\hat{H} = -\frac{\hbar^2}{2m}\nabla^2 + V, \tag{3.8.2}$$

where, as usual,

$$\nabla^2 = \frac{\partial^2}{\partial x_i \partial x_i} \tag{3.8.3}$$

is the *Laplacian operator*. We can now write down the characteristic equation for the energy in this representation; it takes the form

$$\left(-\frac{\hbar^2}{2m}\nabla^2 + V\right)\psi = E\psi, \tag{3.8.4}$$

where $E$ is the eigenvalue for $H$ corresponding to the eigenstate $\psi$. Rearranging this last equation into the form

$$\nabla^2\psi + \frac{2m}{\hbar^2}(E - V)\psi = 0, \tag{3.8.5}$$

we obtain a partial differential equation known as *Schrödinger's Wave Equation*.

As an illustration of the theory, we will solve Schrödinger's equation in two simple cases in the following sections.

If the particle under consideration possesses spin and hence a magnetic moment, then it is possible that its energy $H$ will be dependent upon the spin observables. In these circumstances, $\hat{H}$ will take the form of an $n \times n$ matrix, where $n$ is the number of distinct spin eigenstates, and each element of the matrix will be an operator. Thus, if the particle has spin $\frac{1}{2}\hbar$ and is moving in a uniform magnetic field of strength $B$ directed along the $x_3$-axis, its potential energy in this field is $-B\mu_3$, where $\mu_3$ is the $x_3$-component of its magnetic moment. Now $\mu_3 = \mu_0\sigma_3$, where $\mu_0$ is the Bohr magneton and $\sigma_3$ is the $x_3$-component of the spin axis. It follows that, to allow for the interaction between the spin of the particle and the magnetic field, the expression (3.8.1) for the energy of the particle must be replaced by

$$H = \frac{1}{2m}(p_1^2 + p_2^2 + p_3^2) + V - A\sigma_3. \tag{3.8.6}$$

The first four terms in this expression are independent of the spin and so behave like constants with respect to this observable. But, in any spin representation, a constant $c$ is represented by the matrix $c\mathbf{I}_2$. Hence, employing the $x_1\,x_2\,x_3\,\sigma_3$-representation, the operator $\hat{H}$ is given by

$$\hat{H} = \left(-\frac{\hbar^2}{2m}\nabla^2 + V\right)\mathbf{I}_2 - A\boldsymbol{\sigma}_3, \tag{3.8.7}$$

where $\sigma_3$ is identical with the Pauli matrix $\sigma_z$ in equation (2.2.4). The characteristic equation for the particle's energy is accordingly

$$\begin{pmatrix} -\frac{\hbar^2}{2m}\nabla^2 + V - A & \mathbf{0} \\ \mathbf{0} & -\frac{\hbar^2}{2m}\nabla^2 + V + A \end{pmatrix}\begin{pmatrix}\psi_+ \\ \psi_-\end{pmatrix} = E\begin{pmatrix}\psi_+ \\ \psi_-\end{pmatrix}. \tag{3.8.8}$$

This is equivalent to two Schrödinger equations, viz.

$$\nabla^2\psi_+ + \frac{2m}{\hbar^2}(E + A - V)\psi_+ = 0, \tag{3.8.9}$$

$$\nabla^2\psi_- + \frac{2m}{\hbar^2}(E - A - V)\psi_- = 0. \tag{3.8.10}$$

If $E_1, E_2, \ldots$ are eigenvalues for $E$ in equation (3.8.5) and $\psi_1, \psi_2, \ldots$ are the respective eigenfunctions, then $E_1 - A, E_2 - A, \ldots$ will be eigenvalues for $E$ in equation (3.8.9) with the same set of eigenfunctions. Thus, $E = E_i - A$, $\psi_+ = \psi_i$, is a possible solution of equation (3.8.9). However, the eigenvalues for $E$ in equation (3.8.10) will be $E_1 + A, E_2 + A, \ldots$ and, in general $E_i - A$ will not be included in the sequence; if this is the case, it will be necessary to take $\psi_- = 0$. The solution $E = E_i - A$, $\psi_+ = \psi_i$, $\psi_- = 0$, would then correspond to a particle in the spin eigenstate $\sigma_3 = +1$, whose magnetic P.E. was therefore sharp with value $-A$ and whose remaining energy was sharp with value $E_i$. Other eigenstates exist for which $\psi_+ = 0$, $\sigma_3 = -1$, the magnetic P.E. is $+A$ and the total energy is $E_i + A$. We note that, in this case, the value of the energy completely determines the particle state and hence that this observable alone constitutes a maximal compatible set.

In the absence of a magnetic field, $A = 0$ and the two spinor components $\psi_+, \psi_-$ are seen to satisfy the original Schrödinger equation (3.8.5). The energy must then be associated with a spin observable such as $\sigma_3$ to yield a maximal set.

## 3.9. Particle in a potential well

Throughout this and the succeeding section, we shall employ the coordinate representation and will be considering a particle moving under the action of a conservative force which is always directed parallel to the $x_1$-axis. We shall disregard spin. Then $V$ depends upon $x_1$ alone, since $\partial V/\partial x_2 = \partial V/\partial x_3 = 0$. It is easy to verify that the operators $\hat{p}_2$, $\hat{p}_3$, $\hat{H}$ commute in pairs and hence that the observables $p_2$, $p_3$ and $H$ are compatible. We will investigate states of the particle which are eigenstates with respect to this triad of observables and, in particular, states in which $p_2 = p_3 = 0$, i.e. the particle is moving parallel to the $x_1$-axis. Then $\psi$ must satisfy the characteristic equations

$$\frac{\hbar}{\iota}\frac{\partial \psi}{\partial x_2} = \frac{\hbar}{\iota}\frac{\partial \psi}{\partial x_3} = 0, \tag{3.9.1}$$

i.e. $\psi$ will depend upon $x_1$ alone.

If $H = E$ is the energy eigenvalue, Schrödinger's equation reduces to an ordinary differential equation, viz.

$$\frac{d^2\psi}{dx^2} + \frac{2m}{\hbar^2}(E - V)\psi = 0, \tag{3.9.2}$$

where we have replaced $x_1$ by $x$.

Suppose, now, that the P.E. distribution is in the form of a 'well' determined by the equations

$$\left.\begin{aligned} V &= V_0, \quad x < 0, x > a, \\ &= 0, \quad 0 < x < a, \end{aligned}\right\} \tag{3.9.3}$$

where $V_0 > 0$ is constant. According to classical ideas, if $0 < E < V_0$, the particle will be confined to the region of the well, i.e. $0 < x < a$, since only in this region will its kinetic energy (K.E.) be positive; when the particle arrives at a 'wall' of the well ($x = 0$ or $a$), it will receive an impulse which will cause it to return along its previous path with unaltered speed ($\partial V/\partial x$ is infinite at a wall). Such a state will be termed a *bound state*. However, if $E > V_0$, the particle will move in one direction only from $x = -\infty$ to $x = +\infty$ (or in the opposite sense), its velocity being increased during its passage across the well by the amount necessary to conserve energy. If it is initially in the well, it will escape. A state of this type will be termed a *free state*.

We will only study bound energy eigenstates of the particle for which the eigenvalue $E$ satisfies $0 < E < V_0$. In such a state, $\psi$ has to satisfy the equations

$$\frac{d^2\psi}{dx^2} + \frac{2m}{\hbar^2}(E - V_0)\psi = 0, \quad x < 0, x > a, \tag{3.9.4}$$

$$\frac{d^2\psi}{dx^2} + \frac{2m}{\hbar^2}E\psi = 0, \quad 0 < x < a, \tag{3.9.5}$$

over the ranges indicated. Writing

$$\frac{2m}{\hbar^2}(E-V_0) = -\alpha^2, \quad \frac{2m}{\hbar^2}E = \beta^2, \tag{3.9.6}$$

where $\alpha$, $\beta$ are positive, we easily solve for $\psi$ thus,

$$\psi = A\exp(\alpha x)+B\exp(-\alpha x), \quad x < 0, \tag{3.9.7}$$

$$= C\exp(\iota\beta x)+D\exp(-\iota\beta x), \quad 0 < x < a, \tag{3.9.8}$$

$$= F\exp(\alpha x)+G\exp(-\alpha x), \quad x > a. \tag{3.9.9}$$

But, since $|\psi|^2$ can be interpreted as a probability density, $\psi$ cannot tend to infinity as $x \to \pm\infty$. It follows that we must take $B = F = 0$.

Evidently, the potential discontinuities at $x=0$, $a$ are theoretical idealisations of what must, in practice, be narrow regions over which $V$ changes rapidly from the value 0 to the value $V_0$. In such a region, $\psi$ must continue to satisfy Schrödinger's equation (3.9.2) and it follows that $\psi''$ $(=d^2\psi/dx^2)$ must also change rapidly from one finite value to another as the region is traversed. The change in $\psi'$ from $x=-0$ to $x=+0$ is given by

$$\psi'(+0)-\psi'(-0) = \int_{-0}^{+0} \psi''\,dx = 0, \tag{3.9.10}$$

since $\psi''$ is finite and the length of the interval of integration approaches zero. Thus $\psi'$ is continuous at $x=0$. *A fortiori*, $\psi$ is continuous at this point. Similar considerations show that $\psi$ and $\psi'$ are continuous at $x=a$. Hence, from equations (3.9.7)–(3.9.9), we deduce that

$$C+D = A, \tag{3.9.11}$$

$$C\exp(\iota\beta a)+D\exp(-\iota\beta a) = G\exp(-\alpha a), \tag{3.9.12}$$

$$\iota\beta(C-D) = \alpha A, \tag{3.9.13}$$

$$\iota\beta[C\exp(\iota\beta a)-D\exp(-\iota\beta a)] = -\alpha G\exp(-\alpha a). \tag{3.9.14}$$

Eliminating $A$ and $G$ from the last four equations, we obtain

$$C\exp(-\iota\theta)+D\exp(\iota\theta) = 0, \tag{3.9.15}$$

$$C\exp[\iota(\beta a+\theta)]+D\exp[-\iota(\beta a+\theta)] = 0, \tag{3.9.16}$$

where

$$1+\frac{\iota\beta}{\alpha} = r\exp(\iota\theta). \tag{3.9.17}$$

If $C=D=0$, then $\psi$ is zero identically and there is no particle. But, equations (3.9.15), (3.9.16) possess a non-zero solution in $C$, $D$ if

$$\begin{vmatrix} \exp(-\iota\theta) & \exp(\iota\theta) \\ \exp[\iota(\beta a+\theta)] & \exp[-\iota(\beta a+\theta)] \end{vmatrix} = 0, \tag{3.9.18}$$

i.e. if

$$\exp[2\iota(\beta a+2\theta)] = 1. \tag{3.9.19}$$

Hence,

$$2\beta a+4\theta = 2n\pi, \tag{3.9.20}$$

where $n$ is a positive or negative integer or zero. Substituting for $\theta$ from equation (3.9.17), this condition becomes

$$\tfrac{1}{2}\beta a+\tan^{-1}\frac{\beta}{\alpha} = \tfrac{1}{2}n\pi \tag{3.9.21}$$

or

$$\left.\begin{aligned} \frac{\beta}{\alpha} &= -\tan\tfrac{1}{2}\beta a, \quad n \text{ even}, \\ &= \cot\tfrac{1}{2}\beta a, \quad n \text{ odd}. \end{aligned}\right\} \tag{3.9.22}$$

From equations (3.9.6), it follows that

$$\alpha^2 = \kappa^2-\beta^2, \tag{3.9.23}$$

where

$$\kappa^2 = 2mV_0/\hbar^2, \tag{3.9.24}$$

and thus $\beta$ is required to satisfy one of the equations

$$\left.\begin{aligned} \beta(\kappa^2-\beta^2)^{-\frac{1}{2}} &= -\tan\tfrac{1}{2}\beta a, \\ \beta(\kappa^2-\beta^2)^{-\frac{1}{2}} &= \cot\tfrac{1}{2}\beta a. \end{aligned}\right\} \tag{3.9.25}$$

Graphing the functions $\beta(\kappa^2-\beta^2)^{-\frac{1}{2}}$, $-\tan\frac{1}{2}\beta a$, $\cot\frac{1}{2}\beta a$ against $\beta$, it is easy to see that these equations have solutions $\beta_1, \beta_2, \ldots, \beta_k$ satisfying the inequalities

$$0 < \beta_1 < \frac{\pi}{a} < \beta_2 < \frac{2\pi}{a} < \beta_3 < \frac{3\pi}{a} < \ldots < \beta_k < \frac{k\pi}{a}, \tag{3.9.26}$$

where $k$ is the smallest integer such that

$$k\pi/a > \kappa. \tag{3.9.27}$$

$\beta=0$ is not an acceptable solution since, in this case, $E=0$ and the general solution of equation (3.9.5) is $C+Dx$; this leads to the condition $a\alpha=-2$, which cannot be satisfied, since $a$, $\alpha$ are both positive.

We have shown, therefore, that $\beta$ is quantised. It follows that the spectrum of energy eigenvalues for bound states is discrete, comprising $k$ values in all; these are called the *energy levels* for the system in a bound state. Thus, if the energy of the particle is measured and it is found to be in a bound state, the result of the measurement can only be one of this finite set of eigenvalues. If $E > V_0$, the particle can escape from the well and, as will be shown later (section 4.10), $E$ is not then quantised, i.e. the energy spectrum is continuous.

We shall now calculate the eigenfunctions corresponding to the $k$ energy eigenvalues. From equation (3.9.15), we first deduce that

$$C:D = -\exp(\iota\theta):\exp(-\iota\theta). \tag{3.9.28}$$

Substituting in equation (3.9.8), we find that

$$\psi = \psi_0 \sin(\beta x+\theta), \quad 0 < x < a, \tag{3.9.29}$$

where $\psi_0$ is a constant. Then, from equations (3.9.11), (3.9.12), (3.9.20), it follows that

$$A = \psi_0 \sin\theta, \quad G = (-1)^{n-1}\psi_0 \exp(\alpha a)\sin\theta, \tag{3.9.30}$$

and hence

$$\psi = \psi_0 \exp(\alpha x)\sin\theta, \quad x < 0, \tag{3.9.31}$$

$$= (-1)^{n-1}\psi_0 \exp[\alpha(a-x)]\sin\theta, \quad x > a. \tag{3.9.32}$$

The value of $\psi_0$ will now be chosen to normalise $\psi$ with respect to the $x$-coordinate alone, i.e. such that

$$\int_{-\infty}^{\infty} |\psi|^2\,dx = 1. \tag{3.9.33}$$

It will be found that

$$\psi_0 = \sqrt{\left(\frac{2\alpha}{2+a\alpha}\right)}. \tag{3.9.34}$$

$|\psi|^2 dx$ can now be interpreted physically as the probability of observing the particle with its $x$-coordinate in the interval $(x, x+dx)$, immediately after the particle's energy has been measured to take the energy eigenvalue corresponding to the values of $\alpha$ and $\beta$ occurring in equations (3.9.29)–(3.9.34). It will be observed that, although $\psi \to 0$ as $x \to \pm\infty$, $\psi$ does not vanish anywhere in the regions $x<0$, $x>a$. This implies that there is always a finite probability that the particle will be detected in these regions. However, according to classical ideas, it is impossible for the particle to penetrate into these regions since, if it were to do so, its K.E. would become negative if energy is to be conserved. Nevertheless, the result we have obtained does not imply that the principle of conservation of energy is invalid in quantum mechanics; for the energy and coordinate observables are incompatible ($\hat{H}$ and $\hat{x}$ do not commute) and hence, when $E$ is measured, the particle must not be thought of as occupying any definite position and a subsequent coordinate determination can accordingly result in its being observed at any point on the $x$-axis; if it is observed at a point outside the well, it is not necessary to suppose that its K.E. is then negative, for the process of observation will itself cause an unknown disturbance in the value of $E$, i.e. energy will flow from the observing apparatus to the particle, and hence the particle's energy will cease to be equal to the previously observed eigenvalue; it will, in fact, no longer be known precisely, but will be spread over a spectrum of values with variable probability. Apparently anomalous results, such as the one we have been studying, arise quite frequently in quantum mechanics; they are invariably resolved by remembering that two incompatible observables cannot be sharp simultaneously and consequently that their 'values' can never be inconsistent; if one is measured, a subsequent measurement of the other immediately disturbs the sharp value of the first and so eliminates any possibility of an inconsistency between the values of the observables being detected. Thus, although paradoxical situations arise in quantum mechanics in which it appears that, if we were able to circumvent the uncertainty principle and to measure a particular observable without causing any disturbance to the physical system an impossible result would be found, this principle ensures that no such difficulty can ever arise in practice. Since, to be acceptable, a physical theory has only to conform with observations which can, at least in principle, actually be carried out, the existence of paradoxes of this type is no reason for abandoning the theory of quantum mechanics.

A great simplification is effected in the problem we have been studying if we let $V_0 \to \infty$, i.e. assume that the particle is moving between potential barriers of infinite height which will act as perfectly reflecting walls. Then by equations (3.9.6), (3.9.17), $\alpha \to \infty$ and $\theta \to 0$. Equation (3.9.20) accordingly gives for the eigenvalues of $\beta$,

$$\beta = \frac{n\pi}{a}, \tag{3.9.35}$$

where $n = 1, 2, \ldots$. The corresponding energy eigenvalues follow from equations (3.9.6), viz.,

$$E = E_n = \frac{n^2 h^2}{8ma^2}. \tag{3.9.36}$$

Since, by equation (3.9.24), $\kappa \to \infty$, the inequality (3.9.27) indicates that $n$ is now unbounded.

Within the well, it follows from equation (3.9.29) that the $n$th eigenfunction is

$$\psi = \psi^n = \psi_0 \sin \frac{n\pi x}{a}, \tag{3.9.37}$$

where, by equation (3.9.34),

$$\psi_0 = \sqrt{\left(\frac{2}{a}\right)}. \tag{3.9.38}$$

Outside the well, equations (3.9.31), (3.9.32) confirm our expectation that $\psi$ vanishes.

In the case $n=1$, the particle has smallest energy and is said to be in its *ground state*; then $\psi = \psi_0 \sin(\pi x/a)$ and the most probable position for the particle is midway between the walls. If $n=2$, the particle is in its first *excited state* and its most probable positions are $x = \frac{1}{4}a$, $\frac{3}{4}a$.

It should be noted that only the *magnitude* of the particle's momentum is known in each eigenstate; the sense of the momentum cannot be known, for if it were then, by the uncertainty principle, we could have no knowledge relating to the particle's position, whereas we know that it lies between $x=0$ and $x=a$. This also explains why the zero value for $n$ is not permitted, for with this value $p_x$ would be known precisely (viz. $p_x = 0$) and our knowledge relating to the particle's position would conflict with the uncertainty principle.

## 3.10. The harmonic oscillator

In this section, we will further exemplify the theory by calculating the energy eigenstates of a particle which is constrained to move parallel to the $x$-axis under a force always directed towards the plane $x=0$ and whose magnitude is proportional to the particle's distance from this plane. In classical mechanics, such a system would execute simple harmonic motion and, no matter how great the energy of the motion, the particle would never escape from the centre of attraction. All its states are accordingly bound states and we expect a quantum mechanical treatment to show that its energy is quantised.

Let $m\omega^2 x$ be the magnitude of the attractive force when the particle's coordinate is $x$. Then, if $V$ is the particle's P.E.,

$$\frac{dV}{dx} = m\omega^2 x \tag{3.10.1}$$

and hence

$$V = \tfrac{1}{2}m\omega^2 x^2. \tag{3.10.2}$$

Thus, the particle moves in a potential well of parabolic form. For a state in which the energy takes a sharp value $E$, the wave equation (3.9.2) takes the form

$$\frac{d^2\psi}{dx^2}+\frac{2m}{\hbar^2}(E-\tfrac{1}{2}m\omega^2 x^2)\psi = 0. \tag{3.10.3}$$

Putting

$$x = \sqrt{\left(\frac{\hbar}{m\omega}\right)}\xi, \tag{3.10.4}$$

the wave equation can be written

$$\frac{d^2\psi}{d\xi^2}+(\lambda-\xi^2)\psi = 0, \tag{3.10.5}$$

where

$$\lambda = 2E/\hbar\omega. \tag{3.10.6}$$

To solve equation (3.10.5), we first make a change of dependent variable thus

$$\psi = \exp(-\tfrac{1}{2}\xi^2)\phi. \tag{3.10.7}$$

Upon substitution, the new equation will be found to be

$$\frac{d^2\phi}{d\xi^2}-2\xi\frac{d\phi}{d\xi}+(\lambda-1)\phi = 0. \tag{3.10.8}$$

This equation can be solved by the method of Frobenius. We assume

$$\phi = \sum_{n=0}^{\infty} a_n \xi^{\rho+n} \tag{3.10.9}$$

and then substituting, obtain

$$\sum_{n=0}^{\infty} a_n(\rho+n)(\rho+n-1)\xi^{\rho+n-2}-2\sum_{n=0}^{\infty} a_n(\rho+n)\xi^{\rho+n}+(\lambda-1)\sum_{n=0}^{\infty} a_n\xi^{\rho+n} = 0. \tag{3.10.10}$$

The coefficients of all powers of $\xi$ in the left-hand member of this equation must vanish and the equation will be satisfied identically in $\xi$ provided

$$(\text{coeff. } \xi^{\rho-2}): a_0\,\rho(\rho-1) = 0, \tag{3.10.11}$$

$$(\text{coeff. } \xi^{\rho-1}): a_1(\rho+1)\,\rho = 0, \tag{3.10.12}$$

$$(\text{coeff. } \xi^{\rho+n}): a_{n+2}(\rho+n+2)\,(\rho+n+1)-2a_n(\rho+n)+(\lambda-1)\,a_n = 0, \tag{3.10.13}$$

where $n=0, 1, 2, \ldots$. The equation (3.10.11) is termed the *indicial equation* and is employed to determine $\rho$. Assuming $a_0 \neq 0$, we are forced to take

$$\rho = 0 \text{ or } 1. \tag{3.10.14}$$

In either case, we satisfy equation (3.10.12) by taking

$$a_1 = 0. \tag{3.10.15}$$

There are now two cases to consider: If $\rho=0$, equation (3.10.13) reduces to

$$a_{n+2} = \frac{2n+1-\lambda}{(n+1)\,(n+2)}\,a_n. \tag{3.10.16}$$

Since $a_1=0$, this implies immediately that $a_3=a_5=\ldots=0$. Also, taking $n=0$, we have

$$a_2 = \frac{1-\lambda}{2!}\,a_0. \tag{3.10.17}$$

With $n=2$, we find

$$a_4 = \frac{5-\lambda}{3.4}\,a_2 = \frac{(1-\lambda)\,(5-\lambda)}{4!}\,a_0, \tag{3.10.18}$$

and so on. Taking $a_0=1$, it is seen that the following series is a solution of equation (3.10.8):

$$\phi = 1+\frac{1-\lambda}{2!}\,\xi^2+\frac{(1-\lambda)\,(5-\lambda)}{4!}\,\xi^4+\ldots. \tag{3.10.19}$$

It is easy to prove (e.g. by application of D'Alembert's test) that this series converges for all values of $\lambda$ and $\xi$.

With $\rho=1$, equation (3.10.13) reduces to

$$a_{n+2} = \frac{2n+3-\lambda}{(n+2)\,(n+3)}\,a_n \tag{3.10.20}$$

and we again conclude that $a_1 = a_3 = a_5 = \ldots = 0$. Also, putting $n=0$, we obtain

$$a_2 = \frac{3-\lambda}{3!} a_0 \tag{3.10.21}$$

and then, putting $n=2$,

$$a_4 = \frac{7-\lambda}{4.5} a_2 = \frac{(3-\lambda)(7-\lambda)}{5!} a_0, \tag{3.10.22}$$

and so on. A second solution of equation (3.10.8) is accordingly

$$\phi = \xi + \frac{3-\lambda}{3!}\xi^3 + \frac{(3-\lambda)(7-\lambda)}{5!}\xi^5 + \ldots. \tag{3.10.23}$$

If $\phi_1$, $\phi_2$ represent the two solutions so found, the general solution is then

$$\phi = A\phi_1 + B\phi_2, \tag{3.10.24}$$

where $A$, $B$ are arbitrary constants.

Suppose the series (3.10.19) for $\phi_1$ does not terminate. Then, beyond a certain term of the series, the terms will be all positive or all negative. If they are all negative, substitute $-\phi_1$ for $\phi_1$ in the argument which follows: Writing

$$\phi_1 = \sum_0^\infty u_n (\xi^2)^n, \tag{3.10.25}$$

$$\exp\left(\tfrac{3}{4}\xi^2\right) = \sum_0^\infty v_n (\xi^2)^n, \tag{3.10.26}$$

then $u_n > 0$, $v_n > 0$ for sufficiently large $n$. Also

$$\begin{aligned} \frac{u_n}{u_{n-1}} &= \frac{4n-3-\lambda}{2n(2n-1)}, \\ &= \frac{4 - \dfrac{1}{n}(3+\lambda)}{4n\left(1 - \dfrac{1}{2n}\right)}, \\ &> \frac{3}{4n}, \\ &= \frac{v_n}{v_{n-1}}, \end{aligned} \tag{3.10.27}$$

for sufficiently large $n$. It now follows, by application of the theorem proved in Appendix B, that

$$\phi_1 \geqslant k \exp\left(\tfrac{3}{4}\xi^2\right) + p(\xi^2), \tag{3.10.28}$$

for all $\xi$, where $k$ is a constant and $p$ is a polynomial. Thus, by equation

(3.10.7), the corresponding wave function $\psi_1$ satisfies

$$\psi_1 \geqslant \exp(\tfrac{1}{4}\xi^2)\{k+p\exp(-\tfrac{3}{4}\xi^2)\} \tag{3.10.29}$$

and hence

$$\psi_1 \to +\infty \quad \text{as} \quad \xi \to \pm\infty, \tag{3.10.30}$$

more rapidly than $\exp(\frac{1}{4}\xi^2)$. In the event that it is $-\phi_1$ which satisfies the inequality (3.10.28), we prove similarly that $\psi_1 \to -\infty$ more rapidly than $-\exp(\frac{1}{4}\xi^2)$.

Similarly it may be proved that $|\phi_2/\xi|$ satisfies an inequality of the type (3.10.28) and hence that $|\psi_2|$ approaches $+\infty$ more rapidly than $\xi\exp(\frac{1}{4}\xi^2)$ as $\xi \to \infty$.

We conclude that neither $\psi_1$ nor $\psi_2$ is acceptable as a wave function, since these would imply that the probability of finding the particle with a finite $x$-coordinate is zero.

The only remaining possibility is that the series for $\phi_1$ and/or for $\phi_2$ terminates. Inspection of equations (3.10.19), (3.10.23) shows that this will always happen for one or other of the series if $\lambda$ is an odd positive integer. Thus we are compelled to take

$$\lambda = 2n+1, \tag{3.10.31}$$

where $n = 0, 1, 2$, etc. and it then follows from equation (3.10.6) that the energy $E$ is quantised. Thus

$$E = E_n = \omega\hbar(n+\tfrac{1}{2}). \tag{3.10.32}$$

When $\lambda$ is given by equation (3.10.31), $\phi$ is a polynomial of degree $n$ which is only arbitrary to the extent of a constant multipler. Assuming this multiplier is chosen in such a manner that the coefficient of $\xi^n$ is $2^n$, the polynomial is denoted by $H_n(\xi)$ and is called *Hermite's Polynomial.* The wave amplitude for the energy state $E_n$ is then

$$\psi = \psi^n = A_n \exp(-\tfrac{1}{2}\xi^2) H_n(\xi), \tag{3.10.33}$$

where $A_n$ is a normalisation factor which is calculated in Appendix C to be given by

$$A_n = \left(\frac{2m\omega}{h}\right)^{\frac{1}{4}} \frac{1}{2^{\frac{1}{2}n}\sqrt{(n!)}}. \tag{3.10.34}$$

$\psi$ accordingly approaches zero as $\xi \to \pm\infty$, as is necessary.

The Hermite polynomials of low degree are readily calculated from equations (3.10.19), (3.10.23), to be:

$$\left.\begin{aligned} H_0 &= 1, \quad H_1 = 2\xi, \quad H_2 = 4\xi^2-2, \\ H_3 &= 8\xi^3-12\xi, \quad H_4 = 16\xi^4-48\xi^2+12. \end{aligned}\right\} \tag{3.10.35}$$

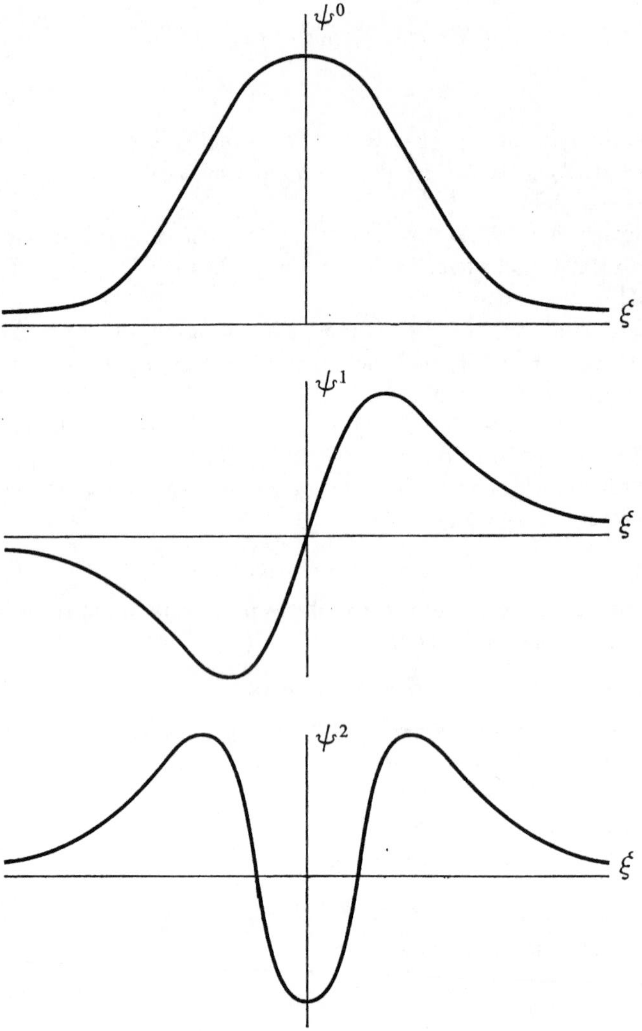


Fig. 3.1. Energy Eigenfunctions for a Harmonic Oscillator.

Further information relating to these polynomials will be found in Appendix C.

The wave amplitudes for the three lowest energy states have been plotted in Fig. 3.1. It will be observed that the most probable position for the particle in its ground state is the origin, but that the probability density possesses peaks on either side of the origin in the excited states. As for the case of a particle moving in a rectangular well, the ground state is not a state of zero energy, since this would imply a precise value for the momentum and a consequent complete lack of information relating to the particle's position; however, it is clear that there is a greater probability of finding the particle in the vicinity of the origin than at a great distance from this point, so that information in regard to its position is not entirely lacking.

The problem of the energy eigenstates of a harmonic oscillator can equally well be solved employing the momentum representation. The Hamiltonian for the system is given by

$$H = \frac{1}{2m}p^2 + \tfrac{1}{2}m\omega^2 x^2 \tag{3.10.36}$$

and hence, in the $p$-representation,

$$\begin{aligned} \hat{H} &= \frac{1}{2m}p^2 + \tfrac{1}{2}m\omega^2\left(-\frac{\hbar}{\iota}\frac{d}{dp}\right)^2, \\ &= \frac{1}{2m}p^2 - \tfrac{1}{2}m\omega^2\hbar^2\frac{d^2}{dp^2}. \end{aligned} \tag{3.10.37}$$

The characteristic equation for the eigenstate in which $H = E$ is accordingly

$$\frac{d^2\phi}{dp^2} + \frac{1}{m^2\omega^2\hbar^2}(2mE - p^2)\phi = 0, \tag{3.10.38}$$

where $\phi(p)$ is the wave function for the eigenstate. Putting

$$p = (m\omega\hbar)^{\frac{1}{2}}\xi, \tag{3.10.39}$$

this equation is reduced to the form of equation (3.10.5). The argument now proceeds along the same lines as before, $E$ is shown to be quantised as at equation (3.10.32) and, in the $n$th eigenstate, $\phi = \phi^n$, where $\phi^n$ is identical with $\psi^n$ as given by equation (3.10.33). It follows, therefore, that in this special case, the probability distributions of particle position and momentum are identical.

## 3.11. Harmonic oscillator in the energy representation

It was demonstrated in the previous section that, when the energy $E$ of a harmonic oscillator is known, its state is completely determined. It follows that the observable $E$ comprises a maximal set of compatible observables for the system and the corresponding eigenstates can accordingly be employed as a basis for a representation of the states and observables of the system. Since $E$ possesses a discrete spectrum, the representation will be in terms of matrices.

In the $x$-representation, the energy eigenstates are determined by the wave functions $\psi^n$ $(n=0,1,\ldots)$ given at equation (3.10.33). It follows from equation (3.4.11) that the matrix $\mathbf{x}$ representing the coordinate of the particle has elements $x_{ij}$ given by

$$x_{ij} = \int_{-\infty}^{\infty} \psi^{i*}\, x\psi^j\, dx,$$

$$= A_i A_j \int_{-\infty}^{\infty} x \exp(-\xi^2)\, H_i(\xi)\, H_j(\xi)\, dx. \tag{3.11.1}$$

Substituting for $x$ from equation (3.10.4) and for $A_i$ from equation (3.10.34), we find that

$$x_{ij} = \frac{1}{\pi}\sqrt{\left(\frac{h}{2^{i+j+1}\, m\omega i!\, j!}\right)} \int_{-\infty}^{\infty} \xi \exp(-\xi^2)\, H_i(\xi)\, H_j(\xi)\, d\xi. \tag{3.11.2}$$

From equations (C.13) and (C.16) of Appendix C, it follows that

$$2\xi H_j = H_{j+1} + 2jH_{j-1}. \tag{3.11.3}$$

Employing the orthonormality condition (C.21), equation (3.11.2) now yields the results

$$x_{i,\,i-1} = x_{i-1,\,i} = \sqrt{\left(\frac{\hbar}{2m\omega}\right)}\, i^{\frac{1}{2}}, \tag{3.11.4}$$

where $i=1, 2, 3, \ldots$. All other elements $x_{ij}$ vanish. Thus the matrix $\mathbf{x}$ is the array

$$\mathbf{x} = \sqrt{\left(\frac{\hbar}{2m\omega}\right)} \begin{pmatrix} 0 & \sqrt{1} & 0 & 0 & \ldots \\ \sqrt{1} & 0 & \sqrt{2} & 0 & \ldots \\ 0 & \sqrt{2} & 0 & \sqrt{3} & \ldots \\ 0 & 0 & \sqrt{3} & 0 & \ldots \\ \ldots & \ldots & \ldots & \ldots & \ldots \end{pmatrix}. \tag{3.11.5}$$

The elements of the matrix representing the momentum component $p_x$ are given by

$$p_{ij} = \frac{\hbar}{\iota} \int_{-\infty}^{\infty} \psi^{i*} \frac{d\psi^j}{dx}\, dx. \tag{3.11.6}$$

It will be found that

$$\mathbf{p} = \frac{1}{\iota}\sqrt{(\tfrac{1}{2}m\omega\hbar)}\begin{pmatrix} 0 & \sqrt{1} & 0 & 0 & \cdots \\ -\sqrt{1} & 0 & \sqrt{2} & 0 & \cdots \\ 0 & -\sqrt{2} & 0 & \sqrt{3} & \cdots \\ 0 & 0 & -\sqrt{3} & 0 & \cdots \\ \cdots & \cdots & \cdots & \cdots & \cdots \end{pmatrix}. \tag{3.11.7}$$

By matrix multiplication, it now follows that

$$\mathbf{x}^2 = \frac{\hbar}{2m\omega}\begin{pmatrix} 1 & 0 & \sqrt{2} & 0 & \cdots \\ 0 & 3 & 0 & \sqrt{6} & \cdots \\ \sqrt{2} & 0 & 5 & 0 & \cdots \\ 0 & \sqrt{6} & 0 & 7 & \cdots \\ \cdots & \cdots & \cdots & \cdots & \cdots \end{pmatrix}. \tag{3.11.8}$$

$$\mathbf{p}^2 = \tfrac{1}{2}m\omega\hbar\begin{pmatrix} 1 & 0 & -\sqrt{2} & 0 & \cdots \\ 0 & 3 & 0 & -\sqrt{6} & \cdots \\ -\sqrt{2} & 0 & 5 & 0 & \cdots \\ 0 & -\sqrt{6} & 0 & 7 & \cdots \\ \cdots & \cdots & \cdots & \cdots & \cdots \end{pmatrix}. \tag{3.11.9}$$

Thus, the matrix representing the energy observable is

$$\begin{aligned}\mathbf{H} &= \frac{1}{2m}\mathbf{p}^2 + \tfrac{1}{2}m\omega^2\mathbf{x}^2, \\ &= \tfrac{1}{2}\omega\hbar\begin{pmatrix} 1 & 0 & 0 & 0 & \cdots \\ 0 & 3 & 0 & 0 & \cdots \\ 0 & 0 & 5 & 0 & \cdots \\ 0 & 0 & 0 & 7 & \cdots \\ \cdots & \cdots & \cdots & \cdots & \cdots \end{pmatrix}.\end{aligned} \tag{3.11.10}$$

This matrix is diagonal as it should be since the eigenstates of $H$ are being employed as the basis. The diagonal elements are the eigenvalues of the energy as already obtained at equation (3.10.32).

## 3.12. Two-particle systems

If the particles are spinless and have coordinates $(x_1, x_2, x_3)$, $(y_1, y_2, y_3)$, then these provide a complete set of compatible observables for the system. Suppose the particles are not identical and that the state of one is specified by the wave function $\psi(x_i)$ and the state of the other is specified by another wave function $\phi(y_j)$. Then $\psi(x_i)\,\phi(y_j)$ is the wave function for the state of the combined system (generalising from equation (1.13.2)).

If the particles have spin, a complete set of compatible observables for the system will include a component of spin for each particle. Suppose the spin component has $m$ eigenvalues for the first particle and $n$ eigenvalues for the second. Then, assuming the particles are not identical, the state of the first particle can be specified by a sequence $\psi_r(x_i)$ $(r=1,2,\ldots,m)$ and the state of the second by a sequence $\phi_s(y_j)$ $(s=1,2,\ldots,n)$. The state of the whole system is then described by the sequence

$$\chi_{rs}(x_i, y_j) = \psi_r(x_i)\,\phi_s(y_j). \tag{3.12.1}$$

This sequence can, of course, be arranged as a column matrix.

Although the wave functions for the system can be separated into two factors as in equation (3.12.1) when the states of the component particles are known individually, it does not follow that such a separation is possible in every state of the system. For example, suppose the particles have no spin and that each is acted upon by a force of attraction due to the other and by no other forces. If $r$ is the distance between the particles, let $V(r)$ be the P.E. due to this attraction. Then, if $m$, $n$ are the masses of the particles, $(p_1, p_2, p_3)$ are the components of momentum of the first particle and $(q_1, q_2, q_3)$ are the components of momentum of the second particle, the classical expression for the energy of the system is

$$H = \frac{1}{2m}(p_1^2+p_2^2+p_3^2)+\frac{1}{2n}(q_1^2+q_2^2+q_3^2)+V. \tag{3.12.2}$$

In the $x_i$, $y_j$-representation, the operator representing $H$ is now seen to be

$$\hat{H} = -\frac{\hbar^2}{2m}\nabla_x^2-\frac{\hbar^2}{2n}\nabla_y^2+V, \tag{3.12.3}$$

where

$$\nabla_x^2 = \frac{\partial^2}{\partial x_i\,\partial x_i}, \quad \nabla_y^2 = \frac{\partial^2}{\partial y_i\,\partial y_i}. \tag{3.12.4}$$

Hence, in a state for which the system as a whole has a sharp total energy $E$, the wave function $\chi(x_i, y_j)$ must satisfy

$$\frac{\hbar^2}{2m}\nabla_x^2\chi+\frac{\hbar^2}{2n}\nabla_y^2\chi+(E-V)\chi = 0. \tag{3.12.5}$$

In general, since $V$ is a function of both sets of coordinates $x_i$, $y_j$, this equation will not possess solutions in the separated form

$$\chi = \psi(x_i)\,\phi(y_j). \tag{3.12.6}$$

However, by making a change of variables, solutions in an alternative separated form can be found and these possess a simple physical significance. We let $(\bar{x}_1, \bar{x}_2, \bar{x}_3)$ be the coordinates of the centre of mass of the particles and $(\xi_1, \xi_2, \xi_3)$ the coordinates of the second particle relative to a parallel non-inertial frame with its origin at the first particle. Then

$$\bar{x}_i = \frac{mx_i + ny_i}{m+n}, \tag{3.12.7}$$

$$\xi_i = y_i - x_i. \tag{3.12.8}$$

Transforming to the new variables $\bar{x}_i$, $\xi_i$, equation (3.12.5) becomes

$$\frac{\hbar^2}{2(m+n)}\nabla^2\chi + \frac{\hbar^2}{2}\left(\frac{1}{m}+\frac{1}{n}\right)\nabla_\xi^2\chi + (E-V)\chi = 0, \tag{3.12.9}$$

where $V = V(\xi_i)$. Since $V$ is a function of the $\xi_i$ alone, separated solutions of the type

$$\chi = \psi(\xi_i)\,\phi(\bar{x}_i) \tag{3.12.10}$$

now exist; for, substituting and dividing by $\chi$, we obtain

$$\frac{1}{\psi}\left[\frac{\hbar^2}{2}\left(\frac{1}{m}+\frac{1}{n}\right)\nabla_\xi^2\psi + (E-V)\psi\right] = -\frac{\hbar^2}{2(m+n)}\frac{1}{\phi}\nabla_{\bar{x}}^2\phi. \tag{3.12.11}$$

The left-hand member of this equation is a function of the $\xi_i$ alone, whereas the right-hand member is a function of the $\bar{x}_i$ alone. It follows by the well-known principle of the separation of variables that both members must be constant. Taking this constant to be $E_0$, we obtain the equations

$$\frac{\hbar^2}{2(m+n)}\nabla_{\bar{x}}^2\phi + E_0\phi = 0, \tag{3.12.12}$$

$$\frac{\hbar^2}{2}\left(\frac{1}{m}+\frac{1}{n}\right)\nabla_\xi^2\psi + (E-E_0-V)\psi = 0. \tag{3.12.13}$$

Equation (3.12.12) is the Schrödinger equation for a particle of mass $(m+n)$, have coordinates $\bar{x}_i$ and moving under no forces with energy $E_0$. Equation (3.12.13) is the Schrödinger equation for a particle of mass $mn/(m+n)$, having coordinates $\xi_i$ and moving with energy $E-E_0$ under an attractive force to the origin derivable from a P.E. function $V(r)$. Thus, there exist energy eigenstates in which a part $E_0$ of the energy may be regarded as being due to the motion of the C.M. and the remaining part $E-E_0$ may be thought of as being due to the motion of the particles relative to one another. The wave function $\psi(\xi_i)$ associated with the relative motion may be thought of as the wave function for the motion of the second particle as observed from the first; this will lead to a Schrödinger equation of the type (3.12.13), provided the mass $n$ of the second particle is replaced by a mass reduced by the factor $m/(m+n)$. The hydrogen atom is an example of the type of system we have been considering, the proton nucleus playing the role of one particle and the orbiting electron of the other. Since the proton is almost 2000 times as massive as the electron, it is natural to think of the proton as being stationary and to consider the motion of the electron relative to it. Our theory indicates that, in calculating the energy eigenstates of the atom, the motion of the proton may be neglected, provided the mass of the electron is reduced by a factor which is approximately 2000/2001. Classical theory also permits us to neglect the motion of one of two mutually attracting bodies by substituting the same 'reduced mass' for the mass of the other.

Returning to the general case, suppose the particles are identical. Then, considerations similar to those which arose in section 2.6 are relevant. Suppose the system is in the state specified by $\chi_{rs}(x_i, y_j)$. If, now, the states of the particles are exchanged, the state of the system is not affected. It follows, as in section 2.6, that $\chi_{rs}(x_i, y_j)$ must be either symmetric or anti-symmetric with respect to exchange of the sets $(x_1, x_2, x_3, r)$, $(y_1, y_2, y_3, s)$. In the case of fermions, $\chi$ is anti-symmetric and in the case of bosons it is symmetric.

Now consider the special case where the states of the individual particles are known. Suppose the first particle is in the state $\psi_r(x_i)$ and the second is in the state $\phi_s(y_j)$ $(r, s = 1, 2, \ldots, n)$. Then equation (3.12.1) would give the wave functions for the system if the particles were distinguishable; this cannot be correct, however, since $\chi_{rs}$ would not then be symmetric or anti-symmetric. Now suppose that the first particle is in the state $\phi_r(x_i)$ and the second is in the state $\psi_s(y_j)$; equation (3.12.1) then gives $\phi_r(x_i)\,\psi_s(y_j)$ for the wave functions of the system. However, this state is not distinct from the first and hence, by the principle of superposition,

$$\chi_{rs} = A\psi_r(x_i)\,\phi_s(y_j) + B\phi_r(x_i)\,\psi_s(y_j), \tag{3.12.14}$$

can also represent the state for all values of $A$ and $B$. Now, by taking $A = B$, we can make $\chi$ symmetric and by taking $A = -B$ we can make $\chi$ anti-symmetric. Thus, for two bosons in the states $\psi$, $\phi$, we take

$$\chi_{rs} = A[\psi_r(x_i)\,\phi_s(y_j) + \phi_r(x_i)\,\psi_s(y_j)]. \tag{3.12.15}$$

For two fermions in these states, we take

$$\chi_{rs} = A[\psi_r(x_i)\,\phi_s(y_j) - \phi_r(x_i)\,\psi_s(y_j)]. \tag{3.12.16}$$

In the case of two fermions in identical states, we note that $\phi = \psi$ and $\chi_{rs}$ vanishes identically. This confirms the exclusion principle, viz. that two fermions cannot exist in identical states.

The generalisation of these ideas to a system of $N$ identical particles presents no difficulty. If $x_i, y_i, z_i, \ldots$ $(i = 1, 2, 3,)$ are the coordinates of the particles and $\psi_r, \phi_s, \theta_t, \ldots$ are the wave functions representing the particle states then, if the particles are fermions, the wave function for the complete system is given by

$$\chi_{rst\ldots}(x_i, y_j, z_k, \ldots) = A \begin{vmatrix} \psi_r(x_i) & \phi_r(x_i) & \theta_r(x_i) & \ldots \\ \psi_s(y_j) & \phi_s(y_j) & \theta_s(y_j) & \ldots \\ \psi_t(z_k) & \phi_t(z_k) & \theta_t(z_k) & \ldots \\ \ldots & \ldots & \ldots & \ldots \end{vmatrix}, \tag{3.12.17}$$

for each term in the expansion of the determinant is a possible wave function for the system and the exchange of two particle states is equivalent to the exchange of two rows of the determinant, which reverses the sign of $\chi$ as required.

If the particles are bosons, equation (3.12.17) remains valid provided all terms in the expansion of the determinant are given positive signs.

## Note on Dirac's notation

This notation has become popular in recent years. It has not been fully employed in this introductory text, since it was felt desirable to relieve the student from the burden of an unfamiliar notation, in order that his attention could be the more fully concentrated upon the essentials of the argument being presented. However, only a slight modification of our notation is needed to convert it into that introduced by Dirac and, to enable the reader to follow calculations expressed in this form, the amendment necessary is described below.

The vector representing the pure state of a system which is prepared by measuring the values of a maximal set of compatible observables to be $a, b, c \ldots$, is denoted in the Dirac notation by

$$|a, b, c, \ldots\rangle,$$

instead of by a single Greek letter (e.g. $\alpha$) as we have been doing. In the case when the observables are quantised, so that the eigenvalues can be arranged in sequences $\{a_i\}$, $\{b_j\}$, $\{c_k\}$, $\ldots$, the symbol $|a_i, b_j, c_k, \ldots\rangle$ representing the eigenstate $a = a_i$, $b = b_j$, $c = c_k$, $\ldots$ is abbreviated to the form

$$|i, j, k, \ldots\rangle.$$

Dirac then introduces the concept of a dual vector space whose elements can also be employed to represent pure states of the system. The vector from this space representing the state $(a, b, c, \dots)$ is denoted by

$$\langle a, b, c, \dots|.$$

If a discrete representation is employed, a vector $|a, b, c, \dots\rangle$ may be taken to correspond to a column matrix and the dual vector $\langle a, b, c, \dots|$ to the conjugate transpose of the column matrix, i.e. a row matrix.

For brevity, in what follows we shall suppress all but the first observable of the maximal set.

Consider the state specified by the vector

$$\alpha|a\rangle + \beta|b\rangle,$$

where $\alpha$, $\beta$ are complex numbers. The corresponding dual vector is taken to be

$$\alpha^*\langle a| + \beta^*\langle b| = \langle a|\,\alpha^* + \langle b|\,\beta^*.$$

Next, suppose we wish to form the scalar product of a pair of vectors $\alpha$, $\beta$ representing two states of the same system. Employing a discrete representation, if $\boldsymbol{\alpha}$, $\boldsymbol{\beta}$ are the corresponding column matrices, we have written this product as

$$\langle\alpha|\beta\rangle = \boldsymbol{\alpha}^\dagger \boldsymbol{\beta}.$$

In the Dirac notation, if the vectors are denoted by $|a\rangle$, $|b\rangle$, their scalar product is written

$$\langle a||b\rangle,$$

i.e. as a product of the vector dual to $|a\rangle$ and the vector $|b\rangle$. The 'double bar' in the centre is then always contracted to a 'single bar' thus

$$\langle a|b\rangle.$$

This, of course, is essentially the notation we have been employing for scalar products and is the reason for our adoption of this notation.

A vector such as $\langle a|$ is called a bra-vector and a vector such as $|b\rangle$ is called a ket-vector. Their product, viz. $\langle a|b\rangle$, is a completed 'bracket' expression (hence the terminology).

In Dirac's notation therefore, if $\hat{x}$ is the operator associated with an observable $x$, the expected value of $x$ in the state $|a\rangle$ will be given by the scalar product of $|a\rangle$ and $\hat{x}|a\rangle$ and this will be written

$$\bar{x} = \langle a|\hat{x}|a\rangle.$$

In our notation, the state would be represented by a vector symbol $\alpha$ and then we would write

$$\bar{x} = \langle \alpha | \hat{x} \alpha \rangle.$$

Now consider the scalar product $P$ of the vectors $|a\rangle$ and $\hat{x}|b\rangle$. This is written as

$$P = \langle a | \hat{x} | b \rangle.$$

Assuming the operator $\hat{x}$ is Hermitian, we know that $P$ is identical with the scalar product $Q$ of the vectors $\hat{x}|a\rangle$ and $|b\rangle$. But, if we denote the dual of the vector $\hat{x}|a\rangle$ by $\langle a|\hat{x}$ (i.e. the operator $\hat{x}$ is being thought of as operating backwards on the dual vector $\langle a|$), then

$$Q = \langle a | \hat{x} | b \rangle$$

and the identity of $P$ and $Q$ becomes a consequence of the notation.

As a final illustration of the notation, consider an eigenvector of the operator $\hat{a}$ corresponding to the eigenvalue $a = a_1$; this will be denoted by $|a_1\rangle$. This vector will satisfy the characteristic equation

$$\hat{a}|a_1\rangle = a_1|a_1\rangle.$$

Taking duals of both sides of this equation, since $a_1$ is real we have

$$\langle a_1|\hat{a} = \langle a_1|a_1.$$

This is the equivalent dual form of the characteristic equation.

CHAPTER 4

# Time variation of states

## 4.1. The Hamiltonian operator

In the previous chapters, the problem of specifying the state of a physical system at a particular instant has been studied and a method for predicting the results of observations made at this instant upon a system in a known state has been developed. In brief, the state of a system has been represented by a vector $\alpha$ in a Hilbert space; then, if $\beta$ is the vector representing an eigenstate in which an observable $q$ takes a sharp value $q'$, the probability that a measurement of $q$ made upon the system in the state $\alpha$ will yield the value $q'$ is $|\langle\alpha|\beta\rangle|^2$. In this chapter, we shall examine the effect of delaying the measurement of $q$ until some later instant, it being assumed that the system is not disturbed by other observations being made upon it in the interim. Thus, knowing the state of the system at some initial instant $t_0$, we wish to be able to predict its state at some later instant $t_1 > t_0$ and thence to calculate the probability that the result of a $q$-observation made at time $t_1$ will be $q'$. We shall accordingly be regarding the state vector $\alpha$ for the system as a function of the time parameter $t$ and will write $\alpha = \alpha(t)$. In Newtonian mechanics it is assumed that, if a system is in a pure state at an initial instant $t_0$, so that the positions and velocities of all its component particles are known, then its state at every later instant is completely determined and may be calculated by application of the principles of the classical theory. In quantum mechanics, we make the corresponding assumption, viz. if a system is in a pure state at $t = t_0$ so that $\alpha(t_0)$ is known, then $\alpha(t)$ is determined for all $t > t_0$ as a smoothly varying vector function of $t$. However, whereas according to the classical view there is no objection to supposing that the system is kept under continuous observation during the period $(t_0, t)$, in quantum mechanics it is necessary to assume that the system is isolated from observing instruments during this interval; for the result of making any measurement on the system is to disturb it into an eigenstate of the observable measured and hence to introduce a discontinuity into the variation of $\alpha(t)$. Our object, therefore, is to determine the manner of variation of $\alpha(t)$ in the interval between two successive sets of observations; the first set of observations at $t_0$ will prepare the system in a certain eigenstate specified by the vector $\alpha(t_0)$; the form of the state vector $\alpha(t)$ at any later instant $t$, prior to the second set of observations, will then be assumed calculable; at $t_1$ the state vector will be $\alpha(t_1)$ and, by applying the methods explained in the earlier chapters, the result of the second set of observations at $t_1$ can then be predicted on a probability basis. However, when these measurements have been performed, the system will have been prepared in a fresh eigenstate $\beta(t_1)$ and this will play the role of an initial state for the subsequent variation of the state vector $\beta(t)$, during the period preceding the next set of observations. Since the disturbing effect an observation has upon a system is, in principle, unknowable, it is impossible to predict *across an instant of observation*; commencing with a system prepared in a specific eigenstate, the best that can be done is to predict the results of the *next* set of observations on a probability basis.

Suppose, then, $\alpha(t)$ represents the state vector at time $t$, where $t_0 < t < t_1$ and $t_0$, $t_1$ are consecutive instants of observation. Knowing $\alpha(t_0)$, it will be assumed that $\alpha(t)$ can be calculated. Thus, we assume the existence of an operator $\hat{T}(t, t_0)$ such that

$$\hat{T}(t, t_0)\,\alpha(t_0) = \alpha(t). \tag{4.1.1}$$

We shall assume that any instant at which $\alpha(t)$ is known can be treated as an initial instant and hence that, for all $t$, $t'$ in the interval $(t_0, t_1)$,

$$\hat{T}(t', t)\,\alpha(t) = \alpha(t'). \tag{4.1.2}$$

Taking $t' = t + \delta t$, we therefore obtain the equation

$$\frac{\alpha(t+\delta t) - \alpha(t)}{\delta t} = \frac{1}{\delta t}[\hat{T}(t+\delta t, t) - 1]\,\alpha(t). \tag{4.1.3}$$

Letting $\delta t \to 0$, we shall assume that

$$\frac{1}{\delta t}[\hat{T}(t+\delta t, t) - 1] \to \hat{S}(t) \tag{4.1.4}$$

and then equation (4.1.3) yields

$$\frac{d\alpha}{dt} = \hat{S}(t)\,\alpha(t). \tag{4.1.5}$$

We now define an operator $\hat{H}$ by the equation

$$\hat{H} = \iota\hbar\hat{S}. \tag{4.1.6}$$

Then equation (4.1.5) becomes

$$\hat{H}\alpha = \iota\hbar\frac{d\alpha}{dt}. \tag{4.1.7}$$

This is the equation of motion for a quantum mechanical system.

It is easy to show that the operator $\hat{H}$ is Hermitian. Taking the scalar product of both members of equation (4.1.7) with $\alpha$, we obtain

$$\langle\alpha|\hat{H}\alpha\rangle = \left\langle\alpha\,\middle|\,\iota\hbar\frac{d\alpha}{dt}\right\rangle = \iota\hbar\left\langle\alpha\,\middle|\,\frac{d\alpha}{dt}\right\rangle, \tag{4.1.8}$$

$$\langle\hat{H}\alpha|\alpha\rangle = \left\langle\iota\hbar\frac{d\alpha}{dt}\,\middle|\,\alpha\right\rangle = -\iota\hbar\left\langle\frac{d\alpha}{dt}\,\middle|\,\alpha\right\rangle. \tag{4.1.9}$$

It now follows that

$$\begin{aligned}\frac{d}{dt}\langle\alpha|\alpha\rangle &= \left\langle\frac{d\alpha}{dt}\,\middle|\,\alpha\right\rangle+\left\langle\alpha\,\middle|\,\frac{d\alpha}{dt}\right\rangle,\\ &= \frac{\iota}{\hbar}[\langle\hat{H}\alpha|\alpha\rangle-\langle\alpha|\hat{H}\alpha\rangle].\end{aligned} \tag{4.1.10}$$

But, assuming $\alpha$ has been normalised, $\langle\alpha|\alpha\rangle=1$ identically. Thus

$$\langle\hat{H}\alpha|\alpha\rangle = \langle\alpha|\hat{H}\alpha\rangle. \tag{4.1.11}$$

Since this equation is valid for arbitrary $\alpha$, it follows that $\hat{H}$ is Hermitian (Ex. 7, Chap. 1).

Suppose that $\alpha(t)$ is the vector representing the state of a system at time $t$, when the system is initially prepared in the state $\alpha(0)$. Let $\beta(t)$ be the vector representing the state of the same system at time $t$, when it is initially prepared in the state $\beta(0)$. Then the Hermitian property of $\hat{H}$ ensures that the scalar product $\langle\alpha|\beta\rangle$ shall be independent of $t$, i.e. the angle between the vectors does not change with the time. This follows since

$$\begin{aligned}\frac{d}{dt}\langle\alpha|\beta\rangle &= \left\langle\frac{d\alpha}{dt}\,\middle|\,\beta\right\rangle+\left\langle\alpha\,\middle|\,\frac{d\beta}{dt}\right\rangle,\\ &= \left\langle-\frac{\iota}{\hbar}\hat{H}\alpha\,\middle|\,\beta\right\rangle+\left\langle\alpha\,\middle|\,-\frac{\iota}{\hbar}\hat{H}\beta\right\rangle,\\ &= \frac{\iota}{\hbar}[\langle\hat{H}\alpha|\beta\rangle-\langle\alpha|\hat{H}\beta\rangle],\\ &= 0.\end{aligned} \tag{4.1.12}$$

The operator $\hat{H}$ will have a characteristic form for every system. For systems which can be described in terms of classical observables alone (i.e. no spin), it will be shown in section 4.3 that $\hat{H}$ is the operator representing the classical Hamiltonian for the system. For more complex systems, it is necessary to guess at the form of $\hat{H}$ and then to compare the theoretical predictions with experiment.

## 4.2. Rates of change of expected values

Let $a$ be any observable for a system whose Hamiltonian operator is $\hat{H}$ and let $\alpha(t)$ be the vector determining the system's state at any time $t$ between two consecutive observations. Then the expected value of $a$ at this instant is given by

$$\bar{a}(t) = \langle \alpha | \hat{a}\alpha \rangle, \tag{4.2.1}$$

where $\hat{a}$ is the operator representing $a$. Differentiating with respect to $t$, we obtain

$$\frac{d\bar{a}}{dt} = \left\langle \frac{d\alpha}{dt} \middle| \hat{a}\alpha \right\rangle + \left\langle \alpha \middle| \frac{d}{dt}(\hat{a}\alpha) \right\rangle. \tag{4.2.2}$$

We shall assume that

$$\frac{d}{dt}(\hat{a}\alpha) = \frac{\partial \hat{a}}{\partial t}\alpha + \hat{a}\frac{d\alpha}{dt}, \tag{4.2.3}$$

where $\partial\hat{a}/\partial t$ is the operator obtained from $\hat{a}$ by formal differentiation with respect to the parameter $t$; in most cases, $\hat{a}$ will not involve this parameter, and then $\partial\hat{a}/\partial t = 0$; equation (4.2.3) then states that the operators $\hat{a}$, $d/dt$ will be assumed to commute. We then have, making use of the equation of motion (4.1.7),

$$\begin{aligned} \frac{d\bar{a}}{dt} &= \left\langle \frac{d\alpha}{dt} \middle| \hat{a}\alpha \right\rangle + \left\langle \alpha \middle| \hat{a}\frac{d\alpha}{dt} \right\rangle + \left\langle \alpha \middle| \frac{\partial \hat{a}}{\partial t}\alpha \right\rangle, \\ &= \left\langle -\frac{\iota}{\hbar}\hat{H}\alpha \middle| \hat{a}\alpha \right\rangle + \left\langle \alpha \middle| -\frac{\iota}{\hbar}\hat{a}\hat{H}\alpha \right\rangle + \left\langle \alpha \middle| \frac{\partial \hat{a}}{\partial t}\alpha \right\rangle, \\ &= \frac{\iota}{\hbar}\left\langle \alpha \middle| \hat{H}\hat{a}\alpha \right\rangle - \frac{\iota}{\hbar}\left\langle \alpha \middle| \hat{a}\hat{H}\alpha \right\rangle + \left\langle \alpha \middle| \frac{\partial \hat{a}}{\partial t}\alpha \right\rangle, \\ &= \langle \alpha | \hat{\dot{a}}\alpha \rangle, \end{aligned} \tag{4.2.4}$$

where the operator $\hat{\dot{a}}$ is defined by the equation

$$\hat{\dot{a}} = \frac{\iota}{\hbar}[\hat{H}, \hat{a}] + \frac{\partial \hat{a}}{\partial t}. \tag{4.2.5}$$

Hence, if $\dot{a}$ is defined to be the observable whose operator is $\hat{\dot{a}}$, equation (4.2.4) may be written

$$\dot{\bar{a}} = \bar{\dot{a}}. \tag{4.2.6}$$

The reader should note that the observable $\dot{a}$ is *defined* by equation (4.2.5) and *not* as the time derivative of the observable $a$. To take

$$\dot{a} = \lim_{\delta t \to 0} \frac{1}{\delta t}[a(t+\delta t)-a(t)], \tag{4.2.7}$$

would imply that a suitable procedure for measuring $\dot{a}$ would be to measure $a$ at the instants $t$, $t+\delta t$ ($\delta t$ very small); however, although this procedure might be expected to yield a satisfactory value for $\dot{a}$ for the system in the state it occupied between the instants $t$, $t+\delta t$, the later observation will disturb the system to such a degree that this value of $\dot{a}$ will no longer be appropriate to the system in the state it occupies subsequent to the instant $t+\delta t$; for this reason, the definition (4.2.7) is unsatisfactory.

Consider the normal case, where $\hat{a}$ is not explicitly dependent on $t$. Then equation (4.2.5) reduces to the form

$$\hat{\dot{a}} = \frac{\iota}{\hbar}[\hat{H}, \hat{a}]. \tag{4.2.8}$$

Hence, if $\hat{H}$ and $\hat{a}$ commute, $\hat{\dot{a}}=0$ and $\dot{\bar{a}}=0$, i.e. $\bar{a}$ remains constant. Also, if $\hat{H}$ and $\hat{a}$ commute, so also do $\hat{H}$ and $\hat{a}^2$, for

$$\hat{H}\hat{a}^2 = \hat{H}\hat{a}\hat{a} = \hat{a}\hat{H}\hat{a} = \hat{a}\hat{a}\hat{H} = \hat{a}^2\hat{H}. \tag{4.2.9}$$

Similarly, it may be shown that $\hat{H}$ commutes with any positive integral power of $\hat{a}$. From this, it follows that $\overline{a^n}$ is constant for all $n$. But $\overline{a^n}$ is the $n$th order moment of $a$ about $a=0$ and the probability distribution of the variate $a$ is completely determined when all its moments about zero are known. The implication is, therefore, that in these circumstances the form of the probability distribution for $a$ is invariant with respect to the time. This leads to an important *conservation principle*. For suppose $a$ is sharp with the eigenvalue $a'$ at some instant; the probability distribution for $a$ is then of a particularly simple form at this instant, viz. $a=a'$ with probability unity and $a$ takes any other value with probability zero. This distribution is maintained during the subsequent motion of the system and it follows, therefore, that if $a$ is measured at any later instant, its value will still be found to be $a'$, i.e. $a$ is conserved. On the other hand, if $\hat{a}$ and $\hat{H}$ do not commute, $a$ may be measured to take the value $a'$ at $t=t_0$, so that $a$ is sharp at this instant, but this simple probability distribution will not be maintained and its value will not be sharp for $t>t_0$. This implies that, if $a$ is to be measured at some subsequent instant $t_1$, the resulting value cannot be predicted with certainty, but only on a probability basis.

In particular, $\hat{H}$ commutes with itself and hence, provided it is not explicitly dependent on $t$, $H$ will be conserved. This suggests that $H$ is then the energy of the system (see next section).

## 4.3. Hamilton's equations

In the case of a system comprising a single particle, the observable corresponding to the operator $\hat{H}$ can be identified with the particle's Hamiltonian. For suppose that $x_i$ $(i=1,2,3)$ are the coordinates of the particle and $p_i$ $(i=1,2,3)$ are its components of linear momentum, all relative to a rectangular cartesian inertial frame. We shall assume that the observable $H$ is expressible in terms of the fundamental observables $x_i$, $p_i$ and the time $t$, thus

$$H = H(x_i, p_i, t). \tag{4.3.1}$$

We shall also assume that $H$ is a polynomial in the $x_i$ and $p_i$ ($H$ can always be approximated to any degree of accuracy desired, by such a polynomial). Then it follows from Ex. 4, Chap. 3 that

$$[\hat{H}, \hat{x}_i] = \frac{\hbar}{\iota}\frac{\partial \hat{H}}{\partial \hat{p}_i}, \quad [\hat{H}, \hat{p}_i] = -\frac{\hbar}{\iota}\frac{\partial \hat{H}}{\partial \hat{x}_i}, \tag{4.3.2}$$

where the partial derivatives have their obvious formal meanings. Since the operators $\hat{x}_i$, $\hat{p}_i$ do not involve the parameter $t$, it follows from equation (4.2.5) that

$$\dot{\hat{x}}_i = \frac{\iota}{\hbar}[\hat{H}, \hat{x}_i], \quad \dot{\hat{p}}_i = \frac{\iota}{\hbar}[\hat{H}, \hat{p}_i]. \tag{4.3.3}$$

Hence, we may conclude that

$$\dot{\hat{x}}_i = \frac{\partial \hat{H}}{\partial \hat{p}_i}, \quad \dot{\hat{p}}_i = -\frac{\partial \hat{H}}{\partial \hat{x}_i}. \tag{4.3.4}$$

But $\partial \hat{H}/\partial \hat{p}_i$, $\partial \hat{H}/\partial \hat{x}_i$ are the operators representing $\partial H/\partial p_i$, $\partial H/\partial x_i$. Thus

$$\dot{x}_i = \frac{\partial H}{\partial p_i}, \quad \dot{p}_i = -\frac{\partial H}{\partial x_i}. \tag{4.3.5}$$

By equation (4.2.6), we now have that

$$\dot{\bar{x}}_i = \overline{\frac{\partial H}{\partial p_i}}, \quad \dot{\bar{p}}_i = -\overline{\frac{\partial H}{\partial x_i}}. \tag{4.3.6}$$

If, therefore, $H$ is identified with the Hamiltonian, Hamilton's classical equations for a particle are validated, on the assumption that these refer to the expected or mean values of the observables involved. But, by the correspondence principle (section 1.9), this is exactly the interpretation we expect to be given to these equations, so that we shall regard our identification of $H$ with the classical Hamiltonian as being strongly indicated in this case. However, our final justification for this step is that it leads to results which are in agreement with observation.

In the case of a system comprising a number of particles, it may be shown similarly that, if the Hamiltonian for the system is expressed in terms of the particles' rectangular cartesian coordinates and the corresponding components of the linear momenta, by identifying $\hat{H}$ with the resulting Hamiltonian operator, we are led to equations of motion for the mean values of the coordinates and momenta, which are identical with Hamilton's equations for the system. We shall accordingly extend our identification of $\hat{H}$ to cover this more general case.

Suppose that $\hat{H}$ does not involve the parameter $t$ explicitly. Then, since $\hat{H}$ commutes with itself, it follows from the results of the previous section that $\overline{H}$ and the probability distribution of $H$ do not vary with $t$. In particular, if $H$ is sharp at one instant, it remains sharp and its value is constant. Now, in these circumstances, the classical Hamiltonian $H$ gives the energy of the system and our result is accordingly the quantum mechanical equivalent of the classical *principle of conservation of energy*.

## 4.4. Steady states

Suppose $\hat{H}$ does not involve the parameter $t$ explicitly and let $H$ be sharp with the value $E$ at $t=t_0$ (e.g. $H$ might be measured to take this value at the instant $t_0$). Then, if $\alpha_0$ is the vector determining the system's state at $t_0$, $\alpha_0$ must satisfy the characteristic equation

$$\hat{H}\alpha_0 = E\alpha_0. \tag{4.4.1}$$

But $H$ remains sharp with the value $E$ for $t > t_0$ and hence, if $\alpha(t)$ represents the state at a later instant $t$, $\alpha$ will also satisfy

$$\hat{H}\alpha = E\alpha. \tag{4.4.2}$$

If now follows from equation (4.1.7) that

$$\frac{d\alpha}{dt} = -\frac{\iota E}{\hbar}\alpha. \tag{4.4.3}$$

Integrating this first order equation under the initial condition $\alpha = \alpha_0$ at $t = t_0$, we obtain

$$\alpha = \alpha_0 \exp[-\iota E(t-t_0)/\hbar]. \tag{4.4.4}$$

This equation indicates the manner in which the state of the system varies with the time. The initial form of $\alpha$, viz. $\alpha_0$, will be referred to as the *vector amplitude* of $\alpha$.

Let $\beta$ be the vector representing a certain state of the system. Then

$$\langle\alpha|\beta\rangle = \exp[\iota E(t-t_0)/\hbar]\langle\alpha_0|\beta\rangle \tag{4.4.5}$$

and hence $|\langle\alpha|\beta\rangle|^2$ is independent of $t$. This implies that the probability of a transition being observed into the state $\beta$ does not change with the time. But $\beta$ is arbitrary. It follows that we have proved that, if $H$ is independent of $t$ and is sharp at some instant, then the probability of any specified transition being observed remains the same for all subsequent instants. We say that the system is in a *steady state*. In particular, if $H$ can be identified with the classical Hamiltonian, then a state of the system in which its energy is sharp is a steady state.

It will be noted from equation (4.4.4) that, when the energy of a system is known to be $E$, then the system's state oscillates sinusoidally with angular frequency $\omega$ (radians/sec) or frequency $\nu$ (cycles/sec) where

$$E = \hbar\omega = h\nu. \tag{4.4.6}$$

This relationship between the energy of a system and an associated frequency was first suggested (1905) by Einstein for the special case of electromagnetic waves. To explain the *photo-electric effect*, he postulated that electromagnetic radiation possessed a corpuscular, as well as an undulatory, nature, so that, in certain circumstances, a beam of light behaved like a stream of particles (*photons*), the energy of each particle being related to the frequency of the radiation by equation (4.4.6).

Suppose we associate with the observable $H$ sufficient other compatible observables to provide a maximal set. Some of these observables may be dependent upon $t$, but we shall assume that $H$ is not. Then, employing the eigenstates of this set of observables at some particular instant as a basis for a representation, the matrix representing $H$ will be diagonal for all $t$ (we shall assume that the spectra of eigenvalues are discrete); this follows since $H$ is independent of $t$ and hence remains a member of the basic set for all $t$. The diagonal elements of $\mathbf{H}$ will be the $H$-eigenvalues, $E_1, E_2, \ldots$. Let $\boldsymbol{\alpha} = \{\alpha_1, \alpha_2, \ldots\}$ be the matrix determining the state of the system at any instant $t$, so that $\alpha_i = \alpha_i(t)$. Then the matrix form of equation (4.1.7) gives

$$\mathbf{H}\boldsymbol{\alpha} = \iota\hbar \frac{d\boldsymbol{\alpha}}{dt}. \tag{4.4.7}$$

Equating corresponding elements from the column matrices forming the two members of this equation, we obtain

$$E_i \alpha_i = \iota\hbar \frac{d\alpha_i}{dt}, \tag{4.4.8}$$

no summation being intended in the left-hand member. This first order differential equation is easily solved to yield

$$\alpha_i = A_i \exp(-\iota E_i t/\hbar), \tag{4.4.9}$$

where the $A_i$ are constants. The manner in which a system's state depends upon $t$ is therefore shown particularly simply in this type of representation.

Since $|\alpha_i|^2 = |A_i|^2$, the probability of observing a transition into the $i$th eigenstate is independent of $t$. This is to be expected, for the probability distribution for $H$ does not change with the time and hence the probability of measuring $H = E_i$ must be independent of $t$.

If all the $A_i$ vanish except $A_1$, the system is in the first eigenstate and $H$ will be sharp with the value $E_1$. Thus, the system will be in a steady state and we shall have

$$\boldsymbol{\alpha} = \begin{pmatrix} A_1 \\ 0 \\ 0 \\ \vdots \end{pmatrix} \exp(-\iota E_1 t/\hbar), \tag{4.4.10}$$

where $|A_1| = 1$. This is a special form of equation (4.4.4). Since any column matrix with elements $\alpha_i$ given by equation (4.4.9) may be expressed as a sum of matrices of the type represented by the right-hand member of equation (4.4.10), it follows that the time variation of the state of a system may be thought of as being due to the superposition of a number of steady states. Thus, we may write

$$\alpha(t) = \sum_i A_i \exp(-\iota E_i t/\hbar). \tag{4.4.11}$$

## 4.5. Time variation of spin

A simple illustration of the general concepts which have been introduced in the previous sections of this chapter, is provided by the problem of the variation in the spin state of a particle caused by the action of an applied field.

We shall assume that the component of the particle's spin in any specified direction possesses two eigenvalues and that the same component of the associated magnetic moment has eigenvalues $\pm\mu_0$. Taking rectangular axes $Ox_1 x_2 x_3$, if $\mu_i$ denotes the component of magnetic moment in the direction $Ox_i$ and $\sigma_i$ represents the corresponding component of the spin axis (eigenvalues $\pm 1$), we have

$$\mu_i = \mu_0 \sigma_i, \quad i = 1,2,3. \tag{4.5.1}$$

To cause a variation in the spin state of the particle, we shall suppose that it is placed in a uniform magnetic field having components $B_i$ ($i=1,2,3$). Then, according to the classical theory of magnetism, the energy of the particle in this field is given by

$$H = -\sum_i \mu_i B_i = -\mu_0 \sum_i \sigma_i B_i. \tag{4.5.2}$$

Since the $B_i$ and $\mu_0$ are constants for the system, the matrix representing this observable is

$$\mathbf{H} = -\mu_0 \sum_i B_i \boldsymbol{\sigma}_i. \tag{4.5.3}$$

Employing the $\sigma_3$-representation, the matrices $\boldsymbol{\sigma}_i$ are the Pauli matrices (equations (2.2.26)) and hence

$$\mathbf{H} = -\mu_0 \begin{pmatrix} B_3 & B_1 - \iota B_2 \\ B_1 + \iota B_2 & -B_3 \end{pmatrix}. \tag{4.5.4}$$

We shall now assume that $H$ can be identified with the Hamiltonian appearing in the equation of motion (4.1.7). Then, if the spin state at any time $t$ is represented by the spinor $\boldsymbol{\alpha}$ (equation (2.2.2)), the variation in this state is governed by the equation

$$\iota\hbar \frac{d\boldsymbol{\alpha}}{dt} = \mathbf{H}\boldsymbol{\alpha}. \tag{4.5.5}$$

This matrix equation is equivalent to the pair of equations

$$\iota\hbar \frac{d\alpha_+}{dt} = -\mu_0[B_3 \alpha_+ + (B_1 - \iota B_2)\alpha_-], \tag{4.5.6}$$

$$\iota\hbar \frac{d\alpha_-}{dt} = -\mu_0[(B_1 + \iota B_2)\alpha_+ - B_3 \alpha_-], \tag{4.5.7}$$

and these completely determine the manner in which the spin state varies after the particle has been prepared in a given state at a certain initial instant.

Since the orientation of the axes $Ox_i$ is completely arbitrary, there is no loss of generality in supposing that the field is directed along the $x_3$-axis. The equations (4.5.6), (4.5.7), then simplify to the form

$$\iota\hbar\frac{d\alpha_+}{dt} = -\mu B\alpha_+, \tag{4.5.8}$$

$$\iota\hbar\frac{d\alpha_-}{dt} = \mu B\alpha_-, \tag{4.5.9}$$

where we have put $\mu_0=\mu$, $B_3=B$. These equations have a general solution

$$\alpha_+ = \alpha_+^0 \exp(\iota\mu Bt/\hbar), \quad \alpha_- = \alpha_-^0 \exp(-\iota\mu Bt/\hbar), \tag{4.5.10}$$

where $\alpha_+^0$, $\alpha_-^0$ are constants of integration. Measuring $t$ from the initial instant, it is clear that $\{\alpha_+^0, \alpha_-^0\}$ is the spinor specifying the initial state.

Referring to equation (4.5.2), we note that, with this orientation of axes, the energy of the system is given by

$$H = -\mu_0\sigma_3 B_3 = -\mu B\sigma_3. \tag{4.5.11}$$

It follows that $H$ and $\sigma_3$ are sharp together. Thus, if $\sigma_3=+1$ at $t=0$, then $H=-\mu B$ and the system is in a steady state. In this case, $\alpha_+^0=1$, $\alpha_-^0=0$ and hence

$$\boldsymbol{\alpha} = \begin{pmatrix}1\\0\end{pmatrix}\exp(\iota\mu Bt/\hbar). \tag{4.5.12}$$

This is a special case of equation (4.4.4). Similarly, if $\sigma_3=-1$ at $t=0$, then

$$\boldsymbol{\alpha} = \begin{pmatrix}0\\1\end{pmatrix}\exp(-\iota\mu Bt/\hbar) \tag{4.5.13}$$

and the system is in a steady state having energy $\mu B$. The general solution (4.5.10) may be obtained by the superposition of two steady states, one of each of the types (4.5.12), (4.5.13).

At any instant, the probabilities of measuring $\sigma_3$ to take the values $+1$ and $-1$ are $|\alpha_+|^2$, $|\alpha_-|^2$ respectively. It is clear from equations (4.5.10) that these are always independent of $t$. The reason for this is that the spin axis may be thought of as precessing steadily about the direction of the applied field, making a constant angle with it and hence that the geometrical relationship between the spin axis and $Ox_3$ never changes.

To prove this, suppose that initially the spin axis is in the direction determined by the spherical polar angles $(\theta, \phi)$ with respect to the frame $Ox_1x_2x_3$, i.e. the component of spin axis in this direction takes the sharp value $+1$ (see Ex. 6, Chap. 2). Then, at $t=0$, $\boldsymbol{\alpha}$ is given by equation (2.3.12). Taking $\gamma = \frac{1}{2}\phi$, we have

$$\alpha_+^0 = \exp(-\tfrac{1}{2}\iota\phi)\cos\tfrac{1}{2}\theta, \quad \alpha_-^0 = \exp(\tfrac{1}{2}\iota\phi)\sin\tfrac{1}{2}\theta. \tag{4.5.14}$$

Substituting in equations (4.5.10), we now obtain

$$\alpha_+ = \exp\left[-\tfrac{1}{2}\iota\left(\phi - \frac{2\mu B}{\hbar}t\right)\right]\cos\tfrac{1}{2}\theta, \tag{4.5.15}$$

$$\alpha_- = \exp\left[\tfrac{1}{2}\iota\left(\phi - \frac{2\mu B}{\hbar}t\right)\right]\sin\tfrac{1}{2}\theta. \tag{4.5.16}$$

Comparing these components of the spinor $\boldsymbol{\alpha}$ with the initial components, it is evident that the spin state has changed only in that $\phi$ has been replaced by $\phi - 2\mu Bt/\hbar$, i.e. the direction of the spin axis has precessed about $Ox_3$ through an angle $2\mu Bt/\hbar$. The angular rate of precession is accordingly constant, viz. $2\mu B/\hbar$.

Next, suppose the component of the particle's spin axis in the direction $(\theta, \phi)$ is measured to be $+1$ at $t=0$ and we desire to know the probability of measuring this component to be $+1$ at a subsequent time $t$. This probability will be given by $|\langle\alpha|\alpha^0\rangle|^2$, where $\alpha^0$ is determined by equations (4.5.14) and $\alpha$ by equations (4.5.15), (4.5.16). Now

$$\begin{aligned}\langle\alpha|\alpha^0\rangle &= \alpha_+^*\alpha_+^0 + \alpha_-^*\alpha_-^0,\\ &= \cos\left(\frac{\mu Bt}{\hbar}\right) - \iota\cos\theta\sin\left(\frac{\mu Bt}{\hbar}\right).\end{aligned} \tag{4.5.17}$$

It follows that the point representing the complex number $\langle\alpha|\alpha^0\rangle$ in the Argand plane, traces out an ellipse with its centre at the origin and its major axis along the real axis; the semi-major axis is of length unity and the semi-minor axis is of length $|\cos\theta|$. The probability $|\langle\alpha|\alpha^0\rangle|^2$ accordingly oscillates between the extreme values 1 and $\cos^2\theta$ (being initially 1 of course). The angular frequency of the oscillation is $2\mu B/\hbar$; this is equal to the angular rate of precession of the spin axis as might be expected (N.B. two complete oscillations correspond to one circuit of the ellipse). The probability of measuring the $(\theta, \phi)$-component of the spin axis to be $-1$ can be found by calculating the scalar product of $\boldsymbol{\alpha}$ with the spinor vector determined by equation (2.3.13); this proves to be

$$\iota\sin\theta\sin\left(\frac{\mu Bt}{\hbar}\right). \tag{4.5.18}$$

The probability is observed to oscillate between the values 0 and $\sin^2\theta$; this it does in such a manner that the net probability that one of the values $\pm 1$ will be obtained remains constant at unity (the reader should check this).

In the special case $\theta = \frac{1}{2}\pi$, where we are measuring the component of spin axis in a direction perpendicular to the field, the probabilities are $\cos^2(\mu Bt/\hbar)$ and $\sin^2(\mu Bt/\hbar)$ respectively. Thus, the system is in the $+1$-eigenstate at $t=0$, in the $-1$-eigenstate at $t=\pi\hbar/2\mu B$ and back again in the $+1$-eigenstate at $t=\pi\hbar/\mu B$; thereafter it repeats this oscillation between the two eigenstates indefinitely. The effect of the field is therefore to cause the particle to flip over rhythmically from one state of spin to the opposite state.

We will next consider the effect of applying an additional field, of constant magnitude, which rotates with angular velocity $\omega$ in a plane at right angles to the uniform field $B$. Thus, we shall take

$$B_1 = \lambda B \cos \omega t, \quad B_2 = \lambda B \sin \omega t, \quad B_3 = B, \tag{4.5.19}$$

$\lambda B$ being the magnitude of the rotating field. Then the equations of motion (4.5.6), (4.5.7), take the form

$$\frac{d\alpha_+}{dt} = \iota\omega_0(\alpha_+ + \lambda\,\alpha_- \exp(-\iota\omega t)), \tag{4.5.20}$$

$$\frac{d\alpha_-}{dt} = \iota\omega_0(\lambda\,\alpha_+ \exp(\iota\omega t) - \alpha_-), \tag{4.5.21}$$

where

$$\omega_0 = \mu B/\hbar. \tag{4.5.22}$$

$2\omega_0$ is the angular velocity of precession of the spin axis in the uniform field.

To solve these equations, we set

$$\alpha_+ = x \exp(\iota\omega_0 t), \quad \alpha_- = y \exp(-\iota\omega_0 t). \tag{4.5.23}$$

Substituting in equations (4.5.20), (4.5.21), it will be found that

$$\frac{dx}{dt} = \iota\omega_0\,\lambda y \exp[-\iota(\omega + 2\omega_0)t], \tag{4.5.24}$$

$$\frac{dy}{dt} = \iota\omega_0\,\lambda x \exp[\iota(\omega + 2\omega_0)t]. \tag{4.5.25}$$

Eliminating $x$ between these equations, it follows that

$$\frac{d^2y}{dt^2} - \iota(\omega + 2\omega_0)\frac{dy}{dt} + \lambda^2\omega_0^2 y = 0. \tag{4.5.26}$$

Assuming that, at $t=0$, the particle is in the eigenstate $\sigma_3=+1$, then $x=1$, $y=0$ at this instant. Thus, $y=0$, $dy/dt=\iota\omega_0\lambda$ at $t=0$ and the appropriate solution of equation (4.5.26) is

$$y = \frac{\iota\lambda\omega_0}{\omega'}\exp\left[\tfrac{1}{2}\iota(\omega+2\omega_0)t\right]\sin\omega' t, \tag{4.5.27}$$

where

$$\omega'^2 = (\omega_0+\tfrac{1}{2}\omega)^2+\lambda^2\omega_0^2. \tag{4.5.28}$$

It now follows that

$$x = \exp\left[-\tfrac{1}{2}\iota(\omega+2\omega_0)t\right]\left[\cos\omega' t+\iota\frac{\omega_0+\frac{1}{2}\omega}{\omega'}\sin\omega' t\right]. \tag{4.5.29}$$

Thus

$$\alpha_+ = \exp(-\tfrac{1}{2}\iota\omega t)\left[\cos\omega' t+\iota\frac{\omega_0+\frac{1}{2}\omega}{\omega'}\sin\omega' t\right], \tag{4.5.30}$$

$$\alpha_- = \frac{\iota\lambda\omega_0}{\omega'}\exp(\tfrac{1}{2}\iota\omega t)\sin\omega' t. \tag{4.5.31}$$

The probability that, at time $t$, a transition into the eigenstate $\sigma_3=-1$ will be found to have occurred, is

$$|\alpha_-|^2 = \frac{\lambda^2\omega_0^2}{\omega'^2}\sin^2\omega' t. \tag{4.5.32}$$

This oscillates between the values 0 and $\lambda^2\omega_0^2/\omega'^2$ with angular frequency $2\omega'$.

Thus, the rotating field causes an oscillation in the spin state of the particle between a state in which $\sigma_3=+1$ with certainty and a state in which $\sigma_3=-1$ with probability $\lambda^2\omega_0^2/\omega'^2$. It is clear from a consideration of equation (4.5.28) that, when $\omega=-2\omega_0$, this probability assumes its maximum value of unity. This is the phenomenon of *spin resonance* and it occurs when the field $\lambda B$ rotates at the same rate and in the same sense as the precession of the spin axis induced by the uniform field. For this value of $\omega$, we have

$$\alpha_+ = \exp(\iota\omega_0 t)\cos\lambda\omega_0 t, \quad \alpha_- = \iota\exp(-\iota\omega_0 t)\sin\lambda\omega_0 t, \tag{4.5.33}$$

and hence

$$|\alpha_+|^2 = \cos^2\lambda\omega_0 t, \quad |\alpha_-|^2 = \sin^2\lambda\omega_0 t. \tag{4.5.34}$$

Thus, the probabilities of finding the particle in the eigenstates $\sigma_3=\pm1$, both oscillate between the values 0 and 1 in such a manner that $|\alpha_+|^2+|\alpha_-|^2=1$.

## 4.6. Conservation of probability

Consider the form taken by the equation of motion (4.1.7) for a system comprising a single spinless particle of mass $m$, moving in a conservative field of force with P.E. $V$. Employing the usual notation and adopting the coordinate representation, $\hat{H}$ is given by equation (3.8.2); equation (4.1.7) accordingly takes the form

$$-\frac{\hbar^2}{2m}\nabla^2\psi + V\psi = i\hbar\frac{\partial\psi}{\partial t}, \tag{4.6.1}$$

where $\psi = \psi(x_1, x_2, x_3, t)$ is the wave function for the particle at time $t$. Taking the complex conjugate of both sides of this equation, we obtain

$$-\frac{\hbar^2}{2m}\nabla^2\psi^* + V\psi^* = -i\hbar\frac{\partial\psi^*}{\partial t}. \tag{4.6.2}$$

Multiplying equation (4.6.1) by $\psi^*$ and equation (4.6.2) by $\psi$ and then subtracting, it will be found that

$$\frac{\hbar^2}{2m}(\psi\nabla^2\psi^* - \psi^*\nabla^2\psi) = i\hbar\frac{\partial}{\partial t}(\psi\psi^*). \tag{4.6.3}$$

But

$$\operatorname{div}(\psi\nabla\psi^*) = \psi\nabla^2\psi^* + \nabla\psi.\nabla\psi^*. \tag{4.6.4}$$

It follows that equation (4.6.3) can be written in the form

$$\frac{\partial\rho}{\partial t} + \operatorname{div}\mathbf{j} = 0, \tag{4.6.5}$$

where

$$\rho = \psi\psi^*, \tag{4.6.6}$$

$$\mathbf{j} = \frac{i\hbar}{2m}(\psi\nabla\psi^* - \psi^*\nabla\psi). \tag{4.6.7}$$

Now $\psi\psi^*$ has been interpreted as a probability density in the $x$-space, i.e. if $d\tau$ is a volume element of this space, $\psi\psi^* d\tau$ is the probability the particle will be located within this element at the instant $t$. The net probability over the whole of the $x$-space is always unity; it follows that probability may be thought of as flowing from one region of space to another in the manner of a fluid, so that although its distribution alters with the time, its overall quantity is conserved. We expect, therefore, that an *equation of continuity*, expressing this conservation, will be deducible from the fundamental equation of motion and that this equation will take the familiar form arising in the theory of fluid flows.

Equation (4.6.5) is exactly of this form, provided we interpret $\mathfrak{j}$ as the *probability current density vector*; this we shall now do. With this interpretation, if $S$ is any closed surface in the $x$-space, the flux of $\mathfrak{j}$ across $S$ in the outwards sense gives the rate at which the probability of locating the particle inside $S$ is decreasing.

In a steady state, $\psi$ depends upon $t$ as indicated in equation (4.4.4). Equations (4.6.6), (4.6.7) then imply that both $\rho$ and $\mathfrak{j}$ are independent of $t$ in such a state.

## 4.7. Free rectilinear motion of a particle. Wave packets

Some of the general concepts introduced in previous sections can be illustrated by reference to the simplest possible mechanical system, comprising a single particle moving parallel to the $x$-axis under no forces. According to classical theory, the particle's motion will be uniform and its momentum will be conserved. In the case of the quantum mechanical system, suppose the particle's momentum is measured precisely at an initial instant $t=t_0$, so that it is set into a pure state, viz. an eigenstate of the momentum observable $p$. Then, by the argument of section 3.7, it follows that the wave function for the particle at this instant is given by

$$\psi = A \exp(\iota px/\hbar), \tag{4.7.1}$$

where $A$ is a constant. Since the particle is moving freely, its total energy $E$ is given by

$$E = \frac{p^2}{2m} + V_0 \tag{4.7.2}$$

where $V_0$ is the constant P.E. of the particle and it follows that, if $p$ is sharp, so is $E$. The results of section 4.4. are accordingly applicable and we deduce that $E$ remains sharp throughout the subsequent motion and that the particle is in a steady state; equation (4.4.4) indicates that the form taken by the wave function at any later instant $t$ is

$$\psi = A \exp[\iota(px - Et)/\hbar], \tag{4.7.3}$$

where a constant factor $\exp(\iota Et_0/\hbar)$ has been absorbed in $A$. It is evident that this represents a complex sinusoidal $\psi$-wave being propagated along the $x$-axis with velocity $E/p$ (the fact that the velocity of propagation is not equal to $v$ has no physical significance, since the $\psi$-wave itself is not directly observable). If $\nu$ is the frequency associated with the wave, then

$$\nu = \frac{E}{2\pi\hbar} \quad \text{or} \quad E = h\nu. \tag{4.7.4}$$

It follows from equation (4.7.3) that $\psi$ satisfies the characteristic equation of the observable $p$ (equation (3.7.3)) for all $t$ and hence that the momentum continues to be sharp with the value $p$. It also follows that $\psi\psi^*$ is constant over the $x$-axis for all $t$, i.e. that all positions for the particle are equally probable; since $p$ is always sharp, this also follows from the uncertainty principle.

Substituting from equation (4.7.3) into equations (4.6.6), (4.6.7), it will be found that

$$\rho = |A|^2, \tag{4.7.5}$$

$$j = \frac{p}{m}|A|^2 = v|A|^2. \tag{4.7.6}$$

By supposing that, instead of a single particle, a very large number of non-interacting particles, each of mass $m$, are in motion in the same state, this pair of equations can be given a straightforward physical meaning. For the number of particles in any region of space will then be proportional to the total probability over this region; hence the particle density will be proportional to the probability density $\rho$ and the particle current density will be proportional to the probability current density $j$. By choosing $|A|$ appropriately, therefore, equations (4.7.5), (4.7.6) can be taken to give the particle density and particle current density respectively. But we have now replaced a single particle moving along $Ox$ with velocity $v$, by a *beam* of such particles; equation (4.7.5) indicates that the particle density in the beam will be uniform (viz. $|A|^2$ particles per unit volume) and equation (4.7.6) gives the resulting flux of particles across a unit area placed normal to the beam. In this manner, a physical meaning can be attached to the constant $A$.

As explained in section 4.4, any other state of the particle can be obtained by a superposition of steady states of the type (4.7.3). Since $p$ possesses a continuous spectrum ranging from $-\infty$ to $+\infty$, this gives a general formula for the wave function of the form

$$\psi(x,t) = \frac{1}{h^{\frac{1}{2}}}\int_{-\infty}^{\infty} \phi(p)\exp[\iota(px-Et)/\hbar]\,dp, \tag{4.7.7}$$

where $E$ is given in terms of $p$ by equation (4.7.2) and we have introduced the factor $h^{-\frac{1}{2}}$ for later convenience. Suppose that at $t=0$, the particle is put in a state for which

$$\psi(x,0) = \sigma^{-\frac{1}{2}}(2\pi)^{-\frac{1}{4}}\exp\left(-\frac{x^2}{4\sigma^2}+\frac{\iota\lambda x}{\hbar}\right). \tag{4.7.8}$$

In this state

$$\rho = \psi\psi^* = \frac{1}{\sigma\sqrt{(2\pi)}}\exp(-x^2/2\sigma^2) \tag{4.7.9}$$

and hence $\psi$ is normalised and the probability distribution of the particle's position along the $x$-axis is of normal error form, with its centre at $x=0$. The variance of the distribution is $\sigma^2$ and thus, if $\sigma$ is small, the particle's position will be known with good accuracy to be in the vicinity of the origin. Putting $t=0$ in equation (4.7.7), we find that $\phi(p)$ satisfies the integral equation

$$h^{-\frac{1}{2}}\int_{-\infty}^{\infty}\phi(p)\exp(\iota px/\hbar)\,dp = \psi(x,0). \tag{4.7.10}$$

Referring to the transformation equations (3.5.29), (3.5.30), it is clear that $\phi(p)$ is the wave function in the $p$-representation for the particle at $t=0$. Employing the inverse transformation, we have that

$$\phi(p) = h^{-\frac{1}{2}}\int_{-\infty}^{\infty}\exp(-\iota px/\hbar)\,\psi(x,0)\,dx. \tag{4.7.11}$$

Substituting for $\psi(x,0)$ from equation (4.7.8) and making use of the standard result

$$\int_{-\infty}^{\infty}\exp(ax^2+bx+c)\,dx = \sqrt{\left(-\frac{\pi}{a}\right)}\exp\left(c-\frac{b^2}{4a}\right), \tag{4.7.12}$$

where $a$ is real and negative, it will be found that

$$\phi(p) = \left(\frac{2}{\pi}\right)^{\frac{1}{4}}\left(\frac{\sigma}{\hbar}\right)^{\frac{1}{2}}\exp[-\sigma^2(p-\lambda)^2/\hbar^2]. \tag{4.7.13}$$

The probability density for $p$ at $t=0$ is accordingly

$$\phi\phi^* = \sqrt{\left(\frac{2}{\pi}\right)}\frac{\sigma}{\hbar}\exp[-2\sigma^2(p-\lambda)^2/\hbar^2]. \tag{4.7.14}$$

The significance of the parameter $\lambda$ is now clear. In the initial state, the probability distribution of the particle's momentum is also of normal error form with variance $\hbar^2/4\sigma^2$ and mean $\lambda$, i.e. $\lambda$ is the expected value of $p$ at $t=0$. It will be observed that, if $\sigma$ is small so that the particle's position is known fairly accurately at $t=0$, then the variance of $p$ is large and the momentum is not known with precision. In fact, if $\sigma_x^2$, $\sigma_p^2$ are variances of $x$, $p$ respectively, then

$$\sigma_x \sigma_p = \sigma . \frac{\hbar}{2\sigma} = \tfrac{1}{2}\hbar, \tag{4.7.15}$$

in agreement with the uncertainty principle (3.6.16).

The wave function for the subsequent motion of the particle can now be found from equation (4.7.7) by substituting for $\phi(p)$ from equation (4.7.13). The result obtained is

$$\psi(x,t) = \sigma^{-\frac{1}{2}}(2\pi)^{-\frac{1}{4}}\mu \exp\left[\mu^2\left\{-\frac{x^2}{4\sigma^2}+\frac{\iota\lambda}{\hbar}\left(x-\frac{\lambda t}{2m}\right)\right\}\right], \tag{4.7.16}$$

where we have taken $V_0=0$ and put

$$\mu = \left(1+\frac{\iota\hbar t}{2m\sigma^2}\right)^{-\frac{1}{2}}. \tag{4.7.17}$$

Thus, the probability density for the position of the particle is given by

$$\psi\psi^* = \frac{1}{\sigma'\sqrt{(2\pi)}}\exp\left[-\frac{1}{2\sigma'^2}\left(x-\frac{\lambda t}{m}\right)^2\right], \tag{4.7.18}$$

where

$$\sigma' = \sigma\left(1+\frac{\hbar^2 t^2}{4m^2\sigma^4}\right)^{\frac{1}{2}}. \tag{4.7.19}$$

This shows that the probability distribution for the particle's position is propagated along the $x$-axis as a normal error pulse. The pulse width (measured by $\sigma'$) increases steadily with $t$ as indicated in equation (4.7.19) and the centre of the pulse moves with uniform velocity $\lambda/m$ (=expected velocity of the particle).

Such a pulse is termed a *wave packet* and is the nearest approximation to a Newtonian particle that quantum mechanics is able to provide. It will be observed that, the smaller the value of $\sigma$, and the more accurately the initial position of the particle is known therefore, the more rapidly the width of the wave packet expands (equation (4.7.19)); this is to be expected since, for small $\sigma$, the momentum is largely uncertain and hence the uncertainty in the position of the particle rapidly increases as time elapses.

It will be left as an exercise for the reader to show from equation (4.7.16) that, in the $p$-representation, the wave function for the particle at time $t$ is given by

$$\phi(p,t) = \left(\frac{2}{\pi}\right)^{\frac{1}{4}}\left(\frac{\sigma}{\hbar}\right)^{\frac{1}{2}}\exp\left[-\frac{\sigma^2}{\hbar^2}(p-\lambda)^2-\frac{\iota p^2 t}{2m\hbar}\right]. \tag{4.7.20}$$

It then follows that the probability density for $p$ is given by

$$\phi\phi^* = \sqrt{\left(\frac{2}{\pi}\right)}\frac{\sigma}{\hbar}\exp\left[-2\sigma^2(p-\lambda)^2/\hbar^2\right]. \qquad (4.7.21)$$

This is identical with the density already calculated for $t=0$ (equation (4.7.14)) and proves that the $p$-distribution does not change with the time. This result may also be derived from the fact that the operators $\hat{p}$ and $\hat{H} = V_0 + \hat{p}^2/2m$ commute and is, in any case, to be expected since no forces act upon the particle.

## 4.8. Freely moving particle incident upon a potential barrier

In this section, we shall continue to study the rectilinear motion of a particle and shall assume that it is moving freely everywhere except at a single point, where it is subjected to an impulsive force. Taking the motion to be parallel to an $x$-axis, suppose the P.E. of the particle is given by the equations

$$\left.\begin{aligned} V &= V_0, \quad x < 0, \\ &= V_1, \quad x > 0. \end{aligned}\right\} \qquad (4.8.1)$$

Then $dV/dx$ vanishes everywhere, except at the origin where, $V$ being discontinuous, its derivative is infinite. If $V_1 > V_0$, $dV/dx = +\infty$ at $x=0$ and the particle will accordingly receive a blow in the negative sense when it arrives at the origin; supposing it to be incident upon the discontinuity from the region $x<0$, the effect will be that of a barrier at $x=0$ which has to be penetrated before the particle can enter the region $x>0$ and resume its free motion.

It will be assumed that the particle is injected into this force field with a sharp value for its total energy $E$; it will accordingly be in a steady state. Before studying the predictions of quantum mechanics with regard to its motion, let us analyse the situation by the methods of Newtonian mechanics. Since the force field is conservative, if the particle penetrates the barrier, energy will be conserved; hence, if $p$ is its momentum in the region $x<0$ and $p'$ is its momentum in the region $x>0$, we shall have

$$\frac{p^2}{2m} + V_0 = E = \frac{p'^2}{2m} + V_1. \qquad (4.8.2)$$

It follows that, if $E < V_1$, $p'$ cannot be real and hence the particle cannot enter the region $x>0$. In these circumstances, if the particle is incident upon the barrier with momentum $p$ from the region $x<0$, it must be reflected by the impulsive force and will return along its previous path with momentum $-p$. If, however, $E > V_1$, such a particle will penetrate the barrier and will subsequently move freely in the region $x > 0$ with momentum $p'$. The predictions of quantum mechanics are rather different.

We will first consider the case where $E > V_0$, $E > V_1$. At the initial instant $t = 0$, when the particle is injected into the force field with energy $E$, its wave function $\psi(x)$ will satisfy the Schrödinger equation (3.8.5). It follows from the equation (4.8.2) that this takes the following forms in the regions $x < 0$, $x > 0$ respectively:

$$\frac{d^2\psi}{dx^2} + \frac{p^2}{\hbar^2}\psi = 0, \tag{4.8.3}$$

$$\frac{d^2\psi}{dx^2} + \frac{p'^2}{\hbar^2}\psi = 0. \tag{4.8.4}$$

The general solutions of these equations are

$$\psi = A\exp(\iota px/\hbar) + B\exp(-\iota px/\hbar), \quad x < 0, \tag{4.8.5}$$

$$\psi = C\exp(\iota p'x/\hbar) + D\exp(-\iota p'x/\hbar), \quad x > 0. \tag{4.8.6}$$

Both $\psi$ and $d\psi/dx$ must be continuous at $x = 0$ (see section 3.9) and hence

$$A + B = C + D, \tag{4.8.7}$$

$$p(A - B) = p'(C - D). \tag{4.8.8}$$

The form of $\psi$ at any later instant can now be obtained from equations (4.8.5) and (4.8.6) by multiplying by an exponential time factor, $\exp(-\iota Et/\hbar)$. Thus, in the region $x < 0$,

$$\psi = A\exp[-\iota(Et - px)/\hbar] + B\exp[-\iota(Et + px)/\hbar], \tag{4.8.9}$$

a result which may be described as the superposition of two trains of complex sinusoidal $\psi$-waves, being propagated in opposite senses along the $x$-axis. The expression for $\psi$ in the region $x > 0$ receives a similar interpretation.

The physical interpretation of this solution for $\psi$ is most easily given in terms of a very large number of non-interacting particles all released under identical initial conditions (i.e. with sharp energy $E$). Since a specification of $E$ does not determine the sense of a particle's motion, we expect to be able to identify two streams of particles, one moving to the left and one to the right. This will be our interpretation of the two trains of waves present in each of the regions $x > 0$, $x < 0$. Thus, in the region

$x<0$, the equation (4.8.9) indicates that there is a stream moving towards the origin having particle density $|A|^2$ and a stream moving away from the origin having particle density $|B|^2$. The current densities (number of particles crossing unit area per second) in these streams are of magnitudes $v|A|^2$, $v|B|^2$ respectively. Similarly, in the region $x>0$, there are two streams with densities $|C|^2$, $|D|^2$ and with current densities $v'|C|^2$, $v'|D|^2$ respectively ($v'=p'/m$).

To maintain the steady state physical situation we have just described, it would be necessary to set up particle sources at $x=-\infty$ and $x=+\infty$. These sources would be provided with a particle emitter and a means of accelerating the particles to an energy $E$, after which they would be beamed parallel to the $x$-axis; each apparatus would also absorb any particles incident upon it. The values of $|A|$ and $|D|$ could then be controlled by monitoring the intensities of the beams generated at $x=-\infty$ and $x=+\infty$ respectively.

Consider the situation brought about when only the apparatus at $x=-\infty$ is emitting particles. Then $D=0$ and $|A|$ may be assumed known. We now have a single beam incident upon the barrier at $x=0$ and this gives rise to a reflected beam and a transmitted beam. The intensities of these beams are determined by $|B|$ and $|C|$ respectively. Solving equations (4.8.7), (4.8.8), we find that

$$B = \frac{p-p'}{p+p'}A, \quad C = \frac{2p}{p+p'}A. \tag{4.8.10}$$

Hence, if $j_0=v|A|^2$ is the current density for the incident beam, it now follows that the current densities for the reflected and transmitted beams are

$$j_r = v|B|^2 = \left(\frac{p-p'}{p+p'}\right)^2 j_0, \tag{4.8.11}$$

$$j_t = v'|C|^2 = \frac{4pp'}{(p+p')^2}j_0, \tag{4.8.12}$$

respectively. It will be found that

$$j_r+j_t = j_0, \tag{4.8.13}$$

as is clearly necessary, since every incident particle will be either transmitted across the barrier or reflected. We can now conclude that the proportion of incident particles which are transmitted is

$$T = j_t/j_0 = \frac{4pp'}{(p+p')^2}; \tag{4.8.14}$$

$T$ is the *transmission coefficient*. The proportion which are reflected is

$$R = j_r/j_0 = \left(\frac{p-p'}{p+p'}\right)^2; \qquad 4.8.15)$$

$R$ is the *reflection coefficient*. For a single particle incident upon the barrier, $T$ is the probability it will be transmitted and $R$ is the probability it will be reflected. It will be noted that, as the height of the barrier is reduced so that $p' \to p$, then $R \to 0$ and $T \to 1$. This implies that there is a continuous relationship between the cause (barrier) and the effect (reflection and transmission of particles), a result which is entirely in accord with Leibniz's principle (section 1.1); according to classical theory, all particles are transmitted.

An unexpected feature of the quantum mechanical solution is that, if $V_1 < V_0$ so that the barrier is replaced by a precipice, then $p' > p$ and $R > 0$, i.e. some particles are reflected. In fact, the deeper the precipice, the larger the value of $R$ and the greater the probability a particle will be reflected.

We will next consider the case where $V_0 < E < V_1$ and, according to classical ideas, particles incident upon the barrier from $x = -\infty$ are all reflected and cannot penetrate into the region $x > 0$. The solutions (4.8.5), (4.8.6) remain valid, except that $p'$ will now be imaginary and we shall write

$$p' = \iota\gamma, \quad \gamma = \sqrt{[2m(V_1 - E)]}. \qquad (4.8.16)$$

The solution for $\psi$ in the region $x > 0$ can then be written

$$\psi = [C\exp(-\gamma x/\hbar) + D\exp(\gamma x/\hbar)]\exp(-\iota Et/\hbar) \qquad (4.8.17)$$

and no longer represents a pair of $\psi$-waves. $C$ and $D$ cannot receive their earlier physical interpretation, therefore. Since $\psi \to \infty$ as $x \to +\infty$ is unrealistic, we must take $D = 0$ as before. Solving equations (4.8.7), (4.8.8), we obtain

$$B = \frac{p - \iota\gamma}{p + \iota\gamma}A, \quad C = \frac{2p}{p + \iota\gamma}A. \qquad (4.8.18)$$

We now calculate that $|B| = |A|$ and hence that

$$j_r = j_0, \qquad (4.8.19)$$

i.e. all particles are reflected. In this respect, the quantum mechanical and classical theories are in accord. However, the particle density in the region $x > 0$ is not zero, but is given by

$$\begin{aligned}\psi\psi^* &= |C|^2 \exp(-2\gamma x/\hbar) = \frac{4p^2}{p^2+\gamma^2}|A|^2 \exp(-2\gamma x/\hbar), \\ &= \frac{4(E-V_0)}{V_1-V_0}|A|^2 \exp(-2\gamma x/\hbar). \end{aligned} \tag{4.8.20}$$

The prediction is that some particles will penetrate into the region $x > 0$, but that the probability of penetrating to a given depth decreases with the depth according to an exponential law. The apparent anomaly that, if energy is to be conserved, a particle having energy $E$ cannot be found in the region $x > 0$, is resolved as in section 3.9.

## 4.9. The tunnel effect

Suppose that a stream of particles moving parallel to $Ox$, each having momentum $p$, is incident upon a potential barrier specified by the equations

$$\left.\begin{aligned} V &= V_0, \quad x < 0, \\ &= V_1, \quad 0 < x < a, \\ &= V_0, \quad x > a, \end{aligned}\right\} \tag{4.9.1}$$

where $V_1 > V_0$. If we suppose that the total energy $E$ of a particle is less than $V_1$, then the prediction of classical mechanics is that no particle will penetrate the barrier into the region $x > a$, and hence all particles will be reflected. However, we know from the analysis carried out in the last section that particles must be assumed to penetrate into the region $0 < x < a$ and hence we expect that a proportion of the incident particles will 'tunnel through' the barrier into the region $x > a$, where they will constitute a transmitted stream.

In the region $x < 0$, neglecting the exponential time factor, we take

$$\psi = A\exp(\iota px/\hbar) + B\exp(-\iota px/\hbar), \tag{4.9.2}$$

where $A$ determines the intensity of the incident stream and $B$ the intensity of the reflected stream. In the region $0 < x < a$, we take

$$\psi = C\exp(\gamma x/\hbar) + D\exp(-\gamma x/\hbar), \tag{4.9.3}$$

where

$$\gamma^2 = 2m(V_1 - E). \tag{4.9.4}$$

In the region $x > a$, we take

$$\psi = F \exp(\iota px/\hbar), \tag{4.9.5}$$

determining the transmitted stream.

The continuity of $\psi$, $d\psi/dx$, across the P.E. discontinuities at $x=0$ and $x=a$, leads to the equations

$$A+B = C+D, \tag{4.9.6}$$

$$A-B = \iota\alpha(D-C), \tag{4.9.7}$$

$$C\exp(\gamma a/\hbar)+D\exp(-\gamma a/\hbar) = F\exp(\iota pa/\hbar), \tag{4.9.8}$$

$$\iota\alpha[D\exp(-\gamma a/\hbar)-C\exp(\gamma a/\hbar)] = F\exp(\iota pa/\hbar), \tag{4.9.9}$$

where

$$\alpha = \frac{\gamma}{p} = \sqrt{\left(\frac{V_1-E}{E-V_0}\right)}. \tag{4.9.10}$$

Eliminating $B$, $C$, $D$ and solving for $F$, it will be found that

$$\frac{F}{A} = \frac{\exp(-\iota pa/\hbar)}{\cosh(\gamma a/\hbar)+\frac{1}{2}\iota(\alpha-1/\alpha)\sinh(\gamma a/\hbar)}. \tag{4.9.11}$$

It now follows that the ratio of the current densities in the transmitted and incident streams is

$$T = \left|\frac{F}{A}\right|^2 = \frac{1}{1+\frac{1}{4}(\alpha+1/\alpha)^2\sinh^2(\gamma a/\hbar)}, \tag{4.9.12}$$

i.e. this proportion of the incident particles will penetrate the barrier.

If $\gamma a$ is small by comparison with $\hbar$, we may approximate by taking $\sinh(\gamma a/\hbar)=\gamma a/\hbar$ and then, to the second order in $\gamma a/\hbar$,

$$T = 1-\left[\tfrac{1}{2}\left(\alpha+\frac{1}{\alpha}\right)\frac{\gamma a}{\hbar}\right]^2. \tag{4.9.13}$$

At the other extreme, if $\gamma a$ is large by comparison with $\hbar$ we can take

$$\sinh(\gamma a/\hbar) = \tfrac{1}{2}\exp(\gamma a/\hbar) \tag{4.9.14}$$

and then

$$T = 16\left(\alpha+\frac{1}{\alpha}\right)^{-2}\exp(-2\gamma a/\hbar) \tag{4.9.15}$$

approximately. The graph of $T$ against $\gamma a/\hbar$ is now easily sketched. Clearly, $T$ is a monotonically decreasing function of $a$ tending to zero as $a$ becomes large. As might be expected, the barrier rapidly becomes 100% effective as its width increases.

The tunnel effect provides explanations for two phenomena which are inconsistent with earlier theories. First there is the cold extraction of electrons from metals by the application of an electric field; the electrons can be attracted away to an anode plate raised to a positive potential and an electric current then flows into the plate. The electrons are trapped in the metal by a potential barrier existing at its surface and, according to classical theory, no electron should be able to leave the metal until this barrier has been sufficiently lowered by the applied field. However, experiment reveals that some current flows into the anode plate even for small applied positive potentials and that this current rises in a continuous manner as the applied potential is increased. This is accounted for by the fact that, even in the presence of the barrier, there is a finite probability that an electron will be able to escape and that, as the barrier is lowered, this probability increases continuously.

Another phenomenon which receives its explanation in terms of the tunnel effect, is the emission of $\alpha$-particles from a radioactive nucleus. Since an $\alpha$-particle is positively charged, when it is outside the nucleus it is repelled by the nucleus according to an inverse square law. However, when it is inside the nucleus this repulsion is completely nullified by the short range, but extremely powerful, nuclear field of attraction. The graph of the P.E. of such a particle in relation to its distance from the centre of the atom is accordingly in the form of a 'well'. An $\alpha$-particle placed in the nucleus with total energy $E$ and executing a one-dimensional motion along a line through the centre would, according to classical ideas, remain trapped in the well indefinitely, being reflected by the walls of the well each time it is incident upon them. However, quantum mechanics predicts that there is a finite probability that the particle will tunnel through the barrier at each impact and hence that, eventually, the particle will escape.

## 4.10. Transmission resonances

In this section, we will consider the problem of a stream of particles incident upon a potential well. Thus, we shall again take $V$ to be given by the equations (4.9.1), but assume that $V_1 < V_0$. Equation (4.9.4) shows that $\gamma$ is now imaginary and is given by

$$\gamma = \iota\sqrt{[2m(E-V_1)]} = \iota p', \tag{4.10.1}$$

where $p'$ is the momentum of a particle of energy $E$ in the well.

From equation (4.9.10), it follows that

$$\alpha = \iota p'/p \tag{4.10.2}$$

and hence equation (4.9.11) is replaced by

$$\frac{F}{A} = \frac{\exp(-\iota pa/\hbar)}{\cos(p'a/\hbar) - \frac{1}{2}\iota\left(\frac{p'}{p} + \frac{p}{p'}\right)\sin(p'a/\hbar)}. \tag{4.10.3}$$

The ratio of the current densities in the transmitted and incident beams is accordingly

$$T = \left|\frac{F}{A}\right|^2 = \left[1 + \tfrac{1}{4}\left(\frac{p}{p'} - \frac{p'}{p}\right)^2 \sin^2\frac{p'a}{\hbar}\right]^{-1}. \tag{4.10.4}$$

If $p' = p$, then $T = 1$ as expected, for then the potential energy is $V_0$ everywhere and there is no well. However, $T = 1$ also provided

$$\sin\frac{p'a}{\hbar} = 0,$$

i.e. provided

$$p' = \frac{nh}{2a}, \tag{4.10.5}$$

where $n$ is a positive integer. This implies that, if the particle energy $E$ takes any one of a set of discrete values enabling condition (4.10.5) to be satisfied, then the beam will be transmitted across the well without attenuation, i.e. the well will be quite transparent to particles having these energies. This is the phenomenon of *transmission resonance*. For particles which are incident upon the well with other energies, there is a finite probability that reflection will take place at one wall or the other, i.e. such particles will be scattered. This result is, of course, quite contrary to the prediction of classical mechanics, according to which a particle will be impulsively accelerated at the first wall and equally retarded at the second wall but will, in all cases, traverse the well without reflection.

This phenomenon can be used to explain the *Ramsauer effect*. The P.E. of an electron within an atom of an inert gas, such as argon, has approximately the form we have been assuming. It follows that such a gas should prove to be relatively transparent to a beam of electrons possessing a critical energy satisfying equation (4.10.5) and hence that, for such a beam, scattering should be small. Ramsauer arrived at this conclusion experimentally.

## 4.11. Heisenberg's representation

Let $\psi^1, \psi^2, \ldots$ be vectors representing a complete set of eigenstates with respect to a maximal set $X$ of compatible observables for a system.

Suppose the system is prepared in the first eigenstate at some initial instant $t=t_0$; then, if $\psi^1(t)$ is the vector representing the state of the system at some later instant $t$, assuming that the system remains undisturbed during the interval $(t_0,t)$, $\psi^1(t)$ will be determined by the equation

$$\hat{H}\psi^1 = \iota\hbar\frac{d\psi^1}{dt}, \tag{4.11.1}$$

where $\hat{H}$ is the Hamiltonian operator for the system. The vectors $\psi^i(t)$ $(i=2,3,\ldots)$ are defined similarly. It must be remarked that, although the $\psi^i$ represent eigenstates with respect to $X$ at the instant $t=t_0$, they do not necessarily represent such eigenstates at a later instant $t$; for example, one of the observables belonging to $X$ might be a component of the linear momentum of one of the particles of the system and, if this is sharp at $t=t_0$, it will not remain sharp unless the same component of the applied force vanishes. Nevertheless, the $\psi^i$ are mutually orthogonal and normalised at $t=t_0$ and, since we have already proved in section 4.1 that the scalar product of two vectors representing states of the same system does not change with $t$, these vectors $\psi^i$ will remain orthogonal for all $t$. It follows that the $\psi^i$ may be employed as a basis for a matrix representation at all times $t$. Such a representation is called a *Heisenberg Representation*.

Previously, we have adopted a fixed orthonormal set as the basis for our representation of a system's state at any instant $t$. The components of the state vector with respect to such a fixed frame naturally varied with $t$ and hence the column matrix specifying the state was a function of $t$. This is termed a *Schrödinger Representation*. In the case of a Heisenberg representation, the frame varies with $t$ and it does so in such a manner that the state vector remains immovable in the frame and its components are constant. This follows from the fact that, if $\alpha(t)$ is the state vector, then $\alpha_i=\langle\psi^i|\alpha\rangle$ is its $i$th component in the frame and this scalar product is independent of $t$. Hence, using a Heisenberg representation, the system's states will be specified by constant column matrices.

However, whereas in a Schrödinger representation observables are represented by matrices which are independent of $t$, in a Heisenberg representation such matrices are functions of $t$. For, if $a$ is any observable and $\hat{a}$ is the operator associated with it, the elements $a_{ij}$ of the matrix representing it are given by equation (1.8.22) and hence, the $\psi^i$ being dependent upon $t$, so are the quantities $a_{ij}$. In particular, suppose the $\psi^i$ represent a set of energy eigenstates at $t=t_0$, $E$ taking the eigenvalue $E$

in the state $\psi^i$. Then, provided $\hat{H}$ is not explicitly dependent upon $t$, $E$ will remain sharp in each of these states and, by equation (4.4.4), we can write

$$\psi^i = \alpha^i \exp(-\iota E_i t/\hbar), \tag{4.11.2}$$

where the vectors $\alpha^i$ are independent of $t$. Substituting in equation (1.8.22), we now obtain

$$a_{ij} = \langle \alpha^i | \hat{a}\alpha^j \rangle \exp[\iota(E_i - E_j)t/\hbar], \tag{4.11.3}$$

showing that the elements of the matrix **a** oscillate as complex sinusoids (except for the elements in the principal diagonal which are seen to be constants). This is the original representation employed by Heisenberg. It is clear that, in this representation, since the basis is a set of energy eigenvectors for all $t$, the Hamiltonian $H$ will be represented by a diagonal matrix, with the energy eigenvalues arranged along the principal diagonal.

The matrix **a** representing an observable $a$ being dependent upon $t$ in a Heisenberg representation, it is pertinent to consider its time rate of change. We have

$$\begin{aligned}\frac{da_{ij}}{dt} &= \frac{d}{dt}\langle \psi^i | \hat{a}\psi^j \rangle, \\ &= \left\langle \psi^i \middle| \left\{ \frac{\iota}{\hbar}[\hat{H}, \hat{a}] + \frac{\partial \hat{a}}{\partial t} \right\} \psi^j \right\rangle,\end{aligned} \tag{4.11.4}$$

the calculation being exactly similar to that which resulted in equation (4.2.4). This last equation asserts that $da_{ij}/dt$ is the $ij$th element of the matrix representing the observable whose associated operator is

$$\frac{\iota}{\hbar}[\hat{H}, \hat{a}] + \frac{\partial \hat{a}}{\partial t}, \tag{4.11.5}$$

i.e. the observable which has been denoted in section (4.2) by $\dot{a}$. But the matrix representing this observable is plainly

$$\frac{\iota}{\hbar}[\mathbf{H}, \mathbf{a}] + \frac{\partial \mathbf{a}}{\partial t}. \tag{4.11.6}$$

It follows that

$$\frac{d\mathbf{a}}{dt} = \frac{\iota}{\hbar}[\mathbf{H}, \mathbf{a}] + \frac{\partial \mathbf{a}}{\partial t}. \tag{4.11.7}$$

This is *Heisenberg's equation of motion* for the observable $a$.

In classical mechanics, if $H = H(p_i, q_i)$, where the $q_i$ are generalised coordinates and $p_i$ are the corresponding components of momentum, if $F = F(p_i, q_i, t)$ is any observable, then

$$\frac{dF}{dt} = \sum_i \left(\frac{\partial F}{\partial q_i}\frac{\partial H}{\partial p_i} - \frac{\partial F}{\partial p_i}\frac{\partial H}{\partial q_i}\right) + \frac{\partial F}{\partial t},$$

$$= [F, H] + \frac{\partial F}{\partial t}, \tag{4.11.8}$$

where $[F, H]$ is called a *Poisson bracket*. Equation (4.11.7) is clearly the quantum mechanical equivalent of the classical equation (4.11.8).

In the special case when the observable $a$ is not explicitly dependent upon $t$ and $\mathbf{H}$, $\mathbf{a}$ commute, equation (4.11.7) reduces to

$$\frac{d\mathbf{a}}{dt} = \mathbf{0}. \tag{4.11.9}$$

This implies that $\mathbf{a}$ is a matrix with constant elements. In this case, as we have already shown, the probability distribution for $a$ is independent of $t$ and we call $a$ a *constant of the motion*.

As an example, consider a system comprising a single particle whose coordinates are $x_i$ $(i = 1,2,3)$ and whose components of momentum are $p_i$ $(i = 1,2,3)$. Then

$$\frac{d\mathbf{x}_i}{dt} = \frac{\iota}{\hbar}[\mathbf{H}, \mathbf{x}_i] = -\frac{\partial \mathbf{H}}{\partial \mathbf{x}_i}, \tag{4.11.10}$$

$$\frac{d\mathbf{p}_i}{dt} = \frac{\iota}{\hbar}[\mathbf{H}, \mathbf{p}_i] = \frac{\partial \mathbf{H}}{\partial \mathbf{p}_i}, \tag{4.11.11}$$

assuming $H$ is a polynomial in its arguments $x_i, p_i$. These are the quantum mechanical counterparts of Hamilton's equations for a particle from the classical theory. It is evident that a Heisenberg representation provides the natural counterparts for quantum mechanics of the equations of motion appearing in the classical theory.

## 4.12. Heisenberg's treatment of the harmonic oscillator problem

As an example of the use of the equations established in the previous section, we will now consider the one-dimensional harmonic oscillator for which the Hamiltonian is given by

$$H = \frac{1}{2m}p^2 + \tfrac{1}{2}m\omega^2 x^2. \tag{4.12.1}$$

Hamilton's equations for the Heisenberg matrices **x**, **p**, take the form

$$\frac{d\mathbf{x}}{dt} = \frac{1}{m}\mathbf{p}, \quad \frac{d\mathbf{p}}{dt} = -m\omega^2\mathbf{x}. \tag{4.12.2}$$

Whence, eliminating **x**, we find that

$$\frac{d^2\mathbf{p}}{dt^2}+\omega^2\mathbf{p} = 0. \tag{4.12.3}$$

It follows that

$$\mathbf{p} = \mathbf{a}\exp(\iota\omega t)+\mathbf{b}\exp(-\iota\omega t), \tag{4.12.4}$$

where **a**, **b** are constant matrices. Hence

$$\iota m\omega\mathbf{x} = \mathbf{a}\exp(\iota\omega t)-\mathbf{b}\exp(-\iota\omega t). \tag{4.12.5}$$

To proceed further, we adopt the special Heisenberg representation based upon a complete set of energy eigenstates. In this representation, **H** is diagonal with diagonal elements $E_0$, $E_1$, $E_2$, ... equal to the energy eigenvalues. The matrix form of equation (4.12.1) accordingly yields the result

$$2mE_i = \sum_{j=0}^{\infty} p_{ij}p_{ji}+m^2\omega^2\sum_{j=0}^{\infty} x_{ij}x_{ji} \geqslant 0, \tag{4.12.6}$$

the inequality following from the fact that the matrices **x**, **p** are Hermitian and hence $p_{ji}=p_{ij}^*$, $x_{ji}=x_{ij}^*$. Thus, the energy eigenvalues are all positive. We shall assume that the basic eigenstates are arranged so that $0<E_0<E_1<E_2<\ldots$ (we also assume that the system is not degenerate with respect to the energy observable).

Now, we know that all the elements of **x** and **p** depend upon $t$ as indicated in equation (4.11.3). It follows from equations (4.12.4), (4.12.5) that

$$\text{if } E_i-E_j \neq \omega\hbar, \quad \text{then } a_{ij} = 0, \tag{4.12.7}$$

and

$$\text{if } E_i-E_j \neq -\omega\hbar, \quad \text{then } b_{ij} = 0. \tag{4.12.8}$$

Suppose the eigenvalue $E_N$ is such that $E_N-\omega\hbar$ is not an eigenvalue. Then $E_N-E_j\neq\omega\hbar$ for any $j$. Hence $a_{Nj}=b_{iN}=0$ for all $i$, $j$. From equations (4.12.4), (4.12.5), we derive the equations

$$2\mathbf{a}\exp(\iota\omega t) = \mathbf{p}+\iota m\omega\mathbf{x}, \quad 2\mathbf{b}\exp(-\iota\omega t) = \mathbf{p}-\iota m\omega\mathbf{x}. \tag{4.12.9}$$

Multiplying these equations together, we obtain

$$\begin{aligned} 4\mathbf{ab} &= \mathbf{p}^2+\omega^2m^2\mathbf{x}^2-\iota\omega m(\mathbf{px}-\mathbf{xp}),\\ &= 2m\mathbf{H}-m\omega\hbar\mathbf{I}, \end{aligned} \tag{4.12.10}$$

where we have made use of the commutation identity (3.6.15). But **H** is diagonal with diagonal elements $E_i$ and hence

$$2mE_i = 4\sum_{j=0}^{\infty} a_{ij}b_{ji}+m\omega\hbar. \tag{4.12.11}$$

In particular, putting $i=N$, this reduces to

$$E_N = \tfrac{1}{2}\omega\hbar. \tag{4.12.12}$$

This shows that only one eigenvalue is such that when $\omega\hbar$ is subtracted the result is not an eigenvalue. But $E_0$ is such an eigenvalue and hence $E_0=\frac{1}{2}\omega\hbar$. The other eigenvalues must form an arithmetical progression with common difference $\omega\hbar$ and thus

$$E_n = E_0+n\omega\hbar = (n+\tfrac{1}{2})\,\omega\hbar, \tag{4.12.13}$$

in agreement with equation (3.10.32).

We have proved, therefore, that

$$E_n-E_{n-1} = \omega\hbar \tag{4.12.14}$$

and it follows from (4.12.7), (4.12.8) that the only non-zero elements of the matrices **a**, **b** are $a_{n,\,n-1}$, $b_{n-1,\,n}$ $(n=1,2,3,\ldots)$. Equation (4.12.11) accordingly yields the result

$$\begin{aligned} 2mE_n &= 4a_{n,\,n-1}\,b_{n-1,\,n}+m\omega\hbar, \\ &= 4a_{n,\,n-1}\,a^*_{n,\,n-1}+m\omega\hbar, \end{aligned} \tag{4.12.15}$$

since, **p** and **x** being Hermitian, it is a consequence of equations (4.12.9) that $a^*_{ij}=b_{ji}$. We now deduce that

$$a_{n,\,n-1} = (\tfrac{1}{2}m\omega\hbar)^{\frac{1}{2}}\,n^{\frac{1}{2}}\exp(\iota\alpha), \tag{4.12.16}$$

where $\alpha$ is arbitrary. Also

$$b_{n-1,\,n} = a^*_{n,\,n-1} = (\tfrac{1}{2}m\omega\hbar)^{\frac{1}{2}}\,n^{\frac{1}{2}}\exp(-\iota\alpha). \tag{4.12.17}$$

Hence, from equations (4.12.4), (4.12.5), we obtain

$$\left.\begin{aligned} x_{n,\,n-1} &= \left(\frac{\hbar}{2m\omega}\right)^{\frac{1}{2}} n^{\frac{1}{2}}\exp[\iota(\omega t+\alpha)], \\ p_{n,\,n-1} &= (-\tfrac{1}{2}m\omega\hbar)^{\frac{1}{2}}\,n^{\frac{1}{2}}\exp[\iota(\omega t+\alpha)], \\ x_{n-1,\,n} &= \left(\frac{\hbar}{2m\omega}\right)^{\frac{1}{2}} n^{\frac{1}{2}}\exp[-\iota(\omega t+\alpha)], \\ p_{n-1,\,n} &= -(-\tfrac{1}{2}m\omega\hbar)^{\frac{1}{2}}\,n^{\frac{1}{2}}\exp[-\iota(\omega t+\alpha)], \end{aligned}\right\} \tag{4.12.18}$$

where $\alpha - \frac{1}{2}\pi$ has been replaced $\alpha$. All other elements of the matrices **x**, **p** are zero. Allowing for the difference of mode of representation, the results (4.12.18) are clearly equivalent to equations (3.11.5), (3.11.7).

## 4.13. Preservation of symmetry and anti-symmetry of system states

It has been explained in section 3.12 that the wave function representing the state of a system of indistinguishable particles must be either symmetric or anti-symmetric with respect to an exchange of the coordinates and spin components of any two particles. If the wave function is symmetric the particles are bosons and if it is anti-symmetric they are fermions. However, if the wave function is known at an initial instant $t=0$, the equation of motion for the system (4.1.7) will determine the form of the wave function at all later instants prior to the next observation. For consistency it is necessary, therefore, that this equation should preserve the properties of symmetry and anti-symmetry of wave functions as time elapses.

Consider the Hamiltonian of a system of identical particles. This will be a function of observables associated with each particle and, since each particle must contribute to it in the same way as every other particle, it will be a symmetric function of these observables; i.e. if the observables associated with two particles are exchanged, the form of the Hamiltonian is unaltered. Thus if $q_i$ represents the set of coordinates and spin components of the $i$th particle, the operator $\hat{H}$ will be a symmetric expression in the $q_i$.

Let $\psi(q_1, q_2, \ldots q_N)$ be the wave function for the system and suppose it is anti-symmetric at time $t$. Then

$$\hat{H}\psi = \iota\hbar \frac{\partial \psi}{\partial t} \tag{4.13.1}$$

and hence the change in $\psi$ during the interval $(t, t+dt)$ is given to the first order by

$$d\psi = -\frac{\iota}{\hbar}\hat{H}\psi\, dt. \tag{4.13.2}$$

But $\hat{H}$ is symmetric and $\psi$ is anti-symmetric in the $q_i$ and therefore it follows from the last equation that $d\psi$ is anti-symmetric. Hence, $\psi + d\psi$ is anti-symmetric and this property of the wave function is accordingly preserved over the interval $dt$. By integration, anti-symmetry will be preserved over any finite time interval.

By a similar argument it may be shown that symmetry of the wave function is also preserved.

CHAPTER 5

# Angular momentum

## 5.1. Angular momentum operators

Suppose that a particle moves relative to an inertial frame determined by rectangular cartesian axes $Oxyz$. If $(p_x, p_y, p_z)$ are the components of its linear momentum in this frame, the classical observable termed the *angular momentum* of the particle about $O$ has components

$$\left.\begin{aligned} L_x &= yp_z - zp_y, \\ L_y &= zp_x - xp_z, \\ L_z &= xp_y - yp_x. \end{aligned}\right\} \tag{5.1.1}$$

Employing the coordinate representation, the corresponding quantum mechanical operators are therefore expected to be

$$\left.\begin{aligned} \hat{L}_x &= \hat{y}\hat{p}_z - \hat{z}\hat{p}_y = \frac{\hbar}{\iota}\left(y\frac{\partial}{\partial z} - z\frac{\partial}{\partial y}\right), \\ \hat{L}_y &= \hat{z}\hat{p}_x - \hat{x}\hat{p}_z = \frac{\hbar}{\iota}\left(z\frac{\partial}{\partial x} - x\frac{\partial}{\partial z}\right), \\ \hat{L}_z &= \hat{x}\hat{p}_y - \hat{y}\hat{p}_x = \frac{\hbar}{\iota}\left(x\frac{\partial}{\partial y} - y\frac{\partial}{\partial x}\right). \end{aligned}\right\} \tag{5.1.2}$$

These operators are clearly Hermitian, for the products $\hat{y}\hat{p}_z$, etc. are all commutative.

Unlike the components of linear momentum, the components of angular momentum are not compatible. This follows since

$$\begin{aligned} [\hat{L}_x, \hat{L}_y] &= -\hbar^2\left(y\frac{\partial}{\partial z} - z\frac{\partial}{\partial y}\right)\left(z\frac{\partial}{\partial x} - x\frac{\partial}{\partial z}\right) + \hbar^2\left(z\frac{\partial}{\partial x} - x\frac{\partial}{\partial z}\right)\left(y\frac{\partial}{\partial z} - z\frac{\partial}{\partial y}\right), \\ &= \hbar^2\left(x\frac{\partial}{\partial y} - y\frac{\partial}{\partial x}\right), \\ &= \iota\hbar\hat{L}_z, \end{aligned} \tag{5.1.3}$$

with similar results for the commutators $[\hat{L}_y, \hat{L}_z]$ and $[\hat{L}_z, \hat{L}_x]$. In general, therefore, only one of the components $L_x$, $L_y$, $L_z$ can be assumed sharp at any instant.†

† If $x = y = z = 0$, the components of linear momentum are unknown; however, in this case, $L_x = L_y = L_z = 0$ and the components of angular momentum are sharp simultaneously. This is an exceptional case.

The identity (5.1.3) explains why it is reasonable to take the eigenvalues of a component of spin, for a particle possessing two spin eigenstates, to be $\pm\frac{1}{2}\hbar$. For suppose the eigenvalues of a spin component are taken, quite generally, to be $\pm\alpha\hbar+\beta$. Then

$$s_x = \alpha\hbar\sigma_x+\beta, \text{ etc.} \tag{5.1.4}$$

Employing any matrix representation, equations (2.2.28) will be valid and it then follows that

$$\begin{aligned}[\mathbf{s}_x, \mathbf{s}_y] &= [\alpha\hbar\boldsymbol{\sigma}_x+\beta\mathbf{I}_2, \alpha\hbar\boldsymbol{\sigma}_y+\beta\mathbf{I}_2], \\ &= \alpha^2\hbar^2[\boldsymbol{\sigma}_x, \boldsymbol{\sigma}_y], \\ &= 2\iota\alpha^2\hbar^2\boldsymbol{\sigma}_z, \\ &= 2\iota\alpha\hbar\mathbf{s}_z-2\iota\alpha\beta\hbar\mathbf{I}_2. \end{aligned} \tag{5.1.5}$$

However, since the spin has been regarded as the intrinsic angular momentum of a particle, it is reasonable to expect that its components will be related by identities of the type (5.1.3). Equation (5.1.5) shows that our expectations will be realised if we take $\alpha=\frac{1}{2}$ and $\beta=0$.

Expressions for the angular momentum operators in terms of spherical polar coordinates $(r,\theta,\phi)$ will next be derived. These coordinates are related to the cartesian coordinates $(x,y,z)$ by the equations

$$x = r\sin\theta\cos\phi, \quad y = r\sin\theta\sin\phi, \quad z = r\cos\theta. \tag{5.1.6}$$

Hence

$$\frac{\partial}{\partial\phi} = \frac{\partial x}{\partial\phi}\frac{\partial}{\partial x}+\frac{\partial y}{\partial\phi}\frac{\partial}{\partial y}+\frac{\partial z}{\partial\phi}\frac{\partial}{\partial z},$$

$$= -r\sin\theta\sin\phi\frac{\partial}{\partial x}+r\sin\theta\cos\phi\frac{\partial}{\partial y}, \tag{5.1.7}$$

$$= x\frac{\partial}{\partial y}-y\frac{\partial}{\partial x}. \tag{5.1.8}$$

Thus

$$\hat{L}_z = \frac{\hbar}{\iota}\frac{\partial}{\partial\phi}. \tag{5.1.9}$$

Also

$$\frac{\partial}{\partial\theta} = \frac{\partial x}{\partial\theta}\frac{\partial}{\partial x}+\frac{\partial y}{\partial\theta}\frac{\partial}{\partial y}+\frac{\partial z}{\partial\theta}\frac{\partial}{\partial z}$$

$$= r\cos\theta\cos\phi\frac{\partial}{\partial x}+r\cos\theta\sin\phi\frac{\partial}{\partial y}-r\sin\theta\frac{\partial}{\partial z}. \tag{5.1.10}$$

Eliminating the operator $\partial/\partial x$ between equations (5.1.7) and (5.1.10), the following result is obtained:

$$\sin\phi\frac{\partial}{\partial\theta}+\cot\theta\cos\phi\frac{\partial}{\partial\phi} = r\cos\theta\frac{\partial}{\partial y}-r\sin\theta\sin\phi\frac{\partial}{\partial z},$$
$$= z\frac{\partial}{\partial y}-y\frac{\partial}{\partial z}. \qquad (5.1.11)$$

This implies that

$$\hat{L}_x = -\frac{\hbar}{\iota}\left(\sin\phi\frac{\partial}{\partial\theta}+\cot\theta\cos\phi\frac{\partial}{\partial\phi}\right). \qquad (5.1.12)$$

Similarly, by eliminating the operator $\partial/\partial y$ between equations (5.1.7) and (5.1.10), it will be found that

$$\hat{L}_y = \frac{\hbar}{\iota}\left(\cos\phi\frac{\partial}{\partial\theta}-\cot\theta\sin\phi\frac{\partial}{\partial\phi}\right). \qquad (5.1.13)$$

If $L$ is the magnitude of the angular momentum vector, then in classical mechanics we have

$$L^2 = L_x^2+L_y^2+L_z^2. \qquad (5.1.14)$$

We shall take the operator $\hat{L}^2$ representing this observable in quantum mechanics to be defined by the equation

$$\hat{L}^2 = \hat{L}_x^2+\hat{L}_y^2+\hat{L}_z^2. \qquad (5.1.15)$$

Since $\hat{L}_x^2$, etc. are Hermitian, so is $\hat{L}^2$.

It may be shown that this observable $L^2$ is compatible with each of the observables $L_x$, $L_y$ and $L_z$. For

$$\begin{aligned}
[\hat{L}^2, \hat{L}_x] &= [\hat{L}_x^2+\hat{L}_y^2+\hat{L}_z^2, \hat{L}_x],\\
&= [\hat{L}_x^2, \hat{L}_x]+[\hat{L}_y^2, \hat{L}_x]+[\hat{L}_z^2, \hat{L}_x],\\
&= [\hat{L}_y\hat{L}_y, \hat{L}_x]+[\hat{L}_z\hat{L}_z, \hat{L}_x],\\
&= \hat{L}_y[\hat{L}_y, \hat{L}_x]+[\hat{L}_y, \hat{L}_x]\hat{L}_y\\
&\quad +\hat{L}_z[\hat{L}_z, \hat{L}_x]+[\hat{L}_z, \hat{L}_x]\hat{L}_z,\\
&= -\iota\hbar\hat{L}_y\hat{L}_z-\iota\hbar\hat{L}_z\hat{L}_y\\
&\quad +\iota\hbar\hat{L}_z\hat{L}_y+\iota\hbar\hat{L}_y\hat{L}_z,\\
&= 0,
\end{aligned} \qquad (5.1.16)$$

having employed the identity given in Ex. 9(d) of Chap. 1 and commutation relations such as (5.1.3). Similarly, it may be shown that

$$[\hat{L}^2, \hat{L}_y] = [\hat{L}^2, \hat{L}_z] = 0. \qquad (5.1.17)$$

These results imply that $L^2$ and any one of the components $L_x$, $L_y$, $L_z$ may be measured simultaneously with precision. However, since the components are incompatible, $L^2$ cannot, in general, be measured simultaneously with a *pair* of components.

Lastly, in this section, we will express $L^2$ in spherical polar coordinates. Allowing the right-hand member of equation (5.1.12) to operate upon itself, it may be calculated that

$$\hat{L}_x^2 = -\hbar^2\left[\sin^2\phi\frac{\partial^2}{\partial\theta^2}+2\cot\theta\sin\phi\cos\phi\frac{\partial^2}{\partial\theta\,\partial\phi}+\cot^2\theta\cos^2\phi\frac{\partial^2}{\partial\phi^2}\right.$$
$$\left.+\cot\theta\cos^2\phi\frac{\partial}{\partial\theta}-(\cot^2\theta+\operatorname{cosec}^2\theta)\sin\phi\cos\phi\frac{\partial}{\partial\phi}\right]. \qquad (5.1.18)$$

Similarly, from equation (5.1.13), it may be found that

$$\hat{L}_y^2 = -\hbar^2\left[\cos^2\phi\frac{\partial^2}{\partial\theta^2}-2\cot\theta\sin\phi\cos\phi\frac{\partial^2}{\partial\theta\,\partial\phi}+\cot^2\theta\sin^2\phi\frac{\partial^2}{\partial\phi^2}\right.$$
$$\left.+\cot\theta\sin^2\phi\frac{\partial}{\partial\theta}+(\cot^2\theta+\operatorname{cosec}^2\theta)\sin\phi\cos\phi\frac{\partial}{\partial\phi}\right]. \qquad (5.1.19)$$

Also, from equation (5.1.9), it follows immediately that

$$\hat{L}_z^2 = -\hbar^2\frac{\partial^2}{\partial\phi^2}. \qquad (5.1.20)$$

Adding equations (5.1.18)–(5.1.20), we find that

$$\hat{L}^2 = -\hbar^2\left(\frac{\partial^2}{\partial\theta^2}+\operatorname{cosec}^2\theta\frac{\partial^2}{\partial\phi^2}+\cot\theta\frac{\partial}{\partial\theta}\right),$$
$$= -\hbar^2\left[\frac{1}{\sin\theta}\frac{\partial}{\partial\theta}\left(\sin\theta\frac{\partial}{\partial\theta}\right)+\frac{1}{\sin^2\theta}\frac{\partial^2}{\partial\phi^2}\right]. \qquad (5.1.21)$$

## 5.2. Eigenfunctions and eigenvalues of angular momentum

The spin (if any) of the particle will be neglected until Section 5.5. Suppose the $z$-component of angular momentum of the particle is observed to take the value $L_z$ at some instant $t$. Then, at this instant, the particle's wave function $\psi$ must satisfy the characteristic equation

$$\hat{L}_z\psi = L_z\psi, \qquad (5.2.1)$$

i.e. in spherical polars

$$\frac{\hbar}{\iota}\frac{\partial\psi}{\partial\phi} = L_z\psi. \qquad (5.2.2)$$

This equation is easily integrated with respect to $\phi$ to yield

$$\psi = \Theta \exp\left(\frac{\iota}{\hbar} L_z \phi\right), \tag{5.2.3}$$

where $\Theta$ is an unknown function of $r$ and $\theta$. But $|\psi|$ must evidently be a single-valued function of position in the $xyz$-space and it may, in fact, be proved that $\psi$ must also be single-valued. Consequently, if $\phi$ is increased by $2\pi$ in the right-hand member of equation (5.2.3), $\psi$ must remain unchanged. This will only be the case if

$$L_z = k\hbar, \tag{5.2.4}$$

where $k$ is a positive or negative integer (or zero). It follows, therefore, that $L_z$ is quantised, i.e. if it is measured, the result must necessarily be one of the values given by equation (5.2.4). This characteristic of an angular momentum component in quantum mechanics contrasts sharply with its essentially continuous nature in classical mechanics. With $L_z$ given by equation (5.2.4), the wave function takes the form

$$\psi = \Theta \exp(\iota k \phi). \tag{5.2.5}$$

Similar results follow for the $x$- and $y$-components, with $\phi$ replaced by the appropriate angles. However, it must be remembered that, in general, neither of these observables can possess a sharply defined value at any instant when $L_z$ is known to possess such a value.

Consider next the square of the magnitude of the angular momentum. If this is measured to be $L^2$ precisely at some instant $t$, the particle's wave function $\psi$ at $t+0$ must satisfy

$$\hat{L}^2 \psi = L^2 \psi. \tag{5.2.6}$$

In spherical polars, this equation is expressible in the form

$$\frac{1}{\sin\theta}\frac{\partial}{\partial\theta}\left(\sin\theta\frac{\partial\psi}{\partial\theta}\right)+\frac{1}{\sin^2\theta}\frac{\partial^2\psi}{\partial\phi^2}+\frac{L^2}{\hbar^2}\psi = 0. \tag{5.2.7}$$

Since $\hat{L}^2$, $\hat{L}_z$ commute, it is permissible to suppose that these observables have been measured simultaneously at the instant $t$. Then, if $L_z = k\hbar$, $\psi$ is given by equation (5.2.5) and substituting in equation (5.2.7) we deduce that

$$\frac{1}{\sin\theta}\frac{\partial}{\partial\theta}\left(\sin\theta\frac{\partial\Theta}{\partial\theta}\right)+\left(\frac{L^2}{\hbar^2}-\frac{k^2}{\sin^2\theta}\right)\Theta = 0. \tag{5.2.8}$$

$\Theta$ is a function of the polar coordinates $r$ and $\theta$. However, we shall regard $r$ as a fixed parameter throughout the remainder of the argument of this section, only $\theta$ being permitted to vary; it will accordingly be appropriate to look upon $\Theta$ as being a function of $\theta$ alone and to replace the operator $\partial/\partial\theta$ by $d/d\theta$.

Changing the independent variable in equation (5.2.8) to $x$, where

$$x = \cos\theta, \tag{5.2.9}$$

this equation can be expressed in the form

$$\frac{d}{dx}\left[(1-x^2)\frac{d\Theta}{dx}\right]+\left(\frac{L^2}{\hbar^2}-\frac{k^2}{1-x^2}\right)\Theta = 0. \tag{5.2.10}$$

We now put

$$\Theta = (1-x^2)^{\frac{1}{2}|k|}\frac{d^{|k|}y}{dx^{|k|}}, \tag{5.2.11}$$

$y$ being a new dependent variable which is clearly arbitrary to the extent of a polynomial of degree $(|k|-1)$ (N.B. $x$, $y$ are *not* cartesian coordinates). Then, from equation (5.2.10), it follows that $y$ must satisfy

$$(1-x^2)\,y^{(|k|+2)}-2(|k|+1)\,xy^{(|k|+1)}+\left(\frac{L^2}{\hbar^2}-|k|-k^2\right)y^{(|k|)} = 0. \tag{5.2.12}$$

By Leibniz's Theorem, this equation may be written

$$\frac{d^{|k|}}{dx^{|k|}}\left[(1-x^2)\,y^{(2)}-2xy^{(1)}+\frac{L^2}{\hbar^2}\,y\right] = 0 \tag{5.2.13}$$

and, integrating $|k|$ times, this leads to the equation

$$(1-x^2)\frac{d^2y}{dx^2}-2x\frac{dy}{dx}+\frac{L^2}{\hbar^2}\,y = 0, \tag{5.2.14}$$

after having neglected a polynomial of degree $(|k|-1)$ in the right-hand member; (it is easy to show that a particular integral appropriate to such a polynomial is a polynomial of the same degree and this makes no contribution to $\Theta$ at equation (5.2.11)). Equation (5.2.14) is, of course, identical with the equation (5.2.10) in the case $k=0$.

The equation (5.2.14) possesses singularities at $x=\pm 1$. Let us transfer the singularity at $x=-1$ to the origin by putting

$$1+x = \xi. \tag{5.2.15}$$

The other singularity is then at $\xi=2$. With the new independent variable $\xi$, equation (5.2.14) becomes

$$\xi(2-\xi)\frac{d^2y}{d\xi^2}+2(1-\xi)\frac{dy}{d\xi}+\lambda(\lambda+1)\,y = 0, \tag{5.2.16}$$

where we have put

$$L^2/\hbar^2 = \lambda(\lambda+1). \tag{5.2.17}$$

There is no loss of generality if we take $\lambda \geqslant 0$. Assuming a solution of equation (5.2.16) in the Frobenius form

$$y = \sum_{n=0}^{\infty} a_n \xi^{\rho+n}, \tag{5.2.18}$$

substituting and equating coefficients of powers of $\xi$ to zero, we find that the indicial equation possesses equal roots $\rho=0$ and that the recurrence relationship for $a_n$ is

$$a_n = \frac{(n+\lambda)(n-\lambda-1)}{2n^2} a_{n-1}. \tag{5.2.19}$$

This leads immediately to a solution

$$y = y_1 = 1-\lambda(\lambda+1)\left(\frac{\xi}{2}\right)+\frac{\lambda(\lambda-1)(\lambda+1)(\lambda+2)}{(2!)^2}\left(\frac{\xi}{2}\right)^2-\dots. \tag{5.2.20}$$

Since $a_n/a_{n-1} \to \frac{1}{2}$ as $n \to \infty$, D'Alembert's test indicates that this series is convergent for $|\xi|<2$. A second solution can be found by the usual method, but this will involve a term $y_1 \log \xi$ and, if this is substituted in equation (5.2.11), it will be found that $\Theta$ becomes infinite as $\xi \to 0$ (i.e. $x \to -1$); since the wave function must be finite for $\theta=\pi$, this solution will not suit our purpose.

Substituting the series $y_1$ in equation (5.2.11) and assuming this series does not terminate, it will be found that

$$\Theta = A(1-\tfrac{1}{2}\xi)^{\frac{1}{2}s}\xi^{\frac{1}{2}s}\sum_{n=0}^{\infty} b_n \xi^n, \tag{5.2.21}$$

where $s=|k|$, $A$ is independent of $\xi$ and

$$b_n = \frac{(s-\lambda)(s-\lambda+1)\dots(s-\lambda+n-1)(s+\lambda+1)(s+\lambda+2)\dots(s+\lambda+n)}{2^n n!(s+1)(s+2)\dots(s+n)}. \tag{5.2.22}$$

It is clear that, for sufficiently large $n$, the coefficients $b_n$ will be all positive or all negative. We shall suppose $b_n$ to be ultimately positive (if $b_n$ is ultimately negative, we consider the series for $-\Theta$). Then

$$\frac{b_n}{b_{n-1}} = \frac{(n+s-\lambda-1)(n+s+\lambda)}{2n(n+s)} > \frac{n+\alpha-1}{2n} \tag{5.2.23}$$

for sufficiently large $n$, where $\frac{1}{2}s < \alpha < s$. [Note: $s=0$ is a special case; in this case, we prove instead that

$$\frac{b_n}{b_{n-1}} > \frac{n-1-r}{2(n-r)},$$

where $r$ is an integer greater than $\lambda(\lambda+1)$, and take as comparison series

$$-(\tfrac{1}{2}\xi)^r \log(1-\tfrac{1}{2}\xi).]$$

Taking as comparison series

$$(1-\tfrac{1}{2}\xi)^{-\alpha} = \sum_{n=0}^{\infty} v_n \xi^n, \tag{5.2.24}$$

the coefficients $v_n$ are all positive and

$$\frac{v_n}{v_{n-1}} = \frac{n+\alpha-1}{2n}. \tag{5.2.25}$$

Hence, by the theorem proved in Appendix B, it follows that

$$\sum b_n \xi^n \geqslant K(1-\tfrac{1}{2}\xi)^{-\alpha} + p(\xi), \tag{5.2.26}$$

where $K$ is a positive constant and $p$ is a polynomial. It now follows from equation (5.2.21) that, since $\alpha > \frac{1}{2}s$, $\Theta$ will become infinite as $\xi \to 2-0$, i.e. $\theta \to 0$. The solution for $\Theta$ derived from equation (5.2.20) is accordingly unsatisfactory at $\theta = 0$.

The only possibility remaining is that the series for $y_1$ should terminate. This it will do if $\lambda$ is a positive integer or zero. Thus we shall take $\lambda = l$ (positive integer or zero) and then $y_1$ will be a polynomial of degree $l$ in $\xi$ and hence also in $x$. If the arbitrary constant multiplier is chosen so that the coefficient of $x^l$ in this polynomial is $(2l)!/2^l(l!)^2$, it is called the *Legendre Polynomial* of degree $l$ and is denoted by $P_l(x)$. The corresponding solution for $\Theta$ obtained from equation (5.2.11) is called the *Associated Legendre Function* of the $k$th order and is denoted by $P_l^k(x)$ ($P_l^k$ and $P_l^{-k}$ are taken to be identical). Some properties of these functions will be found derived in Appendix D. It is clear from equation (5.2.11) that $|k| \leqslant l$, for otherwise $\Theta$ (and hence $\psi$) will be identically zero.

It has been demonstrated, therefore, that if $L_z$ is sharp, $L^2$ is also quantised, its spectrum of eigenvalues being given by equation (5.2.17) with $\lambda = l$, i.e.

$$L^2 = l(l+1)\hbar^2.\dagger \tag{5.2.27}$$

Also, if $L^2$ has the sharp value determined by the last equation and, at the

† Note that $l = \frac{1}{2}, 1$, yield the values of $s^2$ for particles having spins $\frac{1}{2}\hbar, \hbar$.

same instant, $L_z$ has the sharp value given by equation (5.2.4), the wave function at this instant must take the form

$$\psi = RP_l^k(\cos\theta)\exp(\iota k\phi), \tag{5.2.28}$$

where $R$ can depend only upon $r$.

It is usual to adopt the notation

$$Y_{lk}(\theta,\phi) = A_{lk}P_l^k(\cos\theta)\exp(\iota k\phi), \tag{5.2.29}$$

where $A_{lk}$ is chosen so that $Y_{lk}$ is normalised. It is convenient to regard $Y_{lk}$ as having been normalised when the integral of the square of its modulus over the unit sphere $r=1$ is equal to unity. Thus we choose $A_{lk}$ so that

$$A_{lk}^2\int_0^{2\pi} d\phi\int_0^{\pi}[P_l^k(\cos\theta)]^2\sin\theta\, d\theta = 1. \tag{5.2.30}$$

Referring to equation (D.23) of Appendix D, we find that

$$A_{lk}^2 = \frac{2l+1}{4\pi}\cdot\frac{(l-|k|)!}{(l+|k|)!}. \tag{5.2.31}$$

Since $Y_{lk}$ is an eigenfunction of $\hat{L}^2$ corresponding to the eigenvalue $l(l+1)\hbar^2$, it follows that for all $k$

$$\hat{L}^2 Y_{lk} = l(l+1)\hbar^2 Y_{lk}. \tag{5.2.32}$$

Since it is also an eigenfunction of $\hat{L}_z$ for the eigenvalue $k\hbar$, therefore

$$\hat{L}_z Y_{lk} = k\hbar Y_{lk}. \tag{5.2.33}$$

From equations (5.2.4), (5.2.27), it follows that the angular momentum vector of the particle makes an angle $\beta$ with $Oz$ given by the equation

$$\cos\beta = \frac{k}{\sqrt{[l(l+1)]}}. \tag{5.2.34}$$

It will be observed that, since $|k|\leqslant l$, $\beta$ can never equal 0 or $\pi$ and hence that the angular momentum vector cannot be aligned with the $z$-axis. This implies that its direction cannot be known with precision. This also follows from the fact that, if $\beta=0$, then $L_x=L_y=0$ and all components of angular momentum would be sharp simultaneously.

The function $R(r)$ occurring in equation (5.2.28) will become determinate only when a further measurement of the particle's state has been carried out. This must be a measurement of an observable which is compatible with $L_z$ and $L^2$. It will be shown in the next section that the energy is such an observable provided the field of force is of the central type.

### 5.3. Central force fields

Suppose the particle is moving in a conservative field of force with potential energy $V$. In spherical polar coordinates

$$\nabla^2 = \frac{1}{r^2}\frac{\partial}{\partial r}\left(r^2\frac{\partial}{\partial r}\right)+\frac{1}{r^2}\left[\frac{1}{\sin\theta}\frac{\partial}{\partial\theta}\left(\sin\theta\frac{\partial}{\partial\theta}\right)+\frac{1}{\sin^2\theta}\frac{\partial^2}{\partial\phi^2}\right]. \tag{5.3.1}$$

Hence, the Hamiltonian operator $\hat{H}$ may be expressed in the form

$$\hat{H} = -\frac{\hbar^2}{2mr^2}\frac{\partial}{\partial r}\left(r^2\frac{\partial}{\partial r}\right)+\frac{1}{2mr^2}\hat{L}^2+V, \tag{5.3.2}$$

where we have employed equation (5.1.21).

Thus,

$$[\hat{H},\hat{L}_z] = \left[V,\frac{\hbar}{\iota}\frac{\partial}{\partial\phi}\right], \tag{5.3.3}$$

since $\hat{L}_z$ commutes with the first two terms of the right-hand member of equation (5.3.2). Expanding the right-hand member of equation (5.3.3), we find that

$$[\hat{H},\hat{L}_z] = -\frac{\hbar}{\iota}\frac{\partial V}{\partial\phi}. \tag{5.3.4}$$

Applying equation (4.2.5), this result is seen to imply that

$$\frac{d\bar{L}_z}{dt} = -\overline{\frac{\partial V}{\partial\phi}} \tag{5.3.5}$$

or, since

$$-\frac{\partial V}{\partial\phi} = -x\frac{\partial V}{\partial y}+y\frac{\partial V}{\partial x} = xF_y-yF_x, \tag{5.3.6}$$

where $F_x$, $F_y$ are components of force acting upon the particle, equation (5.3.5) is equivalent to

$$\frac{d\bar{L}_z}{dt} = \bar{M}_z, \tag{5.3.7}$$

$M_z$ being the $z$-component of the moment of this force about $O$. Equation (5.3.7) is the quantum mechanical counterpart of the equation of angular momentum for the particle. Similar equations exist for the $x$- and $y$-components.

In the special case when $V$ is independent of $\phi$ so that the force field has axial symmetry about $Oz$, it is clear from equation (5.3.4) that $\hat{H}$ and $\hat{L}_z$ commute. The probability distribution for the observable $L_z$ is then an invariant of the motion and, in particular, if $L_z$ is measured precisely at some instant, it remains with this sharp value throughout the motion; the equation (5.2.5) is then valid for all $t$ and not just for the instant of measurement.

Let us now consider the commutability of the operator $\hat{L}^2$ with $\hat{H}$. Since $\hat{L}^2$ commutes with the first two terms of $\hat{H}$ in equation (5.3.2), wc have

$$[\hat{H}, \hat{L}^2] = [V, \hat{L}^2]. \tag{5.3.8}$$

Clearly, $\hat{H}$ and $\hat{L}^2$ only commute if $V$ is independent of both $\theta$ and $\phi$ and is a function of $r$ and $t$ alone. This is the case when the force acting upon the particle is always directed towards or away from $O$, i.e. a central force field. For motion in such a field $\bar{L}^2$ is conserved and, if $L^2$ is sharp at one instant, it is sharp at every instant with the same value. *A fortiori*, the field is then axially symmetric about $Oz$ and hence $\bar{L}_z$ is also conserved and $L_z$ remains sharp if it is once sharp.

For motion in a central field of force, therefore, if $L_z$ and $L^2$ are measured precisely at some instant, the wave function will be given by equation (5.2.28) at all later instants, until the particle is disturbed. But, since $\hat{H}$ commutes with $\hat{L}_z$ and $\hat{L}^2$, it is now possible for the particle's energy to be sharp also and, provided the force field is steady (i.e. $V$ is independent of $t$), the energy will remain sharp and the particle will be in a steady state. Then, if $E$ is the energy, we know that $t$ must enter into the wave function in the manner indicated in equation (4.4.4). Thus

$$\psi = R(r)\, Y_{lk}(\theta, \phi) \exp(-\iota Et/\hbar). \tag{5.3.9}$$

But $\psi$ must then satisfy the characteristic equation

$$\hat{H}\psi = E\psi. \tag{5.3.10}$$

Substituting for $\hat{H}$ from equation (5.3.2), this gives

$$\frac{1}{r^2}\frac{\partial}{\partial r}\left(r^2 \frac{\partial \psi}{\partial r}\right) - \frac{1}{\hbar^2 r^2}\hat{L}^2\psi + \frac{2m}{\hbar^2}(E-V)\psi = 0. \tag{5.3.11}$$

Substituting for $\psi$ from equation (5.3.9) and employing equation (5.2.32), it will be found that $R$ satisfies the equation

$$\frac{d^2 R}{dr^2} + \frac{2}{r}\frac{dR}{dr} + \left[\frac{2m}{\hbar^2}(E-V) - \frac{l(l+1)}{r^2}\right]R = 0. \tag{5.3.12}$$

As $r \to \infty$, we shall assume $V \to 0$ at least as rapidly as $1/r$ and that $R$ and its derivatives are bounded; then, for large $r$, equation (5.3.12) can be approximated by

$$\frac{d^2 R}{dr^2} - \alpha^2 R = 0, \tag{5.3.13}$$

where

$$\alpha^2 = -\frac{2mE}{\hbar^2}. \tag{5.3.14}$$

By taking $E$ to be negative, we shall limit our discussion to bound states of the particle. Thus, from equation (5.3.13) it follows that

$$R = A \exp(\alpha r) + B \exp(-\alpha r) \tag{5.3.15}$$

for large $r$. But $R$ must not become infinite as $r \to \infty$ and hence $A$ must vanish. We conclude that $R$ behaves like $\exp(-\alpha r)$ at infinity and we shall write accordingly

$$R = \exp(-\alpha r)\chi, \tag{5.3.16}$$

where $\chi$ is a new dependent variable. It is also convenient to change the independent variable to $x$ by putting

$$x = 2\alpha r. \tag{5.3.17}$$

With these changes, equation (5.3.12) assumes the form

$$x\frac{d^2\chi}{dx^2} + (2-x)\frac{d\chi}{dx} - \left[1 + \frac{xV}{4E} + \frac{l(l+1)}{x}\right]\chi = 0. \tag{5.3.18}$$

It can be shown that this equation possesses solutions having the appropriate form at infinity provided $E$ assumes one of a spectrum of eigenvalues. In the next section, these eigenvalues will be calculated for the case when $V$ is the potential energy function for an inverse square law field.

## 5.4. The hydrogen atom

In this section, we will study the motion of an electron having charge $-e$ in the field of a fixed, positively charged nucleus, having charge $Ze$. The case of the hydrogen atom is obtained by taking $Z=1$. Allowance can easily be made for the motion of the nucleus by substituting the 'reduced mass' for the mass of the electron as explained in section (3.12).

Thus, putting

$$V = -\frac{Ze^2}{r} \tag{5.4.1}$$

in equation (5.3.18), this takes the form

$$x\frac{d^2\chi}{dx^2}+(2-x)\frac{d\chi}{dx}-\left[1-\lambda+\frac{l(l+1)}{x}\right]\chi = 0, \tag{5.4.2}$$

where

$$\lambda = \frac{Ze^2}{\hbar}\sqrt{\left(-\frac{m}{2E}\right)}. \tag{5.4.3}$$

This can now be reduced to a form of equation which is associated with the name of Laguerre by putting

$$\chi = x^l y. \tag{5.4.4}$$

The equation for the new independent variable $y$ is

$$xy''+(2l+2-x)y'+(\lambda-l-1)y = 0. \tag{5.4.5}$$

Employing the method of Frobenius, we assume

$$y = \sum_{n=0}^{\infty} a_n x^{\rho+n}. \tag{5.4.6}$$

Upon substituting and equating the coefficients of powers of $x$ to zero, we obtain an indicial equation

$$a_0\rho(\rho+2l+1) = 0 \tag{5.4.7}$$

and a recurrence relationship

$$a_{n+1}(\rho+n+1)(\rho+n+2l+2) = a_n(\rho+n+l+1-\lambda), \tag{5.4.8}$$

valid for $n=0$, 1, 2, etc. The roots of the indicial equation are $\rho=0$, $-(2l+1)$; $l$ being a positive integer or zero, it will be found that the second root leads to the same solution as the first. Taking $\rho=0$, therefore, equation (5.4.8) gives

$$a_{n+1} = \frac{(n+l+1-\lambda)}{(n+1)(n+2l+2)}a_n \tag{5.4.9}$$

and this leads directly to the solution

$$y = y_1 = 1+\frac{l+1-\lambda}{1!(2l+2)}x+\frac{(l+1-\lambda)(l+2-\lambda)}{2!(2l+2)(2l+3)}x^2+\ldots. \tag{5.4.10}$$

A second solution containing a term $y_1 \log x$ can now be found, but is unacceptable for both large and small values of $x$.

If $\lambda$ is not an integer, the terms in the series for $y_1$ are ultimately all positive or all negative for positive $x$. Also

$$\frac{a_n}{a_{n-1}} = \frac{n+l-\lambda}{n(n+2l+1)} > \frac{3}{4n}, \tag{5.4.11}$$

for sufficiently large $n$. Hence, employing the theorem of Appendix B, we conclude that

$$y_1 \geqslant k \exp\left(\tfrac{3}{4}x\right) + p(x) \tag{5.4.12}$$

and $R$ accordingly approaches infinity more rapidly than $kx^l \exp\left(\tfrac{1}{4}x\right)$ as $x \to \infty$. This solution is evidently unsatisfactory.

The only remaining possibility is that $\lambda$ is an integer $n \geqslant l+1$ so that the series terminates; $y_1$ is then a polynomial of degree $(n-l-1)$ and, by comparing equation (5.4.5) with equation (E.1) of Appendix E, it will be seen that $y_1$ is the generalised Laguerre polynomial of degree $(n-l-1)$ and order $(2l+1)$ (except for a numerical factor). Thus the solution for $R$ may be written

$$R = R_{ln} = C_{ln} \exp\left(-\tfrac{1}{2}x\right) x^l L_{n-l-1}^{2l+1}(x), \tag{5.4.13}$$

where $C_{ln}$ is a constant. The complete solution for $\psi$ now follows from equation (5.3.9).

If $\lambda = n$, equation (5.4.3) indicates that the energy $E$ is quantised, the $n$th energy level being given by

$$E = E_n = -\frac{mZ^2 e^4}{2\hbar^2} \cdot \frac{1}{n^2}. \tag{5.4.14}$$

The system is clearly degenerate with respect to the energy observable; for a given energy eigenvalue (i.e. given $n$), $l$ can take any one of the $n$ values 0, 1, 2, ..., $(n-1)$ and, for each such value of $l$, $k$ can assume any one of the $(2l+1)$ values $-l$, $-l+1$, ..., $l$. Thus, if $E = E_n$, there are altogether $n^2$ distinct states of the atom in which $L_z$ and $L^2$ take sharp values; the atom is accordingly $n^2$-fold degenerate with respect to the energy observable. The ground state $n = 1$ is a special case and is not degenerate. If $l = 0$, then $k = 0$ and $\psi$ depends only upon $r$, i.e. the state is spherically symmetric and the electron possesses no angular momentum about the nucleus.

From equations (5.3.14), (5.4.14), we deduce that

$$\alpha = \frac{mZe^2}{n\hbar^2}. \tag{5.4.15}$$

The normalisation condition is that

$$\int_0^\infty dr \int_0^\pi d\theta \int_0^{2\pi} \psi\psi^* r^2 \sin\theta \, d\phi = 1. \tag{5.4.16}$$

Substituting for $\psi$ from equation (5.3.9) and recalling that $Y_{lk}$ has already been normalised, this condition reduces to

$$\int_0^\infty (rR)^2\, dr = 1. \tag{5.4.17}$$

Substituting for $R$ from equation (5.4.13) and employing equation (E.34) of Appendix E, we calculate that

$$C_{ln} = \left(\frac{em^{\frac{1}{2}}Z^{\frac{1}{2}}}{\hbar}\right)^3 \cdot \frac{2}{n^2}\sqrt{\left(\frac{(n-l-1)!}{\{(n+l)!\}^3}\right)}. \tag{5.4.18}$$

Thus, finally,

$$\psi = R_{ln}(x)\, Y_{lk}(\theta, \phi) \exp(-\iota E_n t/\hbar), \tag{5.4.19}$$

where $E_n$ is given by equation (5.4.14) and

$$x = \frac{2mZe^2}{n\hbar^2} r. \tag{5.4.20}$$

We have now demonstrated that any state in which $L_z$, $L^2$ and $H$ are sharp is non-degenerate and hence these observables form a maximal compatible set for the system being considered. It follows that a complete orthonormal set of eigenfunctions for the system is obtained by letting $k$, $l$, $n$ range through all permissible integral values in equation (5.4.19). The wave function for any state of the atom can be expressed as a linear combination of these eigenfunctions and the coefficients in this expansion will determine the probabilities of observing transitions from this state into the eigenstates, as explained in section 3.3.

Our analysis can be employed to explain the spectrum of the light emitted when an electric discharge takes place through hydrogen gas at low pressure. The function of the discharge is to supply energy in a random manner to the hydrogen atoms, thus raising them to energy states above the ground state ($n = 1$). As the atoms fall towards their ground states, either directly or via intermediate states, they discharge energy by emitting bursts of electromagnetic radiation. It was discovered in the nineteenth century that this radiation possessed a discrete spectrum and, at the time, no explanation was forthcoming why this should be so. According to Maxwell's theory of electromagnetism, the orbiting electron should lose energy steadily, emitting radiation having a continuous spectrum, as it spiralled in towards the nucleus. By introducing certain postulates of an *ad hoc* nature relating to the permissible values of the electron's angular momentum about the nucleus, Bohr was able

(1913) to provide an interim explanation of this phenomenon. However, it was not until Schrödinger constructed his theory of wave mechanics (1926), that a thoroughly convincing reason was found. Schrödinger derived this from the fact that the energy of the atom is quantised, as required by equation (5.4.14).

Suppose that an atom is in a steady state corresponding to an energy eigenvalue $E_n$ and that it is then disturbed so that it moves into another steady state for which the energy is $E_{n'}(<E_n)$. The manner in which a small disturbance can cause such a transition will be investigated in section 6.10. As a result of this transition, the atom must discharge a quantity $E_n - E_{n'}$ of energy and Schrödinger assumed that this energy would appear as a photon of electromagnetic radiation. Employing Einstein's equation (4.4.6), he deduced that the frequency of this radiation should be $\nu$, where

$$\nu = \frac{1}{h}(E_n - E_{n'}) = \frac{2\pi^2 me^4}{h^3}\left(\frac{1}{n'^2} - \frac{1}{n^2}\right). \tag{5.4.21}$$

By allowing $n'$, $n$ to range through all integral values, this formula was found to yield the frequencies of all lines in the hydrogen spectrum and had, indeed, been proposed by Balmer, on the basis of actual measurements in 1885.

## 5.5. Particle with spin in a central force field

Two modifications of the theory developed thus far must be made if the particle has spin. Most obviously, the single wave function $\psi$ must be replaced by two, or more, functions, which may be arranged into a column matrix. Also, the Hamiltonian $H$ will, in general, include new terms involving the spin vector. The orbital angular momentum operators $\hat{L}_z$, $\hat{L}^2$, remain unchanged and, since they do not depend upon any spin operators (matrices), when operating upon a column matrix determining the particle's state, they operate upon the elements directly, thus:

$$\hat{L}_z \begin{pmatrix} \psi_1 \\ \psi_2 \\ \vdots \end{pmatrix} = \begin{pmatrix} \hat{L}_z \psi_1 \\ \hat{L}_z \psi_2 \\ \vdots \end{pmatrix}. \tag{5.5.1}$$

As an example, consider an electron moving under the attraction of a positively charged atomic nucleus. The state of the electron at any instant will be specified by a column matrix $\{\psi_+, \psi_-\}$, where $\psi_+$, $\psi_-$ are functions of $(x, y, z, t)$. Since the electron is a magnetic dipole, its Hamiltonian $H$ must include a term representing the energy of this dipole in its motion through the electric field of the nucleus. A magnetic dipole which is stationary in an electrostatic field does not, of course, interact with the field, so that its potential energy is zero. If, however, the dipole is moving, then relative to a frame in which the dipole is stationary the electric field is in motion and this motion will generate a magnetic component of the field; the dipole will interact with this component and the energy of this interaction should be included in $H$. This is the *spin-orbit interaction*.† The magnitude of this effect proves to be comparatively small and we shall first neglect it. With this approximation, the operators $\hat{H}$, $\hat{L}^2$, $\hat{L}_z$ are all unaltered by the presence of spin and accordingly continue to commute; this guarantees the continued existence of states for which the corresponding observables are sharp. However, a complete specification of an eigenstate in terms of the eigenvalues of these observables alone, can no longer be effected; a further observable involving the spin must clearly be introduced to complete a maximal set of compatible observables.

Suppose, then, we search for eigenstates in which the observables $H$, $L^2$, $L_z$, $\sigma_z$ are all sharp. Such a state is possible, since $\hat{\sigma}_z$ commutes with each of the operators $\hat{H}$, $\hat{L}^2$, $\hat{L}_z$. That this is so can be seen thus: Employing the $xyz\sigma_z$-representation, we have

$$\hat{H}\hat{\sigma}_z\psi = \hat{H}\begin{pmatrix}1 & 0\\ 0 & -1\end{pmatrix}\begin{pmatrix}\psi_+\\ \psi_-\end{pmatrix} = \hat{H}\begin{pmatrix}\psi_+\\ -\psi_-\end{pmatrix} = \begin{pmatrix}\hat{H}\psi_+\\ -\hat{H}\psi_-\end{pmatrix}, \tag{5.5.2}$$

$$\hat{\sigma}_z\hat{H}\psi = \begin{pmatrix}1 & 0\\ 0 & -1\end{pmatrix}\begin{pmatrix}\hat{H}\psi_+\\ \hat{H}\psi_-\end{pmatrix} = \begin{pmatrix}\hat{H}\psi_+\\ -\hat{H}\psi_-\end{pmatrix}, \tag{5.5.3}$$

$\psi$ being an arbitrary vector. Now, eigenstates for $H$ must be represented by vectors satisfying the characteristic equation

$$\hat{H}\psi = E_n\psi, \tag{5.5.4}$$

i.e.

$$\hat{H}\begin{pmatrix}\psi_+\\ \psi_-\end{pmatrix} = E_n\begin{pmatrix}\psi_+\\ \psi_-\end{pmatrix}. \tag{5.5.5}$$

Equating like elements from the two members of this equation, we obtain the equations

$$\hat{H}\psi_+ = E_n\psi_+, \quad \hat{H}\psi_- = E_n\psi_-, \tag{5.5.6}$$

showing that $\psi_+$, $\psi_-$ are energy eigenfunctions corresponding to the eigenvalue $E_n$ as for the case when spin was neglected. Similarly, it can

† There is also an interaction between the magnetic moments of the electron and nucleus. However, this effect proves to be very small and will be neglected. It is responsible for what is termed the *hyperfine* splitting of the energy levels of the atom.

be shown that $\psi_+, \psi_-$ are eigenfunctions of $L^2$ and $L_z$ as for the case of no spin. Finally, for the eigenstate $\sigma_z = \pm 1$, we must have

$$\hat{\sigma}_z \psi = \psi, \tag{5.5.7}$$

i.e.

$$\begin{pmatrix} 1 & 0 \\ 0 & -1 \end{pmatrix} \begin{pmatrix} \psi_+ \\ \psi_- \end{pmatrix} = \begin{pmatrix} \psi_+ \\ \psi_- \end{pmatrix}. \tag{5.5.8}$$

Thus $\psi_- = 0$ and we have proved that, in the eigenstate $H = E_n$, $L^2 = l(l+1)\hbar^2$, $L_z = k\hbar$, $\sigma_z = +1$, the state vector is given by

$$\boldsymbol{\psi} = \boldsymbol{\psi}_{kln+} = \begin{pmatrix} R_{ln}(x)\, Y_{lk}(\theta, \phi) \\ 0 \end{pmatrix} \exp(-\iota E_n t/\hbar). \tag{5.5.9}$$

It is left as an exercise for the reader to obtain the corresponding result $\boldsymbol{\psi}_{kln-}$ for the eigenstate in which $\sigma_z = -1$.

If spin-orbit interaction is taken into account, it may be proved that an additional term

$$\frac{e^2 Z}{2m^2 c^2} \frac{1}{r^3} \mathfrak{L} . \mathfrak{s} \tag{5.5.10}$$

must be added to $H$ ($c$ is the velocity of light). Since

$$\mathfrak{L} . \mathfrak{s} = L_x s_x + L_y s_y + L_z s_z, \tag{5.5.11}$$

$\hat{H}$ continues to commute with $\hat{L}^2$, but it no longer commutes with $\hat{L}_z$ or $\hat{\sigma}_z$. This implies that $H, L^2, L_z, \sigma_z$ cannot all be sharp together. Instead it is necessary to replace $L_z$, $\sigma_z$ by $J_z$, $J^2$, where $\mathfrak{J}$ is the sum of the orbital and spin angular momenta of the electron. We shall prove that $\hat{H}$, $\hat{L}^2$, $\hat{J}^2, \hat{J}_z$ commute in pairs and hence constitute a maximal set of compatible observables.

Thus, we define $\mathfrak{J}$, the overall angular momentum, by

$$\mathfrak{J} = \mathfrak{L} + \mathfrak{s}. \tag{5.5.12}$$

Then

$$\hat{J}_z = \hat{L}_z + \hat{s}_z, \text{ etc.} \tag{5.5.13}$$

and, if $J$ is the magnitude of $\mathfrak{J}$,

$$\begin{aligned} \hat{J}^2 &= \hat{J}_x^2 + \hat{J}_y^2 + \hat{J}_z^2, \\ &= \hat{L}^2 + \tfrac{3}{4}\hbar^2 + 2(\hat{L}_x \hat{s}_x + \hat{L}_y \hat{s}_y + \hat{L}_z \hat{s}_z), \end{aligned} \tag{5.5.14}$$

since $\hat{s}^2 = \frac{3}{4}\hbar^2$ by equation (2.2.34) and the components of $\mathfrak{L}$ commute with those of $\mathfrak{s}$. But $\hat{L}^2$ commutes with the components of $\mathfrak{L}$ and $\mathfrak{s}$. It follows that $\hat{L}^2$ commutes with both $\hat{J}_z$ and $\hat{J}^2$.

Further,

$$[\hat{J}_x, \hat{J}_y] = [\hat{L}_x, \hat{L}_y] + [\hat{s}_x, \hat{L}_y] + [\hat{L}_x, \hat{s}_y] + [\hat{s}_x, \hat{s}_y],$$
$$= \iota\hbar(\hat{L}_z + \hat{s}_z),$$
$$= \iota\hbar\hat{J}_z, \tag{5.5.15}$$

by the commutation relations for $\mathfrak{L}$ and $\mathfrak{s}$. An argument similar to that employed to establish equation (5.1.16), now proves that $\hat{J}^2$ commutes with $\hat{J}_z$.

We have proved, therefore, that $\hat{L}^2, \hat{J}^2, \hat{J}_z$ commute in pairs. It remains to show that they all commute with $\hat{H}$. Consider first the Hamiltonian without the spin-orbit interaction term; this is given by equation (5.3.2). This commutes with $\hat{L}^2$ and, since it also commutes with all components of $\mathfrak{L}$ and $\mathfrak{s}$, it follows from equation (5.5.14) that it commutes with $\hat{J}^2$ and from equation (5.5.13) that it commutes with $\hat{J}_z$. Finally, we shall take account of the spin-orbit interaction term (5.5.10). By equation (5.5.14), the operator corresponding to this term is equivalent to

$$\frac{e^2 Z}{4m^2 c^2} \frac{1}{r^3} (\hat{J}^2 - \hat{L}^2 - \tfrac{3}{4}\hbar^2). \tag{5.5.16}$$

It now follows from what has already been proved that this operator commutes with $\hat{L}^2$, $\hat{J}^2$ and $\hat{J}_z$.

We now seek a state in which $H$, $L^2$, $J^2$, $J_z$ are sharp simultaneously. Suppose, first, that $J_z$ is sharp with an eigenvalue $q\hbar$, so that the state vector satisfies

$$\hat{J}_z \psi = q\hbar\psi. \tag{5.5.17}$$

Employing the $xyzs_z$-representation, this equation can be written

$$\left[\hat{L}_z + \tfrac{1}{2}\hbar \begin{pmatrix} 1 & 0 \\ 0 & -1 \end{pmatrix}\right] \begin{pmatrix} \psi_+ \\ \psi_- \end{pmatrix} = q\hbar \begin{pmatrix} \psi_+ \\ \psi_- \end{pmatrix}, \tag{5.5.18}$$

which leads to the pair of equations

$$\hat{L}_z \psi_+ = (q - \tfrac{1}{2})\hbar\psi_+, \tag{5.5.19}$$

$$\hat{L}_z \psi_- = (q + \tfrac{1}{2})\hbar\psi_-. \tag{5.5.20}$$

These equations can now be solved for $\psi_+$, $\psi_-$ by a method identical to that employed with equation (5.2.1). It is found that both $q - \frac{1}{2}$ and $q + \frac{1}{2}$ must be integers and hence that $q$ must be of the form

$$q = \text{integer} + \tfrac{1}{2}. \tag{5.5.21}$$

With such a value for $q$,

$$\psi_+ = A\exp[\iota(q - \tfrac{1}{2})\phi], \quad \psi_- = B\exp[\iota(q + \tfrac{1}{2})\phi] \tag{5.5.22}$$

where $A$, $B$ are functions of $r$ and $\theta$ and include the usual steady state time factor. We note that the eigenvalues for $J_z$ belong to the sequence $\ldots -\frac{5}{2}\hbar, -\frac{3}{2}\hbar, -\frac{1}{2}\hbar, \frac{1}{2}\hbar, \frac{3}{2}\hbar, \frac{5}{2}\hbar, \ldots$.

Next, suppose that $L^2 = \lambda\hbar^2$, so that $\psi$ satisfies

$$\hat{L}^2\psi = \lambda\hbar^2\psi. \tag{5.5.23}$$

This is equivalent, in the chosen representation, to the pair of equations

$$\hat{L}^2\psi_+ = \lambda\hbar^2\psi_+, \quad \hat{L}^2\psi_- = \lambda\hbar^2\psi_-. \tag{5.5.24}$$

From previous theory (section 5.2), we conclude that $\lambda = l(l+1)$, where $l$ is a positive integer or zero and that

$$\psi_+ = A' P_l^{q-\frac{1}{2}}(\cos\theta)\exp[\iota(q-\tfrac{1}{2})\phi] = AY_{l,\,q-\frac{1}{2}}(\theta,\phi), \tag{5.5.25}$$

$$\psi_- = B' P_l^{q+\frac{1}{2}}(\cos\theta)\exp[\iota(q+\tfrac{1}{2})\phi] = BY_{l,\,q+\frac{1}{2}}(\theta,\phi), \tag{5.5.26}$$

where $A$, $B$ are now functions of $r$ and $t$ alone. It also follows from our earlier analysis that $\psi_+$ is identically zero unless

$$|q-\tfrac{1}{2}| \leqslant l, \tag{5.5.27}$$

and that $\psi_-$ is identically zero unless

$$|q+\tfrac{1}{2}| \leqslant l. \tag{5.5.28}$$

Hence, for a given value of $l$, the eigenvalues of $q$ form the sequence $-l-\frac{1}{2}, -l+\frac{1}{2}, \ldots, l-\frac{1}{2}, l+\frac{1}{2}$. However, if $q$ takes the extreme value $-l-\frac{1}{2}$, the inequality (5.5.27) does not hold and hence $\psi_+ = 0$; this implies that $s_z = -\frac{1}{2}\hbar$ with certainty. Again, if $q$ takes the extreme value $l+\frac{1}{2}$, the inequality (5.5.28) is not satisfied and then $\psi_- = 0$, i.e. $s_z = +\frac{1}{2}\hbar$ with certainty. For the other values of $q$, the values of $s_z$ are determined on a probability basis only; the probabilities will be calculated later in this section.

If $J^2 = j(j+1)\hbar^2$, then $\psi$ must also satisfy the equation

$$\hat{J}^2\psi = j(j+1)\hbar^2\psi. \tag{5.5.29}$$

The operator $\hat{J}^2$ is given by equation (5.5.14). Substituting Pauli matrices for the operators $\hat{\sigma}_x$, $\hat{\sigma}_y$, $\hat{\sigma}_z$, this equation will be found to be equivalent to the equation

$$\left[\hat{L}^2 + \tfrac{3}{4}\hbar^2 + \hbar\begin{pmatrix}\hat{L}_z & \hat{L}_x - \iota\hat{L}_y \\ \hat{L}_x + \iota\hat{L}_y & -\hat{L}_z\end{pmatrix}\right]\begin{pmatrix}\psi_+ \\ \psi_-\end{pmatrix} = j(j+1)\hbar^2\begin{pmatrix}\psi_+ \\ \psi_-\end{pmatrix}. \tag{5.5.30}$$

Substituting for $\psi_+$, $\psi_-$ from equations (5.5.25), (5.5.26) and making use of equations (5.2.32), (5.2.33), and the results given in Ex. 4, Chap. 5, it will be found that $A$ and $B$ must satisfy the equations

$$[l(l+1)-j(j+1)+q+\tfrac{1}{4}]A-\sqrt{[(l+\tfrac{1}{2})^2-q^2]}\,B = 0, \tag{5.5.31}$$

$$-\sqrt{[(l+\tfrac{1}{2})^2-q^2]}\,A+[l(l+1)-j(j+1)-q+\tfrac{1}{4}]B = 0. \tag{5.5.32}$$

If $A$, $B$ are not both to vanish, it is necessary that

$$\begin{vmatrix} l(l+1)-j(j+1)+q+\tfrac{1}{4} & -\sqrt{[(l+\tfrac{1}{2})^2-q^2]} \\ -\sqrt{[(l+\tfrac{1}{2})^2-q^2]} & l(l+1)-j(j+1)-q+\tfrac{1}{4} \end{vmatrix} = 0. \tag{5.5.33}$$

Since $l \geqslant 0$ and $j \geqslant 0$, this equation possesses two relevant solutions only, viz.

$$j = l-\tfrac{1}{2} \quad \text{and} \quad j = l+\tfrac{1}{2}. \tag{5.5.34}$$

In the case $j=l-\frac{1}{2}$ (this case cannot arise if $l=0$), it will be found that

$$A:B = \sqrt{(l-q+\tfrac{1}{2})}:\sqrt{(l+q+\tfrac{1}{2})} \tag{5.5.35}$$

and hence that

$$\psi = \begin{pmatrix} \sqrt{(l-q+\tfrac{1}{2})}\, Y_{l,\,q-\frac{1}{2}}(\theta,\phi) \\ \sqrt{(l+q+\tfrac{1}{2})}\, Y_{l,\,q+\frac{1}{2}}(\theta,\phi) \end{pmatrix} X(r)\exp(-\iota Et/\hbar). \tag{5.5.36}$$

In the case $j=l+\frac{1}{2}$, it will be found that

$$A:B = -\sqrt{(l+q+\tfrac{1}{2})}:\sqrt{(l-q+\tfrac{1}{2})} \tag{5.5.37}$$

and hence that

$$\psi = \begin{pmatrix} \sqrt{(l+q+\tfrac{1}{2})}\, Y_{l,\,q-\frac{1}{2}}(\theta,\phi) \\ -\sqrt{(l-q+\tfrac{1}{2})}\, Y_{l,\,q+\frac{1}{2}}(\theta,\phi) \end{pmatrix} X(r)\exp(-\iota Et/\hbar). \tag{5.5.38}$$

However, we have proved already that if $q=l+\frac{1}{2}$, then $B=0$ and if $q=-l-\frac{1}{2}$, then $A=0$. We see from equations (5.5.36), (5.5.38), that this can only be so in the case $j=l+\frac{1}{2}$. Thus, if $j=l+\frac{1}{2}$, $q$ is permitted to range over the values $-l-\frac{1}{2}, \ldots, l+\frac{1}{2}$, but if $j=l-\frac{1}{2}$, $q$ is confined to the range $-l+\frac{1}{2}, \ldots, l-\frac{1}{2}$.

In the case $j=l+\frac{1}{2}$, the orbital and spin angular momentum vectors may be thought of as being aligned and in the same sense, so that their contributions to the overall angular momentum are additive. In the case $j=l-\frac{1}{2}$, the two vectors may be regarded as being aligned, but in opposite senses, so that their contributions must be subtracted. In reality, of course, neither of these vectors has a precise direction and, further, the magnitude of the orbital angular momentum is not $l\hbar$. The view that has been suggested has only the status of a mnemonic therefore and must not be adopted as a basis for exact calculations. In the case when the spin and orbital angular momenta are aligned, as might be expected, $J_z$ is able to assume the extreme values $\pm(l+\frac{1}{2})\hbar$. When the two angular momenta are opposed, however, the extreme values of $J_z$ are smaller in magnitude, viz. $\pm(l-\frac{1}{2})\hbar$.

If, finally, $\psi$ is to be an eigenvector of the Hamiltonian, it must satisfy an equation

$$\hat{H}\psi = E\psi. \tag{5.5.39}$$

This will determine $X(r)$ and the energy eigenvalues. Taking the spin-orbit interaction term in the form (5.5.16), we have

$$\hat{H} = -\frac{\hbar^2}{2mr^2}\frac{\partial}{\partial r}\left(r^2\frac{\partial}{\partial r}\right)+\frac{1}{2mr^2}\hat{L}^2-\frac{Ze^2}{r}+\frac{e^2Z}{4m^2c^2}\frac{1}{r^3}(\hat{J}^2-\hat{L}^2-\tfrac{3}{4}\hbar^2). \tag{5.5.40}$$

Thus, substituting for $\psi$ from equation (5.5.38) into equation (5.5.39) and taking advantage of the fact that $\psi$ is an eigenvector of both $\hat{L}^2$ and $\hat{J}^2$ (equations (5.5.23), (5.5.29)), it will be found that $X(r)$ satisfies the equation

$$\frac{d^2X}{dr^2}+\frac{2}{r}\frac{dX}{dr}+FX = 0, \tag{5.5.41}$$

where

$$F = \frac{2mE}{\hbar^2}+\frac{2mZe^2}{\hbar^2}\frac{1}{r}-\frac{l(l+1)}{r^2}-\frac{e^2Z}{2mc^2}\frac{1}{r^3}\{j(j+1)-l(l+1)-\tfrac{3}{4}\}. \tag{5.5.42}$$

Thus, in the case $j=l-\frac{1}{2}$,

$$F = \frac{2mE}{\hbar^2}+\frac{2mZe^2}{\hbar^2}\frac{1}{r}-\frac{l(l+1)}{r^2}+\frac{e^2Z}{2mc^2}\frac{l+1}{r^3}. \tag{5.5.43}$$

In the case $j=l+\frac{1}{2}$,

$$F = \frac{2mE}{\hbar^2}+\frac{2mZe^2}{\hbar^2}\frac{1}{r}-\frac{l(l+1)}{r^2}-\frac{e^2Z}{2mc^2}\frac{l}{r^3}. \tag{5.5.44}$$

Equation (5.5.41) is accordingly identical with the equation (5.3.12) obtained for the case of a particle having no spin, except for the addition of an extra term representing the spin-orbit coupling.

On account of the presence of the factor $1/c^2$ in the spin-orbit coupling term of $F$, this term proves to be very small by comparison with the others. A good approximation to the solution of equation (5.5.41) can therefore be found, in the first instance, by neglecting this term entirely; a correction to this approximation, allowing for the presence of the spin-orbit coupling term, will then be found in section 6.4 by the method of perturbations.

With this approximation, equation (5.5.41) becomes identical with equation (5.3.12) and hence

$$X = BR_{ln}, \tag{5.5.45}$$

where $R_{ln}$ is given by equations (5.4.13), (5.4.18) and $B$ is a normalisation factor. The energy eigenvalues are again given by equation (5.4.14). Thus, for the case $j = l - \frac{1}{2}$,

$$\psi = \begin{pmatrix} \sqrt{(l-q+\frac{1}{2})}\, Y_{l,\,q-\frac{1}{2}}(\theta, \phi) \\ \sqrt{(l+q+\frac{1}{2})}\, Y_{l,\,q+\frac{1}{2}}(\theta, \phi) \end{pmatrix} BR_{ln} \exp(-\iota E_n t/\hbar). \tag{5.5.46}$$

$\psi$ is normalised if

$$\int_0^\infty dr \int_0^\pi d\theta \int_0^{2\pi} (|\psi_+|^2 + |\psi_-|^2)\, r^2 \sin\theta\, d\phi = 1. \tag{5.5.47}$$

Since the functions $R_{ln}\, Y_{l,\,q-\frac{1}{2}}$ and $R_{ln}\, Y_{l,\,q+\frac{1}{2}}$ are already normalised, this condition is easily seen to be equivalent to

$$(l-q+\tfrac{1}{2})\, B^2 + (l+q+\tfrac{1}{2})\, B^2 = 1, \tag{5.5.48}$$

or

$$B = 1/\sqrt{(2l+1)}. \tag{5.5.49}$$

Hence, finally,

$$\psi = \begin{pmatrix} \sqrt{\left(\dfrac{l-q+\frac{1}{2}}{2l+1}\right)}\, Y_{l,\,q-\frac{1}{2}}(\theta, \phi) \\ \sqrt{\left(\dfrac{l+q+\frac{1}{2}}{2l+1}\right)}\, Y_{l,\,q+\frac{1}{2}}(\theta, \phi) \end{pmatrix} R_{ln} \exp(-\iota E_n t/\hbar), \tag{5.5.50}$$

where $l = 1, 2, \ldots, (n-1)$, $q = -l+\frac{1}{2}, -l+\frac{3}{2}, \ldots, l-\frac{1}{2}$.

For the case $j = l + \frac{1}{2}$, we calculate similarly that

$$\psi = \begin{pmatrix} \sqrt{\left(\dfrac{l+q+\frac{1}{2}}{2l+1}\right)}\, Y_{l,\,q-\frac{1}{2}}(\theta, \phi) \\ -\sqrt{\left(\dfrac{l-q+\frac{1}{2}}{2l+1}\right)}\, Y_{l,\,q+\frac{1}{2}}(\theta, \phi) \end{pmatrix} R_{ln} \exp(-\iota E_n t/\hbar), \tag{5.5.51}$$

where $l = 0, 1, 2, \ldots, (n-1)$, $q = -l-\frac{1}{2}, -l+\frac{1}{2}, \ldots, l+\frac{1}{2}$.

It now follows from equation (5.5.50) that, in the case when the spin and orbital angular momenta have opposite senses, the probabilities of measuring $s_z = \frac{1}{2}\hbar$ and $-\frac{1}{2}\hbar$ are

$$\frac{l-q+\frac{1}{2}}{2l+1}, \quad \frac{l+q+\frac{1}{2}}{2l+1}, \tag{5.5.52}$$

respectively. In the case when the spin and orbital angular momenta are aligned, these probabilities are found from equation (5.5.51) to be

$$\frac{l+q+\frac{1}{2}}{2l+1}, \quad \frac{l-q+\frac{1}{2}}{2l+1}, \tag{5.5.53}$$

respectively.

## 5.6. Charged particle in an electromagnetic field

Thus far, it has been assumed that the field of force to which the moving particle is subject is conservative, so that the force it exerts upon the particle can be derived from a scalar potential energy function $V$. This assumption is no longer valid when the particle is electrically charged and moves in an electromagnetic field which possesses a magnetic component as well as an electric component in the inertial frame we happen to be employing; for the magnetic field will be responsible for a force whose magnitude and direction depend upon the velocity of the particle in addition to its position. Thus, if $\mathfrak{E}$, $\mathfrak{B}$ are the electric and magnetic field intensities respectively at the point occupied by the particle when its velocity vector is $\mathfrak{v}$, the force $\mathfrak{F}$ exerted upon the particle is given by Lorentz's formula, viz.

$$\mathfrak{F} = e\mathfrak{E} + \frac{e}{c}\mathfrak{v} \times \mathfrak{B}, \tag{5.6.1}$$

where $e$ is the charge on the particle in electrostatic units and $c$ is the velocity of light (cm/sec) (units for $\mathfrak{E}$ and $\mathfrak{B}$ are assumed to be Gaussian). The electric and magnetic components of any electromagnetic field can be derived from a vector potential function $\mathfrak{A}(x,y,z,t)$ and a scalar potential $\Phi(x,y,z,t)$ by means of the equations

$$\mathfrak{E} = -\operatorname{grad}\Phi - \frac{1}{c}\frac{\partial \mathfrak{A}}{\partial t}, \tag{5.6.2}$$

$$\mathfrak{B} = \operatorname{curl}\mathfrak{A}. \tag{5.6.3}$$

It can now be demonstrated that a Hamiltonian function can be constructed from which the particle's motion can be derived in the usual way. Thus, if we take

$$H = \frac{1}{2m}\left(\mathfrak{p} - \frac{e}{c}\mathfrak{A}\right)^2 + e\Phi, \tag{5.6.4}$$

it is shown in Appendix F that Hamilton's equations yield the correct equation of motion for the particle. If $\mathfrak{A}$ vanishes so that there is no magnetic field, this clearly reduces to the form of equation (3.8.1) with $V = e\Phi$, the P.E. of the particle in the electric field.

The quantum mechanical operator representing the classical observable $H$ as given by equation (5.6.4) is presumably

$$\begin{aligned} \hat{H} &= \frac{1}{2m}\left(\frac{\hbar}{\iota}\nabla - \frac{e}{c}\mathfrak{A}\right)^2 + e\Phi, \\ &= -\frac{\hbar^2}{2m}\nabla^2 - \frac{e}{2mc}\frac{\hbar}{\iota}(\mathfrak{A}.\nabla + \nabla.\mathfrak{A}) + e\Phi + \frac{e^2}{2mc^2}\mathfrak{A}^2, \end{aligned} \tag{5.6.5}$$

where we have replaced the operator $\mathfrak{A}.\nabla$ (in general non-Hermitian) by its symmetrised product in the usual way and have neglected, for the present, any spin the particle may have.

The argument of section 4.6 will now be repeated for this, more general, case. The equation of motion (4.1.7) takes the form

$$-\frac{\hbar^2}{2m}\nabla^2\psi - \frac{e}{2mc}\frac{\hbar}{\iota}\{\mathfrak{A}.\nabla\psi + \operatorname{div}(\mathfrak{A}\psi)\} + \left(e\Phi + \frac{e^2}{2mc^2}\mathfrak{A}^2\right)\psi = \iota\hbar\frac{\partial\psi}{\partial t}. \tag{5.6.6}$$

Taking the complex conjugate of this equation and proceeding as after equation (4.6.2), we now obtain the equation

$$\frac{\partial}{\partial t}(\psi\psi^*) + \operatorname{div}\left[\frac{\iota\hbar}{2m}(\psi\nabla\psi^* - \psi^*\nabla\psi) - \frac{e}{mc}\psi\psi^*\mathfrak{A}\right] = 0. \tag{5.6.7}$$

This is the modified form of the continuity equation for probability in the case when the particle is subjected to magnetic forces (cf. equation (4.6.5)). The probability current density is given by

$$\mathfrak{j} = \frac{\iota\hbar}{2m}(\psi\nabla\psi^* - \psi^*\nabla\psi) - \frac{e}{mc}\psi\psi^*\mathfrak{A}. \tag{5.6.8}$$

A point to be noted is that the vector potential $\mathfrak{A}$ defined by equation (5.6.3) is indeterminate to the extent that the gradient of an arbitrary function of $(x, y, z.t)$ may be added to it without affecting this equation. If this is done, the scalar potential $\Phi$ and the wave function $\psi$ are changed (the latter by a factor of unit modulus having no physical significance), but it is found that the value of $\mathfrak{j}$ given by equation (5.6.8) is not affected when all these changes are taken into account.

## 5.7. The Normal Zeeman Effect

Consider an electron of mass $m$ and electric charge $-e$, which moves under the action of a conservative force having potential energy function $V$. Suppose the electron is also acted upon by a uniform steady magnetic field of strength $B$, whose direction is parallel to the $z$-axis. A vector potential $\mathfrak{A}$ for this field is easily verified to be given by

$$A_x = -\tfrac{1}{2}By, \quad A_y = \tfrac{1}{2}Bx, \quad A_z = 0. \tag{5.7.1}$$

An appropriate Hamiltonian operator for the system is accordingly given by equation (5.6.5) to be

$$\hat{H} = -\frac{\hbar^2}{2m}\nabla^2 + \frac{eB}{2mc}\frac{\hbar}{\iota}\left(x\frac{\partial}{\partial y} - y\frac{\partial}{\partial x}\right) + \frac{e^2B^2}{8mc^2}(x^2+y^2) + V. \tag{5.7.2}$$

However, this expression for $\hat{H}$ neglects the spin of the electron. We shall allow for this approximately, by including a term representing the P.E. of the dipole moment of the electron in the magnetic field $\mathfrak{B}$, i.e.

$$-\mathfrak{B}.(-\mu_0\mathfrak{z}) = \frac{e\hbar B}{2mc}\sigma_z, \tag{5.7.3}$$

where $\mu_0$ is the Bohr magneton given by equation (2.1.1). Due to the electron's motion, the magnetic field it 'sees' will differ from $\mathfrak{B}$, but if we assume that its velocity is small by comparison with $c$, the error of our approximation will be small. Then, assuming the magnetic field to be weak so that terms involving $B^2$ can be neglected and employing the $xyz\sigma_z$-representation, the Hamiltonian operator is given by

$$\hat{H} = -\frac{\hbar^2}{2mr^2}\frac{\partial}{\partial r}\left(r^2\frac{\partial}{\partial r}\right) + \frac{1}{2mr^2}\hat{L}^2 + V + \frac{eB}{2mc}(\hat{L}_z + \hbar\hat{\sigma}_z), \tag{5.7.4}$$

where $\hat{\sigma}_z$ is determined by the matrix given at equation (2.2.4) and spherical polars have replaced cartesian coordinates. The spin-orbit interaction is being neglected.

Consider the case of an electron in a hydrogen-type atom; then $V$ is a function of $r$ alone. Due to the presence of the magnetic field, the Hamiltonian, which was formerly given by equation (5.3.2), has been modified by the addition of a term $eB(\hat{L}_z + \hbar\hat{\sigma}_z)/2mc$. This extra term commutes with $\hat{L}_z$, $\hat{L}^2$ and $\hat{\sigma}_z$ and it follows that eigenstates in which the observables $H$, $L_z$, $L^2$, $\sigma_z$ are all sharp together, still exist. The argument given in the first part of section 5.5. is accordingly still relevant, except that $\{\psi_+, \psi_-\}$ has now to satisfy a modified characteristic equation for energy which leads to the pair of equations

$$\left[\hat{H}_0+\frac{eB}{2mc}(\hat{L}_z+\hbar)\right]\psi_+ = E\psi_+, \tag{5.7.5}$$

$$\left[\hat{H}_0+\frac{eB}{2mc}(\hat{L}_z-\hbar)\right]\psi_- = E\psi_-, \tag{5.7.6}$$

where $\hat{H}_0$ is the Hamiltonian operator without the magnetic field term.

Suppose the electron is in the eigenstate for which $H=E$, $L_z=k\hbar$, $L^2=l(l+1)\hbar^2$ and $\sigma_z=+1$. Then

$$\psi_+ = RY_{lk}(\theta,\phi)\exp(-\iota Et/\hbar), \quad \psi_- = 0. \tag{5.7.7}$$

Substituting in equation (5.7.5) for $\psi_+$, we find that $R$ satisfies the equation

$$\frac{d^2R}{dr^2}+\frac{2}{r}\frac{dR}{dr}+\left[\frac{2m}{\hbar^2}(E-V)-\frac{l(l+1)}{r^2}-\frac{eB}{c\hbar}(k+1)\right]R = 0. \tag{5.7.8}$$

This equation reduces to equation (5.3.12) when the magnetic field is absent. However, even in the presence of this field, if we put

$$E = E_0+\frac{eB}{2mc}(k+1)\hbar \tag{5.7.9}$$

in equation (5.7.8), it becomes identical with equation (5.3.12) after $E$ has been replaced by $E_0$. This implies that our previous calculation of $R$ continues to be valid provided we substitute $E_0$ for $E$.

Thus $E_0$ will be quantised as required by equation (5.4.14) and hence $E$ will be quantised according to the equation

$$E = E_{nk} = -\frac{mZ^2e^4}{2\hbar^2}\cdot\frac{1}{n^2}+\frac{eB}{2mc}(k+1)\hbar. \tag{5.7.10}$$

The eigenstate corresponding to the quantum numbers $k$, $l$, $n$ and for which $\sigma_z=+1$, is then given by

$$\psi_+ = R_{ln}Y_{lk}\exp(-\iota E_{nk}t/\hbar), \quad \psi_- = 0. \tag{5.7.11}$$

Similarly, the eigenstate corresponding to the quantum numbers $k$, $l$, $n$ and for which $\sigma_z=-1$, is determined by

$$\psi_+ = 0, \quad \psi_- = R_{ln}Y_{lk}\exp(-\iota E'_{nk}t/\hbar), \tag{5.7.12}$$

where

$$E'_{nk} = -\frac{mZ^2e^4}{2\hbar^2}\cdot\frac{1}{n^2}+\frac{eB}{2mc}(k-1)\hbar. \tag{5.7.13}$$

Suppose a transition occurs between a state $(k,l,n)$ and a state $(k',l',n')$. It will be shown in section 6.13 that certain selection rules require that $\Delta k = k' - k = 0$, 1 or $-1$, $\Delta l = l' - l = 1$ or $-1$ and that the spin component is unchanged. In the absence of the magnetic field, the frequency of the radiation emitted (or absorbed) by the atom will be determined by the energy change $E_n - E_{n'}$. For given values of $n$ and $n'$, this leads to a single line in the atom's spectrum. However, in the presence of the magnetic field, the possible energy changes are

$$E_{nk} - E_{n'k'} = E_n - E_{n'} + \frac{eB}{2mc}(k - k')\hbar, \tag{5.7.14}$$

if $\sigma_z = +1$ and

$$E'_{nk} - E'_{n'k'} = E_n - E_{n'} + \frac{eB}{2mc}(k - k')\hbar, \tag{5.7.15}$$

if $\sigma_z = -1$. Since $k' = k$, $k+1$ or $k-1$, these transitions yield a triplet of spectrum lines, the middle one of which will be in the same position as the line resulting from the same transitions in the absence of the magnetic field. The effect of the field upon the spectrum of the atom is to cause its lines to split into three components; this is the *Normal Zeeman Effect*. If the spin-orbit interaction is allowed for, it is found that the pattern of the splitting of the energy levels is more complex and leads to the *Anomalous Zeeman Effect*.

Equations (5.7.10), (5.7.13) indicate that, when the magnetic field is brought into existence, the electron has its energy altered by an amount

$$\frac{eB}{2mc}(L_z + 2s_z). \tag{5.7.16}$$

This is precisely the energy change we would expect if the atom were a magnetic dipole of moment

$$-\frac{e}{2mc}(\mathfrak{L} + 2\mathfrak{s}). \tag{5.7.17}$$

The second term in this expression is accounted for by the magnetic moment of the electron due to its spin. The first term is due to the magnetic moment associated with the electron's orbital motion and a very simple atomic model can be used to make this identification seem very plausible.

Suppose the electron is rotating about the nucleus in a circular orbit of radius $a$ with speed $v$. The number of circuits it makes in unit time is then $v/2\pi a$ and, hence, the charge flowing past a fixed point on the orbit per unit time is $ev/2\pi a$. We can therefore think of the nucleus as being surrounded by a current of strength $ev/2\pi a$ and enclosing an area $\pi a^2$. Such a current behaves classically like a dipole of moment

$$\frac{1}{c}\cdot\frac{ev}{2\pi a}\cdot\pi a^2 = \frac{eva}{2c}\cdot \tag{5.7.18}$$

Since $L = mva$, this leads to an expected moment of $eL/2mc$.

CHAPTER 6

# Perturbation methods

## 6.1. Perturbed steady states

Suppose that $\hat{H}$ is the Hamiltonian operator for a system subject to a steady applied force field, so that $\hat{H}$ does not depend upon $t$ explicitly. We shall assume that the eigenvalues for the energy $E$ of the system form a discrete spectrum $E_1, E_2, \ldots$ and that the system is non-degenerate with respect to each such eigenvalue (the effect of degeneracy will be considered in section 6.2). Let $\psi^1 \exp(-\iota E_1 t/\hbar), \psi^2 \exp(-\iota E_2 t/\hbar), \ldots$ be the normalised eigenvectors corresponding to the steady states having energies $E_1, E_2, \ldots$ respectively; then the vector amplitudes $\psi^1, \psi^2, \ldots$ form a complete orthonormal set. The problem we shall study is to examine the effect on the eigenvalues $E_i$ and the vector amplitudes $\psi^i$ of a *small* change in the applied force field to which the system is subject. For example, the system may be an electron moving under the Coulomb attraction towards a positively charged nucleus and the change in the force field may be brought about by the application of a uniform magnetic field (cf. the Zeeman effect, section 5.7) or by the application of a uniform electric field (the Stark effect, section 6.4). If such an additional field component is weak by comparison with the strength of the field originally present, it is referred to as a *perturbation* of the original field. Let the perturbed Hamiltonian for the particle be given by

$$\hat{H}' = \hat{H} + \epsilon \hat{K}, \tag{6.1.1}$$

where $\epsilon$ is a numerical multiplier, whose value will be taken to be sufficiently small to validate our later arguments. We shall assume that $\hat{K}$ (like $\hat{H}$) is not time dependent.

Let $E_n$ denote a particular one of the energy eigenvalues and let $\psi^n$ be the corresponding normalised vector amplitude. We shall suppose that the effect of the perturbation is to shift this energy level to a neighbouring value $E_n'$, where

$$E_n' = E_n + \epsilon F_n + \epsilon^2 G_n + \ldots, \tag{6.1.2}$$

and that a vector amplitude corresponding to $E_n'$ is given by

$$\psi^{n\prime} = \psi^n + \epsilon \phi^n + \epsilon^2 \chi^n + \ldots. \tag{6.1.3}$$

This vector amplitude $\psi^{n\prime}$ will not be supposed normalised, but its arbitrary multiplier will be determined by the condition

$$\langle\psi^n|\psi^{n\prime}\rangle = 1. \tag{6.1.4}$$

Substituting for $\psi^{n\prime}$ in this condition from equation (6.1.3), it will be seen that it is satisfied for all sufficiently small $\epsilon$ if, and only if,

$$\langle\psi^n|\phi^n\rangle = \langle\psi^n|\chi^n\rangle = \ldots = 0. \tag{6.1.5}$$

Since $\psi^{n\prime}$ is an eigenvector for $\hat{H}'$, we know that

$$\hat{H}'\psi^{n\prime} = E'_n\psi^{n\prime}. \tag{6.1.6}$$

(The summation convention is suspended throughout this chapter.) Substituting for $\hat{H}'$, $E'_n$, $\psi^{n\prime}$ from equations (6.1.1)–(6.1.3) and equating coefficients of like powers of $\epsilon$ in the two members of equation (6.1.6), we obtain the equations

$$\hat{H}\psi^n = E_n\psi^n, \tag{6.1.7}$$

$$\hat{H}\phi^n+\hat{K}\psi^n = E_n\phi^n+F_n\psi^n, \tag{6.1.8}$$

$$\hat{H}\chi^n+\hat{K}\phi^n = E_n\chi^n+F_n\phi^n+G_n\psi^n, \tag{6.1.9}$$

etc.

Equation (6.1.7) provides no new information. Taking the scalar product of both members of equation (6.1.8) with $\psi^i$, there are two cases to consider. If $i \neq n$, then $\psi^i$, $\psi^n$ are orthogonal and we find that

$$\langle\psi^i|\hat{H}\phi^n\rangle+\langle\psi^i|\hat{K}\psi^n\rangle = E_n\langle\psi^i|\phi^n\rangle. \tag{6.1.10}$$

But $\hat{H}$ is Hermitian and hence

$$\langle\psi^i|\hat{H}\phi^n\rangle = \langle\hat{H}\psi^i|\phi^n\rangle = E_i\langle\psi^i|\phi^n\rangle. \tag{6.1.11}$$

It follows that

$$\langle\psi^i|\phi^n\rangle = \frac{1}{E_n-E_i}\langle\psi^i|\hat{K}\psi^n\rangle. \tag{6.1.12}$$

But, referring to equation (1.6.18), it is clear that the left-hand member of this equation is the coefficient of $\psi^i$ in the expansion of $\phi^n$ in terms of these eigenvectors. We shall denote this coefficient by $a_i$. Also, the scalar product in the right-hand member is the *in*th element of the matrix representing $\hat{K}$ when we employ the eigenvectors $\psi^i$ as a basis. Denoting this element by $K_{in}$, we have

$$a_i = \frac{K_{in}}{E_n-E_i}\cdot \tag{6.1.13}$$

The coefficient of $\psi^n$ in this eigenvector expansion of $\phi^n$ is zero by the first of equations (6.1.5).

In the case when $i=n$, employing the first of equations (6.1.5), we obtain

$$\langle\psi^n|\hat{H}\phi^n\rangle+\langle\psi^n|\hat{K}\psi^n\rangle = F_n\langle\psi_n|\psi_n\rangle = F_n. \tag{6.1.14}$$

Since $\hat{H}$ is Hermitian,

$$\langle\psi^n|\hat{H}\phi^n\rangle = \langle\hat{H}\psi^n|\phi^n\rangle = E_n\langle\psi^n|\phi^n\rangle = 0. \tag{6.1.15}$$

Hence,

$$F_n = \langle\psi^n|\hat{K}\psi^n\rangle = K_{nn}, \tag{6.1.16}$$

where $K_{nn}$ denotes the diagonal element of the matrix representing $\hat{K}$, in the position corresponding to the $n$th eigenvector.

Equation (6.1.16) gives the first order term in the expansion (6.1.2). The first order term in the expansion (6.1.3) is obtained from equation (6.1.13) in the form

$$\phi^n = \sum_{i=1}^{\infty}{}' \frac{K_{in}\psi^i}{E_n-E_i}, \tag{6.1.17}$$

the prime attached to the summation operator indicating that the term for which $i=n$ is to be omitted.

The second order terms are calculable from equation (6.1.9). Taking the scalar product of both members with $\psi^i$, in the case $i\neq n$, we obtain

$$E_i\langle\psi^i|\chi^n\rangle+\langle\psi^i|\hat{K}\phi^n\rangle = E_n\langle\psi^i|\chi^n\rangle+F_n\langle\psi^i|\phi^n\rangle. \tag{6.1.18}$$

But,

$$b_i = \langle\psi^i|\chi^n\rangle \tag{6.1.19}$$

is the coefficient of $\psi^i$ in the eigenvector expansion of $\chi^n$ and thus equation (6.1.18) shows that

$$\begin{aligned} b_i &= \frac{1}{E_n-E_i}\langle\psi^i|(\hat{K}-F_n)\phi^n\rangle, \\ &= \frac{1}{E_n-E_i}\sum_{j=1}^{\infty}{}' \frac{K_{jn}}{E_n-E_j}\langle\psi^i|(\hat{K}-F_n)\psi^j\rangle, \\ &= \frac{1}{E_n-E_i}\left[\sum_{j=1}^{\infty}{}' \frac{K_{jn}K_{ij}}{E_n-E_j}-\frac{K_{nn}K_{in}}{E_n-E_i}\right]. \end{aligned} \tag{6.1.20}$$

The second of equations (6.1.5) shows that the coefficient of $\psi^n$ in the expansion of $\chi^n$ is zero.

In the case when $i=n$, equation (6.1.9) yields

$$\langle\psi^n|\hat{K}\phi^n\rangle = F_n\langle\psi^n|\phi^n\rangle+G_n. \tag{6.1.21}$$

Substituting for $\phi^n$ from equation (6.1.17) and for $F_n$ from equation (6.1.16), we find that

$$G_n = \sum_{i=1}^{\infty}{}' \frac{|K_{in}|^2}{E_n-E_i}, \tag{6.1.22}$$

since $K_{ni}=K_{in}^*$.

Higher order terms can now be evaluated by a similar method if required, but the working soon becomes excessively laborious.

It will be observed that the perturbed energy eigenvalues and eigenvectors can be written down when the matrix representing $\hat{K}$ has been calculated. As an example, consider the perturbed harmonic oscillator for which the Hamiltonian is given by

$$H = \frac{p^2}{2m}+\tfrac{1}{2}m\omega^2x^2+\epsilon x^4. \tag{6.1.23}$$

In this case $\hat{K}=x^4$ and, employing the $x$-representation, the eigenfunctions forming the basis are given at equations (3.10.33), (3.10.34). Thus,

$$K_{ij} = \frac{1}{\pi^{5/2}}\left(\frac{h}{2m\omega}\right)^2\frac{1}{2^{\frac{1}{2}(i+j)}\sqrt{(i!)}\sqrt{(j!)}}\int_{-\infty}^{\infty}\xi^4\exp(-\xi^2)H_i(\xi)H_j(\xi)\,d\xi. \tag{6.1.24}$$

Employing the recurrence relationship (3.11.3), we find that

$$\xi^2H_i = \tfrac{1}{4}H_{i+2}+\tfrac{1}{2}(2i+1)H_i+i(i-1)H_{i-2}. \tag{6.1.25}$$

The integral in equation (6.1.24) can now be expressed as the sum of a number of integrals of the type appearing in the identity (C.21) and hence can be evaluated. The results are

$$K_{ii} = \tfrac{3}{4}\left(\frac{\hbar}{m\omega}\right)^2(2i^2+2i+1), \tag{6.1.26}$$

$$K_{i,\,i+2} = \tfrac{1}{2}\left(\frac{\hbar}{m\omega}\right)^2\sqrt{[(i+1)(i+2)]}(2i+3), \tag{6.1.27}$$

$$K_{i,\,i+4} = \tfrac{1}{4}\left(\frac{\hbar}{m\omega}\right)^2\sqrt{[(i+1)(i+2)(i+3)(i+4)]}, \tag{6.1.28}$$

$$K_{i,\,i-2} = \tfrac{1}{2}\left(\frac{\hbar}{m\omega}\right)^2 \sqrt{[i(i-1)]}\,(2i-1), \tag{6.1.29}$$

$$K_{i,\,i-4} = \tfrac{1}{4}\left(\frac{\hbar}{m\omega}\right)^2 \sqrt{[i(i-1)(i-2)(i-3)]}, \tag{6.1.30}$$

all other elements $K_{ij}$ being zero.

Substituting in equations (6.1.16), (6.1.17), we now find that, to the first order in $\epsilon$,

$$E'_n = \omega\hbar(n+\tfrac{1}{2}) + \tfrac{3}{4}\epsilon\left(\frac{\hbar}{m\omega}\right)^2 (2n^2+2n+1), \tag{6.1.31}$$

giving the perturbed energy levels and

$$\psi^{n\prime} = \psi^n + \frac{\epsilon}{\omega\hbar}(\tfrac{1}{4}K_{n-4,\,n}\psi^{n-4} + \tfrac{1}{2}K_{n-2,\,n}\psi^{n-2} - \tfrac{1}{2}K_{n+2,\,n}\psi^{n+2} - \tfrac{1}{4}K_{n+4,\,n}\psi^{n+4}), \tag{6.1.32}$$

giving the perturbed energy eigenfunctions.

## 6.2. The degenerate case

If the unperturbed system is degenerate with respect to its energy, we shall suppose that further observables are associated with the energy to form a maximal set of compatible observables. The $\psi^i$ will then denote a complete orthonormal set of eigenvectors with respect to these observables and the $E_i$ will denote the corresponding energy eigenvalues (no longer all distinct).

Suppose the original system is $s$-fold degenerate with respect to a particular energy eigenvalue $E$. Then, for convenience, we will take the orthonormal set of $s$ eigenvectors corresponding to $E$ to be $\psi^1, \psi^2, \ldots, \psi^s$. When the perturbation is applied to the system, these $s$ states will, in general, be affected differently and we shall assume that the degenerate level $E$ splits into $s$ distinct levels $E'_1, E'_2, \ldots, E'_s$ as given by equation (6.1.2) (with $E_1 = E_2 = \ldots = E_s = E$). Applying the argument of the previous section to the particular level $E_j$, we obtain equation (6.1.12) in the form

$$\langle\psi^i|\hat{K}\psi^j\rangle = (E_j - E_i)\langle\psi^i|\phi^j\rangle, \tag{6.2.1}$$

where $i \neq j$. However, if $i, j \leqslant s$, $E_i = E_j$ and thus

$$K_{ij} = \langle\psi^i|\hat{K}\psi^j\rangle = 0, \tag{6.2.2}$$

i.e. our argument can only be valid if the first $s$ rows and columns of the matrix $\mathbf{K}$ form a diagonal sub-matrix. We must accordingly try to choose the orthonormal set $\psi^1, \psi^2, \ldots, \psi^s$, so that this shall be the case. If this can be done, equation (6.1.16) will yield the values of $F_1, F_2, \ldots, F_s$ which determine the perturbed energy levels. Inspection of equation (6.1.16) reveals that the $F_i$ are the diagonal elements of the diagonalised sub-matrix $(K_{ij})$, $i, j \leqslant s$.

Suppose $\theta^1, \theta^2, \ldots, \theta^s$ is our first choice of an orthonormal set of eigenvectors corresponding to the eigenvalue $E$. Employing these as a basis, the sub-matrix $\boldsymbol{\varkappa} = (K_{ij})$ will not, in general, be diagonal. Consider the set of all vectors which are linearly dependent upon the $\theta^i$; all such vectors will be eigenvectors for the energy eigenvalue $E$. If $\alpha$ is such a vector and

$$\alpha = \sum_{i=1}^{s} a_i \theta^i, \tag{6.2.3}$$

the $a_i$ are the components of $\alpha$ in the directions of the $\theta^i$ and the set $(a_1, a_2, \ldots, a_s)$ completely determines the vector $\alpha$. Thus the set of all vectors $\alpha$ constitutes a complex vector space of the type introduced in section 1.5. This space has $s$ dimensions and the $\theta^i$ form a set of base vectors.

Since it is constructed from a sub-set of the complete set of energy eigenvectors, it is called a *sub-space* of the space of vectors which represent all possible states of the system. Now, as explained in section 1.6, any set of $s$ mutually orthogonal and normalised vectors in this sub-space can be accepted as a set of base vectors. Let $\psi^i$ $(i = 1, 2, \ldots, s)$ be such a set. Then the $\psi^i$ are all eigenvectors for the energy eigenvalue $E$ and might have been chosen instead of the $\theta^i$. The effect of a change from the orthonormal set $\theta^i$ to the orthonormal set $\psi^i$ upon a matrix such as $\boldsymbol{\varkappa}$ has already been calculated at equation (1.8.16); thus if $\mathbf{U}$ is the unitary matrix of the transformation which transforms the components of a vector relative to the $\theta^i$-frame into its components relative to the $\psi^i$-frame, i.e. if

$$u_{ij} = \langle \psi^i | \theta^j \rangle, \tag{6.2.4}$$

then $\boldsymbol{\varkappa}$ is transformed into $\boldsymbol{\varkappa}'$, where

$$\boldsymbol{\varkappa}' = \mathbf{U}\boldsymbol{\varkappa}\mathbf{U}^{-1}. \tag{6.2.5}$$

It is shown in texts devoted to linear algebra that a unitary matrix $\mathbf{U}$ can always be found having the property that it reduces an arbitrary Hermitian matrix $\boldsymbol{\varkappa}$ to diagonal form by a transformation of the type (6.2.5); methods by which the matrix $\mathbf{U}$ can be calculated in any particular case, will also be found described in such works, but we shall not require to employ these techniques in this book. Choosing the matrix **U** to be of this form, the new orthonormal set may be calculated from the equation

$$\psi^i = \sum_{j=1}^{s} u_{ij}^* \theta^j. \tag{6.2.6}$$

It remains to calculate the diagonal elements of $\boldsymbol{\varkappa}'$ which, as has already been explained, give the values of the first order changes in the energy level, viz. $F_1, F_2, \ldots, F_s$.

Consider the equation for the characteristic roots of the diagonal matrix $\boldsymbol{\varkappa}'$, viz.

$$|\boldsymbol{\varkappa}' - \lambda\mathbf{I}| = 0. \tag{6.2.7}$$

This can be written

$$(F_1 - \lambda)(F_2 - \lambda)\ldots(F_s - \lambda) = 0 \tag{6.2.8}$$

and it follows that $F_1, F_2, \ldots, F_s$ are the characteristic roots. Now

$$\begin{aligned} |\boldsymbol{\varkappa}' - \lambda\mathbf{I}| &= |\mathbf{U}\boldsymbol{\varkappa}\mathbf{U}^{-1} - \lambda\mathbf{U}\mathbf{U}^{-1}|, \\ &= |\mathbf{U}(\boldsymbol{\varkappa} - \lambda\mathbf{I})\,\mathbf{U}^{-1}|, \\ &= |\mathbf{U}|\,|\boldsymbol{\varkappa} - \lambda\mathbf{I}|\,|\mathbf{U}|^{-1}, \\ &= |\boldsymbol{\varkappa} - \lambda\mathbf{I}|, \end{aligned} \tag{6.2.9}$$

since $|\mathbf{U}| = 1$. Thus, equation (6.2.7) is equivalent to the equation for the characteristic roots of the matrix $\boldsymbol{\varkappa}$, viz.

$$|\boldsymbol{\varkappa} - \lambda\mathbf{I}| = 0. \tag{6.2.10}$$

Thus, to calculate the values of the $F_i$, we have only to determine the characteristic roots of the matrix representing the perturbation operator $\hat{K}$ employing any basis.

The analysis of the degenerate case will not be carried further in this book.

## 6.3. The Zeeman Effect

This has already been considered in section 5.7. We will here derive the result of that section by the method of perturbations, regarding the weak uniform magnetic field $B$ as a small disturbance of the system comprising a point charge $-e$ moving in a central field $V$. Neglecting terms $O(B^2)$, the expression (5.7.4) for the Hamiltonian yields

$$\hat{H} = -\frac{\hbar^2}{2m}\nabla^2 + V, \quad \hat{K} = \hat{L}_z + \hbar\hat{\sigma}_z, \tag{6.3.1}$$

where we have taken

$$\epsilon = \frac{eB}{2mc}. \tag{6.3.2}$$

Associating the compatible observables $L^2$, $L_z$, $\sigma_z$ with $H$ to form a maximal set, eigenvectors for the unperturbed system, which is degenerate with respect to $H$, take the form shown in equation (5.5.9) (and a similar equation for the cases where $\sigma_z = -1$). Then, the elements of the matrix $\mathbf{K}$ corresponding to the energy eigenvalue $E_n$ are given by

$$K_{klns,\,k'l'ns'} = \int \psi^\dagger_{klns} \hat{K} \psi_{k'l'ns'}\, d\tau, \tag{6.3.3}$$

where the subscripts $s$, $s'$ take the values $+$ and $-$. Since $\psi_{klns}$ is an eigenvector for $\hat{L}_z$ and $\hat{\sigma}_z$, it is also an eigenvector for $\hat{K}$; thus

$$\hat{K}\psi_{klns} = (k \pm 1)\hbar\psi_{klns}, \tag{6.3.4}$$

the positive sign being taken when $s$ is $+$ and the negative sign when $s$ is $-$. But the vectors $\psi_{klns}$ form an orthonormal set and hence

$$K_{klns,\,k'l'ns'} = (k \pm 1)\hbar\delta_{kk'}\,\delta_{ll'}\,\delta_{ss'}. \tag{6.3.5}$$

Thus, the sub-matrix of $\mathbf{K}$ corresponding to $E_n$ is already diagonal when simultaneous eigenfunctions of the observables $L^2$, $L_z$, $H$, $\sigma_z$ are employed as a basis and its diagonal elements are $(k \pm 1)\hbar$ (positive sign if $s$, $s'$ are both $+$ and negative sign if they are both $-$).

Thus, as a result of the perturbation, the energy level $E_n$ for which $L^2 = l(l+1)\hbar^2$, $L_z = k\hbar$, $\sigma_z = +1$, becomes the level

$$E_n + \frac{eB}{2mc}(k+1)\hbar, \tag{6.3.6}$$

and the level $E_n$ for which $L^2 = l(l+1)\hbar^2$, $L_z = k\hbar$, $\sigma_z = -1$, becomes the level

$$E_n + \frac{eB}{2mc}(k-1)\hbar. \tag{6.3.7}$$

These results are in agreement with those already found in section 5.7.

It is evident that this increase in the number of distinct energy levels implies that the order of the degeneracy of the system with respect to the energy observable has been reduced by the perturbation. The system is no longer degenerate with respect to the quantum number $k$. However, degeneracy with respect to the quantum number $l$ is still present, since the energy levels do not depend upon this number.

## 6.4. Spin-orbit coupling

Since the term introduced into the Hamiltonian (see (5.5.10)) to allow for the interaction between the spin of the electron in a hydrogen-type atom and its orbital motion, is small by comparison with the remaining terms, it may be treated as a perturbation. Thus, taking this term in the form given at (5.5.16), we put

$$\epsilon = \frac{e^2 Z}{4m^2 c^2}, \quad \hat{K} = \frac{1}{r^3}(\hat{J}^2 - \hat{L}^2 - \tfrac{3}{4}\hbar^2). \tag{6.4.1}$$

We now consider the splitting of the energy level $E_n$ caused by this perturbation. A complete set of orthonormal eigenvectors for this energy level, in the absence of the perturbation, has been calculated in section 5.5 and the members of this set are given at equations (5.5.50), (5.5.51). Quantum numbers by which these eigenstates may be distinguished are provided by $q$, $l$, $n$, $j$ and the corresponding eigenvector will be denoted by $\psi_{qlnj}$. $j$ can take only two values, viz., $l+\frac{1}{2}$ and $l-\frac{1}{2}$; these will be denoted by $j_+$ and $j_-$ respectively. Then $\psi_{qlnj_-}$ is given by equation (5.5.50) and $\psi_{qlnj_+}$ is given by equation (5.5.51).

Since $\hat{K}$ commutes with $\hat{J}_z$, $\hat{L}^2$ and $\hat{J}^2$ (but not with $\hat{H}$, the Hamiltonian without the perturbation term), and $\psi_{qlnj}$ are simultaneous eigenvectors for $\hat{J}_z$, $\hat{L}^2$ and $\hat{J}^2$, the part of the matrix representing $\hat{K}$ corresponding to a given value of $n$, when these eigenvectors are employed as a basis, will be diagonal. Thus,

$$\begin{aligned} K_{q'l'nj'qlnj} &= \langle \psi_{q'l'nj'} | \hat{K}\psi_{qlnj} \rangle, \\ &= \left\langle \psi_{q'l'nj'} \middle| \frac{\hbar^2}{r^3}[j(j+1) - l(l+1) - \tfrac{3}{4}]\psi_{qlnj} \right\rangle, \\ &= \hbar^2[j(j+1) - l(l+1) - \tfrac{3}{4}]\left\langle \psi_{q'l'nj'} \middle| \frac{1}{r^3}\psi_{qlnj} \right\rangle, \\ &= \hbar^2[j(j+1) - l(l+1) - \tfrac{3}{4}]\int \frac{1}{r^3}\psi^\dagger_{q'l'nj'}\psi_{qlnj}\,d\tau, \end{aligned} \tag{6.4.2}$$

and it follows from the fact that the functions $\psi_{qlnj}$ are orthogonal that this integral vanishes unless $q=q'$, $l=l'$, $j=j'$. Taking $q'=q$, $l'=l$, $j'=j$, we obtain the diagonal elements of the submatrix of **K** corresponding to the energy level $E_n$ in the form

$$\begin{aligned} K_{qlnj,qlnj} &= l\hbar^2 I, \quad \text{if} \quad j = j_+, \\ &= -(l+1)\hbar^2 I, \quad \text{if} \quad j = j_-, \end{aligned} \tag{6.4.3}$$

where

$$I = \int \frac{1}{r^3} \psi^{\dagger}_{qlnj} \psi_{qlnj} \, d\tau,$$

$$= \int \frac{1}{r^3} \{|Y_{l,\, q-\frac{1}{2}}|^2 + |Y_{l,\, q+\frac{1}{2}}|^2\} R^2_{ln} r^2 \sin\theta \, dr \, d\theta \, d\phi,$$

$$= \int_0^\infty \frac{1}{r} R^2_{ln} \, dr, \tag{6.4.4}$$

since the $Y_{lk}$ are normalised.

Substituting in equation (6.4.4) for $R_{ln}$ from equation (5.4.13), we find that

$$I = \left(\frac{e^2 mZ}{\hbar^2}\right)^3 \cdot \frac{4}{n^4} \cdot \frac{(n-l-1)!}{\{(n+l)!\}^3} \int_0^\infty \exp(-x) \, x^{2l-1} \, (L^{2l+1}_{n-l-1})^2 \, dx. \tag{6.4.5}$$

The integral can now be calculated by making use of equation (E.39) in Appendix E. The final result is

$$I = \left(\frac{e^2 mZ}{\hbar^2}\right)^3 \cdot \frac{1}{n^3 \, l(l+1)\,(l+\frac{1}{2})}. \tag{6.4.6}$$

The perturbed energy levels into which the formerly degenerate level $E_n$ splits, can now be calculated. Corresponding to the quantum numbers $q$, $l$, $n$, $j_+$, we have the level

$$E = E_n \left[1 - \frac{(\alpha Z)^2}{n(l+1)\,(2l+1)}\right], \tag{6.4.7}$$

where $E_n$ is given by equation (5.4.14) and $\alpha$ is the dimensionless *fine-structure constant* given by

$$\alpha = \frac{e^2}{c\hbar}. \tag{6.4.8}$$

Corresponding to the quantum numbers $q$, $l$, $n$, $j_-$, there is the level

$$E = E_n \left[1 + \frac{(\alpha Z)^2}{nl(2l+1)}\right]. \tag{6.4.9}$$

It will be observed that energy degeneracy with respect to the quantum number $q$, still persists.

The formulae (6.4.7), (6.4.9) do not lead to results which are in accord with observation. The reason for this is that the correction which allows for the special relativity variation of the electron mass with its velocity proves to be of the same order as the correction required by the spin-orbit coupling. Thus, according to the special theory of relativity, the Hamiltonian for the electron is given by the formula

$$H = mc^2\left(1+\frac{p^2}{m^2c^2}\right)^{\frac{1}{2}} - mc^2 + V, \tag{6.4.10}$$

where $m$ is the rest mass. If we assume that $v$ is small by comparison with the velocity of light, this gives approximately

$$H = \frac{p^2}{2m} - \frac{p^4}{8m^3c^2} + V. \tag{6.4.11}$$

To this degree of approximation, therefore, the Hamiltonian for the system should be corrected by the subtraction of a further term $p^4/8m^3c^2$. We shall treat this term as a further perturbation which is responsible for an additional shift in the energy level corresponding to the quantum numbers $q, l, n, j$.

Thus, we take

$$\epsilon = -\frac{1}{2mc^2}, \quad \hat{K} = \left(\frac{1}{2m}\hat{p}^2\right)^2 = (\hat{H}-V)^2, \tag{6.4.12}$$

where $H$ denotes the unperturbed Hamiltonian, and $V = -Ze^2/r$ is the P.E. of the electron in the field of the nucleus. Then

$$\begin{aligned} \hat{K}\psi_{qlnj} &= (\hat{H}-V)^2\psi_{qlnj}, \\ &= (E_n - V)^2\psi_{qlnj}, \end{aligned} \tag{6.4.13}$$

since $\psi_{qlnj}$ is an eigenfunction of $\hat{H}$ associated with the eigenvalue $E_n$.

Proceeding now as before equation (6.4.2), we show that the submatrix of $\mathbf{K}$ corresponding to the energy level $E_n$ is diagonal and the diagonal elements are given by

$$\begin{aligned} K_{qlnj,\,qlnj} &= \int (E_n - V)^2 \psi^\dagger_{qlnj}\psi_{qlnj}\,d\tau, \\ &= \int_0^\infty r^2(E_n-V)^2 R_{ln}^2\,dr \\ &= E_n^2\int_0^\infty r^2R_{ln}^2\,dr + 2Ze^2E_n\int_0^\infty rR_{ln}^2\,dr \\ &\quad + Z^2e^4\int_0^\infty R_{ln}^2\,dr. \end{aligned} \tag{6.4.14}$$

The first integral is unity since $R_{ln}$ has been normalised. The second integral is

$$\frac{e^2 mZ}{\hbar^2}\cdot\frac{(n-l-1)!}{n^2\{(n+l)!\}^3}\int_0^\infty \exp(-x)\,x^{2l+1}(L_{n-l-1}^{2l+1})^2\,dx \tag{6.4.15}$$

and this can be calculated from equation (E.30) of Appendix E to give

$$\frac{e^2 mZ}{\hbar^2}\cdot\frac{1}{n^2}. \tag{6.4.16}$$

The third integral is

$$2\left(\frac{e^2 mZ}{\hbar^2}\right)^2\cdot\frac{(n-l-1)!}{n^3\{(n+l)!\}^3}\int_0^\infty \exp(-x)\,x^{2l}(L_{n-l-1}^{2l+1})^2\,dx \tag{6.4.17}$$

and this can be calculated from equation (E. 38) of Appendix E to give

$$\left(\frac{e^2 mZ}{\hbar^2}\right)^2\cdot\frac{1}{n^3(l+\frac{1}{2})}. \tag{6.4.18}$$

From equation (6.4.14), we now find that

$$K_{qlnj,\,qlnj} = E_n^2\left[\frac{4n}{l+\frac{1}{2}}-3\right] \tag{6.4.19}$$

and it follows that the required correction to $E_n$ is

$$E_n(\alpha Z)^2\frac{1}{n^2}\left[\frac{n}{l+\frac{1}{2}}-\frac{3}{4}\right]. \tag{6.4.20}$$

Applying this correction to equation (6.4.7), we find that the perturbed energy level associated with the quantum numbers $q$, $l$, $n$, $j_+$ is

$$E = E_n\left[1+\left(\frac{\alpha Z}{n}\right)^2\left(\frac{n}{l+1}-\frac{3}{4}\right)\right]. \tag{6.4.21}$$

Applying the correction to equation (6.4.9), we find that the perturbed energy level associated with the quantum numbers $q$, $l$, $n$, $j_-$ is

$$E = E_n\left[1+\left(\frac{\alpha Z}{n}\right)^2\left(\frac{n}{l}-\frac{3}{4}\right)\right]. \tag{6.4.22}$$

This last pair of equations is easily seen to be equivalent to the single equation

$$E = E_n\left[1+\left(\frac{\alpha Z}{n}\right)^2\left(\frac{n}{j+\frac{1}{2}}-\frac{3}{4}\right)\right]. \tag{6.4.23}$$

This equation is found to be in good agreement with observation.

By taking the Hamiltonian in the exact relativistic form (6.4.10), Dirac replaced the Schrödinger characteristic equation for the energy by a new equation which now bears his name and was thus able to make full allowance for relativistic effects. Employing this equation, it will be shown in section 7.9 that the exact equation for the energy level associated with the quantum numbers $q$, $l$, $n$, $j$, is

$$E = mc^2\left[1+\left\{\frac{\alpha Z}{n-j-\frac{1}{2}+\sqrt{\{(j+\frac{1}{2})^2-\alpha^2 Z^2\}}}\right\}^2\right]^{-\frac{1}{2}}. \tag{6.4.24}$$

If the left hand member of this equation is expanded in ascending powers of the fine-structure constant $\alpha$ as far as the terms in $\alpha^4$, it will be found to yield equation (6.4.23) (with an additional rest energy term $mc^2$).

## 6.5. The Stark Effect

As an example of a perturbation calculation in the degenerate case which does not lead immediately to a sub-matrix for $\hat{K}$ which is diagonal, we will consider the effect produced on the energy levels of a hydrogen-type atom when it is placed in a uniform electric field. Thus, the unperturbed system comprises a single electron having charge $-e$ moving in a Coulomb field of force, for which the Hamiltonian operator is

$$\hat{H} = -\frac{\hbar^2}{2m}\nabla^2 - \frac{Ze^2}{r}. \tag{6.5.1}$$

The spin-orbit coupling will be neglected. The perturbing field, assumed to be of uniform intensity $F$ in the direction of the $z$-axis, contributes an additional term

$$eFz = eFr\cos\theta. \tag{6.5.2}$$

Thus, in this case, we take

$$\epsilon = eF, \quad \hat{K} = r\cos\theta. \tag{6.5.3}$$

The unperturbed system is degenerate with respect to the energy, a complete system of eigenvectors being given in the $xyz\sigma_z$-representation by the spinors $\psi_{kln+}$, $\psi_{kln-}$, defined at (5.5.9). The elements of the sub-matrix of $\mathbf{K}$ corresponding to the energy level $E = E_n$ are given by

$$K_{k'l'ns',\,klns} = \int \psi^{\dagger}_{k'l'ns'}\, r\cos\theta\, \psi_{klns}\, d\tau, \tag{6.5.4}$$

where the subscript $s$ can be $+$ or $-$. Substituting for $\psi_{klns}$ from equation (5.5.9), we find that

$$K_{k'l'ns',\,klns} = \delta_{ss'}\int_0^\infty r^3 R_{ln} R_{l'n}\, dr \times \int_0^{2\pi} d\phi \int_0^{\pi} Y^*_{l'k'}\, Y_{lk} \sin\theta\cos\theta\, d\theta. \tag{6.5.5}$$

Now

$$\int_0^{2\pi} d\phi \int_0^{\pi} Y^*_{l'k'}\, Y_{lk} \sin\theta \cos\theta \, d\theta$$

$$= A_{l'k'} A_{lk} \int_0^{2\pi} \exp[\iota(k-k')\phi]\, d\phi \times \int_0^{\pi} P^{k'}_{l'} P^k_l \sin\theta \cos\theta \, d\theta,$$

$$= 2\pi \delta_{kk'} A_{lk} A_{l'k} \int_0^{\pi} P^k_{l'} P^k_l \sin\theta \cos\theta \, d\theta, \tag{6.5.6}$$

where $A_{lk}$ is given by equation (5.2.31). Also

$$\int_0^{\pi} P^k_{l'} P^k_l \sin\theta \cos\theta \, d\theta = \int_{-1}^{1} P^k_l(\mu) P^k_{l'}(\mu)\, \mu \, d\mu, \tag{6.5.7}$$

where $\mu = \cos\theta$. This integral has been calculated at equation (D.27) of Appendix D and, using this result, we obtain

$$\left.\begin{aligned} \int_0^{2\pi} d\phi \int_0^{\pi} Y^*_{l'k'}\, Y_{lk} \sin\theta \cos\theta \, d\theta &= \left(\frac{l^2-k^2}{4l^2-1}\right)^{\frac{1}{2}} \delta_{kk'}, \quad \text{if } l = l'+1, \\ &= \left(\frac{l'^2-k^2}{4l'^2-1}\right)^{\frac{1}{2}} \delta_{kk'}, \quad \text{if } l' = l+1, \end{aligned}\right\} \tag{6.5.8}$$

and is zero otherwise.

Suppose $l = l'+1$. Then

$$\int_0^{\infty} r^3 R_{ln} R_{l'n}\, dr$$

$$= \frac{\hbar^2}{4mZe^2} \cdot \frac{(n-l-1)!}{\{(n+l)!\}^3} \sqrt{[(n-l)(n+l)^3]} \int_0^{\infty} \exp(-x)\, x^{2l+2} L^{2l+1}_{n-l-1} L^{2l-1}_{n-l}\, dx \tag{6.5.9}$$

where the relationship between $x$ and $r$ is given by equation (5.4.20). This integral has been calculated in Appendix E (equation (E.41)). Employing this result, we find that

$$\int_0^{\infty} r^3 R_{ln} R_{l'n}\, dr = -\frac{3\hbar^2}{2mZe^2} n \sqrt{(n^2-l^2)}. \tag{6.5.10}$$

It now follows from equation (6.5.5) that, if $l = l'+1$, $k = k'$, $s = s'$,

$$K_{k'l'ns',klns} = -\frac{3\hbar^2 n}{2mZe^2} \left[\frac{(l^2-k^2)(n^2-l^2)}{4l^2-1}\right]^{\frac{1}{2}}, \tag{6.5.11}$$

and if $l'=l+1$, $k=k'$, $s=s'$,

$$K_{k'\,l'\,ns',\,klns} = -\frac{3\hbar^2 n}{2mZe^2}\left[\frac{(l'^2-k^2)(n^2-l'^2)}{4l'^2-1}\right]^{\frac{1}{2}}. \tag{6.5.12}$$

All other elements of the sub-matrix of **K** vanish.

For a given $n$, we shall arrange the elements of the sub-matrix as follows: The triples of subscripts $(k,l,s)$ are first arranged in the order shown below:

$$\begin{array}{lllll}
(0,0,+), & (0,1,+), & (0,2,+), & \ldots, & (0,n-1,+),\\
(0,0,-), & (0,1,-), & (0,2,-), & \ldots, & (0,n-1,-),\\
(1,1,+), & (1,2,+), & (1,3,+), & \ldots, & (1,n-1,+),\\
(1,1,-), & (1,2,-), & (1,3,-), & \ldots, & (1,n-1,-),\\
(-1,1,+), & (-1,2,+), & (-1,3,+), & \ldots, & (-1,n-1,+),\\
(-1,1,-), & (-1,2,-), & (-1,3,-), & \ldots, & (-1,n-1,-),\\
(2,2,+), & (2,3,+), & (2,4,+), & \ldots, & (2,n-1,+),\\
\ldots & \ldots & \ldots & \ldots, & (-n+1,n-1,-).
\end{array}$$

I.e. $k$ takes the values $0$, $\pm 1$, $\pm 2$, $\ldots$, $\pm(n-1)$ and $s$ the values $+$, $-$ successively and, for each pair of values of $k$ and $s$, $l$ ranges over the values $|k|$, $|k|+1$, $\ldots$, $n-1$. There are $2n^2$ terms in the sequence and this, therefore, is the order of the sub-matrix. Then, if the triple $(k,l,s)$ stands in the $r$th position in this sequence, the element $K_{k'\,l'\,ns',\,klns}$ is placed in the $r$th column. Similarly, if the triple $(k',l',s')$ stands in the $r'$th position in this sequence, the element is placed in the $r'$th row. Since the elements of the sub-matrix are zero unless $k=k'$, $s=s'$, the resulting matrix will comprise non-zero square matrix blocks arranged along the principal diagonal, each block corresponding to equal fixed values of $k$ and $k'$ and of $s$ and $s'$, and zero elements everywhere else. Thus, the first block along the diagonal will be formed from the elements $K_{0l'\,n+,\,0ln+}$, where $l'$ refers to the row and $l$ to the column in which the element is placed. The characteristic equation (6.2.10) for the sub-matrix will now involve a determinant which can be expanded as the product of the characteristic determinants for the separate blocks. This implies that the set of characteristic roots of the sub-matrix will be the sum of the sets of characteristic roots for the blocks.

Consider, therefore, a particular block for which $k$ and $k'$ take the same value $k$. Then $l$, $l'$ range over the values $|k|$, $|k|+1, \ldots, n-1$, but only the elements for which $l=l'+1$ or $l'=l+1$ are non-zero. We shall write $k$ for

$|k|$ throughout the remainder of the argument. Then the block takes the form below,

$$\begin{pmatrix} 0 & a_1 & 0 & \dots & 0 & 0 \\ a_1 & 0 & a_2 & \dots & 0 & 0 \\ 0 & a_2 & 0 & \dots & 0 & 0 \\ \dots & \dots & \dots & \dots & \dots & \dots \\ \dots & \dots & \dots & \dots & 0 & a_{n-k-1} \\ \dots & \dots & \dots & \dots & a_{n-k-1} & 0 \end{pmatrix}, \quad (6.5.13)$$

all elements except those lying on either side of the principal diagonal being zero. It follows from equations (6.5.11), (6.5.12), that

$$a_i = -\frac{3\hbar^2 n}{2mZe^2}\left[\frac{\{(k+i)^2-k^2\}\{n^2-(k+i)^2\}}{4(k+i)^2-1}\right]^{\frac{1}{2}}, \quad (6.5.14)$$

where $i = 1, 2, \dots, (n-k-1)$. The characteristic equation for this block can now be written

$$\begin{vmatrix} \alpha & b_1 & 0 & \dots & 0 & 0 \\ b_1 & \alpha & b_2 & \dots & 0 & 0 \\ 0 & b_2 & \alpha & \dots & 0 & 0 \\ \dots & \dots & \dots & \dots & \dots & \dots \\ \dots & \dots & \dots & \dots & \alpha & b_{n-k-1} \\ \dots & \dots & \dots & \dots & b_{n-k-1} & \alpha \end{vmatrix} = 0, \quad (6.5.15)$$

where

$$\alpha = \frac{2mZe^2}{3\hbar^2 n}\lambda, \quad (6.5.16)$$

and

$$b_i^2 = \frac{i(2k+i)(n-k-i)(n+k+i)}{(2k+2i-1)(2k+2i+1)}. \quad (6.5.17)$$

The determinant occurring in equation (6.5.15) can be calculated by Schlaff's method as described in Appendix G. The result is that it factorises thus:

$$[\alpha^2-(n-k-1)^2][\alpha^2-(n-k-3)^2][\alpha^2-(n-k-5)^2]\dots, \quad (6.5.18)$$

the final factor being $\alpha^2-1$ or $\alpha$ according as $(n-k)$ is even or odd respectively.

The characteristic roots are now easily seen to be given by

$$\alpha = \pm(n-k-1), \pm(n-k-3), \dots \quad (6.5.19)$$

and hence, from equation (6.5.16),

$$\lambda = \frac{3\hbar^2 nr}{2mZe^2}, \tag{6.5.20}$$

where $r = -(n-k-1),\ -(n-k-3),\ \ldots,\ (n-k-3),\ (n-k-1)$. Since $k$ can take all integral values from 0 to $(n-1)$, we obtain all the characteristic roots of the sub-matrix originally considered by letting $r$ range over the integral values from $-(n-1)$ to $+(n-1)$ (many of these roots will be repeated).

Thus the energy eigenstate $E_n$ splits into $(2n-1)$ distinct states given by

$$E = E_n + eF\lambda = E_n + \frac{3\hbar^2 Fnr}{2mZe}. \tag{6.5.21}$$

## 6.6. Transition theory

In the previous sections of this chapter, we have considered the effect of a small perturbing force upon the steady energy states of a system; this analysis is referred to as *time-independent perturbation theory*. In this and later sections, we shall attempt to follow the evolution in time of a system from an initial steady state, after a small perturbing force has been applied; this approach is termed *time-dependent perturbation theory*. We shall suppose that the perturbation is switched on at $t=0$, when the system is in an unperturbed steady state having vector amplitude $\psi^n$ and energy $E_n$. At some later instant $t$, the perturbation is supposed discontinued and the energy of the system is again measured; in general, its new energy $E_m$ will be found to be different from its original energy $E_n$, so that a *transition* will have been caused by the perturbing force. The new energy $E_m$ will, of course, belong to the spectrum of energy eigenvalues for the unperturbed system, but we shall not be able to predict with precision the energy change which will be found to have occurred; such a prediction can only be made on a probability basis. Our main object will therefore be to calculate the probability of every possible transition from the initial steady state $E_n$ to a final state $E_m$.

We shall take the perturbed Hamiltonian for the system to be given by equation (6.1.1) and shall no longer assume $\hat{K}$ to be independent of $t$. Assuming the $\psi^i$ form a complete orthonormal set, any state $\psi$ can be expanded at some instant $t$ in terms of these functions thus:

$$\psi = \sum_{i=1}^{\infty} c_i \psi^i \exp(-\iota) E_i t/\hbar. \tag{6.6.1}$$

If $\psi$ represents a state of the unperturbed system, we know that the $c_i$ will be constants (see equation (4.4.11)); however, if $\psi$ refers to the perturbed system, the $c_i$ will vary with $t$. At $t=0$, we know that $\psi=\psi^n$ and hence

$$\left.\begin{aligned} c_i &= 0, \quad i \neq n, \\ &= 1, \quad i = n. \end{aligned}\right\} \tag{6.6.2}$$

At any later instant $t$, if the perturbation is discontinued and $E$ is measured, we know that it will be found to take the value $E_i$ with probability $|c_i|^2$. Our problem, therefore, is to calculate the coefficients $c_i$ as functions of $t$.

To perform this calculation, we make use of the fact that $\psi$ must satisfy the equation of motion

$$(\hat{H}+\epsilon\hat{K})\,\psi = \iota\hbar\frac{\partial\psi}{\partial t}. \tag{6.6.3}$$

Substituting in this equation from equation (6.6.1) and remembering that $\hat{H}\psi^i=E_i\psi^i$, we obtain

$$\epsilon\sum_{i=1}^{\infty} c_i \exp\left(-\iota E_i t/\hbar\right)\hat{K}\psi^i = \iota\hbar\sum_{i=1}^{\infty}\frac{dc_i}{dt}\psi^i \exp\left(-\iota E_i t/\hbar\right). \tag{6.6.4}$$

(We have assumed $\hat{K}f(t)=f(t)\hat{K}$.) Taking the scalar product of both members of this equation with $\psi^m$, since the $\psi^i$ are orthogonal, we find that

$$\iota\hbar\frac{dc_m}{dt}\exp\left(-\iota E_m t/\hbar\right) = \epsilon\sum_{i=1}^{\infty} c_i K_{mi}\exp\left(-\iota E_i t/\hbar\right), \tag{6.6.5}$$

where $K_{mi}$ is an element of the matrix representing $\hat{K}$ when the $\psi^i$ are employed as basis. Writing

$$\omega_{mi} = (E_m - E_i)/\hbar, \tag{6.6.6}$$

equation (6.6.5) takes the form

$$\iota\hbar\frac{dc_m}{dt} = \epsilon\sum_{i=1}^{\infty} c_i K_{mi}\exp\left(\iota\omega_{mi}t\right), \tag{6.6.7}$$

where $m=1, 2, \ldots$. These equations, together with the initial conditions (6.6.2), determine the coefficients $c_m$ as functions of $t$.

It is clear from equations (6.6.7), that all derivatives $dc_m/dt$ are small with $\epsilon$. It follows that all the coefficients $c_m$ are initially small of order $\epsilon$, with the exception of $c_n$, which is then approximately unity. Hence, for values of $t$ near to $t=0$, the equations (6.6.7) can be approximated by

$$\iota\hbar\frac{dc_m}{dt} = \epsilon K_{mn}\exp(\iota\omega_{mn}t). \tag{6.6.8}$$

Integrating for the cases when $m\neq n$, we obtain

$$c_m = -\frac{\iota\epsilon}{\hbar}\int_0^t K_{mn}\exp(\iota\omega_{mn}t)\,dt. \tag{6.6.9}$$

Note that, in general, the matrix elements $K_{mn}$ are functions of $t$. Provided the integral in equation (6.6.9) remains bounded as $t\to\infty$, $c_m$ will continue to be of the first order in $\epsilon$ and our approximation will then be valid for all values of $t$. If, however, the modulus of the integral increases without limit, the approximation must ultimately break down.

In the case $m=n$, equation (6.6.7) may be approximated by

$$\iota\hbar\frac{dc_n}{dt} = \epsilon K_{nn}c_n \tag{6.6.10}$$

which, upon integration, yields

$$c_n = \exp\left(-\frac{\iota\epsilon}{\hbar}\int_0^t K_{nn}\,dt\right). \tag{6.6.11}$$

Now $K_{nn}$ is real, since $\hat{K}$ is Hermitian. Hence $|c_n|=1$ for all $t$. It follows that all our approximations are valid provided the $c_i$ remain small for $i\neq n$.

We will next consider two particular classes of perturbations, (i) those which are independent of $t$ and (ii) those which vary sinusoidally with $t$, for both of which this condition for the validity of our approximation is satisfied.

## 6.7. Time-independent perturbing force

If $\hat{K}$, and therefore the $K_{mn}$, are independent of $t$, equation (6.6.9) yields

$$c_m = \frac{\epsilon K_{mn}}{\hbar\omega_{mn}}[1-\exp(\iota\omega_{mn}t)]. \tag{6.7.1}$$

We have here assumed that $\omega_{mn}\neq 0$ and hence that the system is non-degenerate. If the system is degenerate, some of the $\omega_{mn}$ will vanish and it then follows from equation (6.6.9) that the corresponding $c_m$ will increase linearly with $t$ and hence that our approximation must ultimately fail. However, if the system is not degenerate, equation (6.7.1) indicates that the $c_m$ will remain of the first order in $\epsilon$. In this case

$$|c_m|^2 = \left(\frac{\epsilon|K_{mn}|}{\hbar}\cdot\frac{\sin\frac{1}{2}\omega_{mn}t}{\frac{1}{2}\omega_{mn}}\right)^2, \tag{6.7.2}$$

yielding the probability of a transition from the initial state $E_n$ to another state $E_m$ being observed at time $t$. It will be noted that this probability oscillates sinusoidally between the values 0 and $(2\epsilon|K_{mn}|/\hbar\omega_{mn})^2$.

Thus, as soon as the perturbing force comes into operation, the system may be thought of as commencing to move into every one of its other possible energy states simultaneously, although the trends towards such states are soon reversed, so that the probabilities of finding the system in any one of these states remains always small. It will be further observed from equation (6.7.2), that the probability of finding the system in a state whose energy differs greatly from the initial energy $E_n$ is smaller than the probability of finding it in a neighbouring energy state; this follows since $\omega_{mn}$ is large when the difference between $E_m$ and $E_n$ is large. On the other hand, if the system is degenerate, the probability of finding the system in a state differing from the initial state, but having the same energy, can become large. These results also follow from the principle of energy conservation for, since the Hamiltonian for the system is independent of $t$, the probability distribution of its energy cannot change; but, initially, this distribution is strongly 'peaked' near the value $E_n$† and hence there must always be a high probability of measuring $E$ to have this value $E_n$.

If the energy $E$ of the system is not quantised (e.g. the case of a particle not in a bound state) so that the system is perturbed into a continuum of energy states, by modification of the boundary conditions a very large number of discrete states in close proximity to one another may sometimes be substituted for the continuum and then our previous results become applicable. An example will be given in the next section. Suppose that, in such a case, $\rho(E)\,dE$ denotes the number of such discrete states within the energy interval $(E, E+dE)$. Then, the probability $dP$ that a transition from an initial energy state $E_0$ to a final state in the

† In the presence of the perturbing force, $E_n$ is not an eigenvalue for the energy. However, there will be a perturbed eigenvalue close to this and this value will be assumed by $E$, initially, with high probability.

range $(E, E+dE)$ will be found to have taken place at time $t$ after switching on the perturbation, is given by equation (6.7.2) to be

$$dP = 4\epsilon^2 |K_{E_0 E}|^2 \frac{\sin^2\{(E-E_0)t/2\hbar\}}{(E-E_0)^2} \rho(E)\, dE. \tag{6.7.3}$$

Hence, the total probability $P_{12}$ that a transition from the state $E_0$ into a state having energy $E$ lying in the interval $(E_1, E_2)$ has taken place at time $t$ is given by

$$P_{12} = \frac{2\epsilon^2 t}{\hbar} \int_{x_1}^{x_2} |K_{E_0 E}|^2 \rho(E) \frac{\sin^2 x}{x^2}\, dx, \tag{6.7.4}$$

where

$$x = (E-E_0)t/2\hbar, \quad x_1 = (E_1-E_0)t/2\hbar, \quad x_2 = (E_2-E_0)t/2\hbar. \tag{6.7.5}$$

If, now, $E_0 < E_1 < E_2$ and we suppose $|K_{E_0 E}|^2 \rho(E)$ to be bounded so that $|K_{E_0 E}|^2 \rho(E) < M$, then

$$P_{12} < \frac{2\epsilon^2 t}{\hbar} M \int_{x_1}^{x_2} \frac{dx}{x^2} = \frac{4\epsilon^2 M(E_2-E_1)}{(E_1-E_0)(E_2-E_0)}. \tag{6.7.6}$$

Thus $P_{12}$ remains small of order $\epsilon^2$ as $t \to \infty$. The same is true of $P_{12}$ if $E_1 < E_2 < E_0$. However, if $E_1 < E_0 < E_2$, then $x_1 \to -\infty$ and $x_2 \to +\infty$ as $t \to \infty$. Hence, for large values of $t$, we have approximately

$$P_{12} = \frac{2\epsilon^2 t}{\hbar} \int_{-\infty}^{\infty} |K_{E_0 E}|^2 \rho(E) \frac{\sin^2 x}{x^2}\, dx. \tag{6.7.7}$$

But, for values of $x$ outside an interval such as $(-4\pi, 4\pi)$, the integrand in this integral is small and it follows that such values of $x$ contribute only a small amount to its value. Also, over such an interval of $x$, if $t$ is large, it follows from the first of equations (6.7.5) that $E$ will differ very little from $E_0$. Hence, for large $t$, $P_{12}$ can be approximated by

$$\begin{aligned} P_{12} &= \frac{2\epsilon^2 t}{\hbar} |K_{E_0 E_0}|^2 \rho(E_0) \int_{-\infty}^{\infty} \frac{\sin^2 x}{x^2}\, dx, \\ &= \frac{2\pi}{\hbar} \epsilon^2 t |K_{E_0 E_0}|^2 \rho(E_0). \end{aligned} \tag{6.7.8}$$

We have proved, therefore, that the probability of a transition of the energy from $E_0$ into any interval not containing $E_0$ remains always small, but that the probability of a transition into any interval, however small, containing $E_0$, is independent of the length of the interval and increases linearly with $t$. Thus, the probability rate of transition from energy $E_0$ to energy lying within the interval $(E, E+dE)$ is

$$w = \frac{2\pi}{\hbar}\epsilon^2|K_{E_0 E_0}|^2 \rho(E_0)\, \delta(E-E_0)\, dE. \tag{6.7.9}$$

In other words, transitions take place at a constant rate and in such a way that energy is conserved.

## 6.8. Scattering by a centre of force

Suppose that a particle, moving uniformly under no forces, approaches a centre of force. Ultimately, the particle will recede from the centre and its motion will again be uniform; however, in general, its line of motion will have been deflected by the encounter and we say that the particle has been scattered. In this section, we shall calculate the probability that the line of motion will be deflected through an angle $\alpha$.

Let $\mathfrak{r}$ be the position vector for the particle, the origin being at the centre of force. Taking the Hamiltonian for the particle to be $p^2/2m + V(r)$, where $V(r)$ is the potential energy of the particle in the field of force, we shall treat $V$ as a small perturbation brought into play at $t=0$. Prior to this instant, we shall suppose that the particle is moving uniformly with momentum $\mathfrak{k}\hbar$. At time $t$, the centre of force is supposed annihilated and the particle's momentum is again measured and found to be $\mathfrak{k}_0\hbar$. Since, as explained in the last section, energy will be conserved, it follows that $|\mathfrak{k}| = |\mathfrak{k}_0| = k$. Our problem is to calculate the probability distribution of the angle of deflection between $\mathfrak{k}_0$ and $\mathfrak{k}$.

Consider a particle moving freely with momentum $\mathfrak{k}\hbar$ relative to rectangular axes $Oxyz$. With the object of replacing a continuous momentum spectrum by a discrete one and of permitting the normalisation of the wave function, we shall suppose that our system incorporates a device which is capable of transferring the particle, upon arrival at the plane $x=L$, instantaneously to the point having the same $y$- and $z$-coordinates in the plane $x=0$ without any change in momentum; the device will also be supposed capable of transferring the particle between these planes in the opposite sense (i.e. from $x=0$ to $x=L$), between the planes $y=0$, $y=L$ in either sense and between the planes $z=0$, $z=L$ in either sense.

The effect is to confine the particle to a box of volume $L^3$. The wave amplitude for the particle in this box is given by

$$\psi = A \exp(\iota \mathfrak{k}.\mathfrak{r}). \tag{6.8.1}$$

The boundary conditions are that $\psi$ and its first partial derivatives with respect to $x$, $y$, $z$ must be identical at corresponding points of opposite faces of the box. These conditions are easily shown to require that

$$k_x = 2n_x \pi/L, \quad k_y = 2n_y \pi/L, \quad k_z = 2n_z \pi/L, \tag{6.8.2}$$

where $n_x$, $n_y$, $n_z$ are positive or negative integers (not all zero) and $\mathfrak{k} = (k_x, k_y, k_z)$. It is clear that, if $L$ is large, the discrete momentum spectrum specified by equations (6.8.2) constitutes a good approximation to the continuous spectrum. Since $|\psi| = |A|$ inside the box and $|\psi| = 0$ outside, $\psi$ is normalised if we take $A = 1/L^{3/2}$. Thus

$$\psi = L^{-3/2} \exp(\iota \mathfrak{k}.\mathfrak{r}). \tag{6.8.3}$$

Taking $k_x$, $k_y$, $k_z$ to be rectangular cartesian coordinates of a point in $\mathfrak{k}$-space, the various particle states permitted by the equations (6.8.2) are represented by points forming a lattice; the number of points per unit volume is $(L/2\pi)^3$. Consider the states which are represented by points lying within a solid angle $d\omega$ with its apex at the origin and between concentric spheres of radii $k$, $k+dk$, with centres at the origin. The volume of this region is $k^2 d\omega dk$ and hence the number of such states is $k^2 d\omega dk (L/2\pi)^3$. If $E$ is the energy of the particle when its momentum is of magnitude $k\hbar$, then

$$E = \frac{k^2 \hbar^2}{2m} \tag{6.8.4}$$

and it follows that

$$dE = \frac{k\hbar^2}{m} dk. \tag{6.8.5}$$

Hence, the number of particle states for which the momentum direction lies within the solid angle $d\omega$ and for which the energy lies in the interval $(E, E+dE)$ is

$$\rho\, dE = \left(\frac{L}{2\pi}\right)^3 \frac{mk}{\hbar^2} d\omega\, dE \tag{6.8.6}$$

and thus

$$\rho = \left(\frac{L}{2\pi}\right)^3 \frac{mk}{\hbar^2} d\omega. \tag{6.8.7}$$

The probability rate of transition from an initial momentum state $\mathfrak{k}_0$ into a state $\mathfrak{k}$ whose direction lies within the solid angle $d\omega$ can now be calculated from equation (6.7.9).

We have

$$\epsilon K_{E_0 E_0} = \frac{1}{L^3} \int \exp(-\iota \mathfrak{k} . \mathfrak{r}) V(r) \exp(\iota \mathfrak{k}_0 . \mathfrak{r})\, d\tau$$

$$= \frac{1}{L^3} \int \exp[\iota(\mathfrak{k}_0 - \mathfrak{k}) . \mathfrak{r}] V(r)\, d\tau. \qquad (6.8.8)$$

Choosing the $z$-axis in the direction of the vector $\mathfrak{k}_0 - \mathfrak{k}$ and introducing spherical polar coordinates $(r, \theta, \phi)$, since

$$(\mathfrak{k}_0 - \mathfrak{k}) . \mathfrak{r} = |\mathfrak{k}_0 - \mathfrak{k}|\, r \cos\theta = 2k \sin\tfrac{1}{2}\alpha\, r \cos\theta, \qquad (6.8.9)$$

we obtain

$$\epsilon K_{E_0 E_0} = \frac{1}{L^3} \int\int\int \exp(\iota a r \cos\theta)\, V(r)\, r^2 \sin\theta\, dr\, d\theta\, d\phi, \qquad (6.8.10)$$

where $a = 2k \sin\frac{1}{2}\alpha$. The integrations with respect to $\phi(0 < \phi < 2\pi)$ and $\theta(0 < \theta < \pi)$ are easily performed and yield the result

$$\epsilon K_{E_0 E_0} = \frac{4\pi}{aL^3} \int_0^\infty r V \sin ar\, dr. \qquad (6.8.11)$$

It now follows from equations (6.7.9), (6.8.7), (6.8.11), that the probability rate of transition is given by

$$w = \frac{m\, d\omega}{L^3 \hbar^3 k \sin^2 \frac{1}{2}\alpha} \left( \int_0^\infty r V \sin ar\, dr \right)^2. \qquad (6.8.12)$$

But the probability current density for the initial state $\mathfrak{k}_0$ is calculated from equation (4.6.7) to be $\mathfrak{j} = \hbar \mathfrak{k}_0 / m L^3$. Thus, equation (6.8.12) can be written

$$w = j \left( \frac{m}{\hbar^2 k \sin\frac{1}{2}\alpha} \int_0^\infty r V \sin ar\, dr \right)^2 d\omega. \qquad (6.8.13)$$

If we now substitute an incident stream of particles each having momentum $\mathfrak{k}_0 \hbar$ for the single particle and interpret $j$ as the current density of this stream, we can interpret $w$ as the current density for the scattered stream whose momenta have directions lying within a solid angle $d\omega$

making an angle $\alpha$ with the direction of the incident stream. Writing $w = j\sigma\,d\omega$, $\sigma$ is termed the *differential scattering cross-section* and we have therefore shown that

$$\sigma = \left(\frac{m}{\hbar^2 k \sin \frac{1}{2}\alpha} \int_0^\infty rV \sin ar\,dr\right)^2. \tag{6.8.14}$$

In the case of a positively charged atomic nucleus shielded by a cloud of orbital electrons, the potential energy of an approaching electron is given approximately by a formula of the type

$$V = -\frac{Ze^2}{r} \exp(-r/r_0). \tag{6.8.15}$$

In this case,

$$\sigma = \left[\frac{2mZe^2 r_0^2}{\hbar^2(1+4r_0^2 k^2 \sin^2 \frac{1}{2}\alpha)}\right]^2. \tag{6.8.16}$$

By letting $r_0 \to \infty$, we approach the case of scattering of electrons by an inverse square law centre of attraction. This gives

$$\sigma = \left(\frac{mZe^2}{2\hbar^2 k^2 \sin^2 \frac{1}{2}\alpha}\right)^2 = \left(\frac{Ze^2}{4E \sin^2 \frac{1}{2}\alpha}\right)^2, \tag{6.8.17}$$

which is the classical Rutherford formula.

## 6.9. Oscillatory perturbing force

In this section, we will consider a particular example of a time-dependent perturbation, viz. when $\hat{K}$ oscillates sinusoidally with respect to $t$. We shall take

$$\hat{K} = \hat{F} \exp(-\iota\omega t) + \hat{G} \exp(\iota\omega t), \tag{6.9.1}$$

where $\hat{F}$, $\hat{G}$ are time-independent operators. The matrix elements for $\hat{K}$, $\hat{F}$, $\hat{G}$ are related by the equation

$$K_{mn} = F_{mn} \exp(-\iota\omega t) + G_{mn} \exp(\iota\omega t), \tag{6.9.2}$$

but, although $\hat{K}$ must be Hermitian, since the multipliers $\exp(\pm\iota\omega t)$ are complex, the operators $\hat{F}$, $\hat{G}$ are not Hermitian and, hence, neither are the matrices $(F_{mn})$, $(G_{mn})$. However, $K_{nm}^* = K_{mn}$ requires that

$$F_{nm}^* \exp(\iota\omega t) + G_{nm}^* \exp(-\iota\omega t) = F_{mn} \exp(-\iota\omega t) + G_{mn} \exp(\iota\omega t) \tag{6.9.3}$$

for all values of $t$. Thus

$$F^*_{nm} = G_{mn}, \quad G^*_{nm} = F_{mn}. \tag{6.9.4}$$

These conditions are identical and enable us to write

$$K_{mn} = F_{mn} \exp(-\iota\omega t) + F^*_{nm} \exp(\iota\omega t). \tag{6.9.5}$$

Substituting for $K_{mn}$ in equation (6.6.9), we calculate that

$$c_m = \frac{\epsilon F_{mn}\{1-\exp[\iota(\omega_{mn}-\omega)t]\}}{\hbar(\omega_{mn}-\omega)} + \frac{\epsilon F^*_{nm}\{1-\exp[\iota(\omega_{mn}+\omega)t]\}}{\hbar(\omega_{mn}+\omega)}, \tag{6.9.6}$$

provided $m \neq n$ and $\omega \neq \pm\omega_{mn}$. Clearly, $c_m$ oscillates and remains of order $\epsilon$, so that our approximations are valid. As for the time-independent perturbation, the trend of the system away from the eigenstate $E_n$ towards the eigenstate $E_m$ is soon reversed and the probability of finding that the system has made a transition always remains small. If, however, $\omega = \omega_{mn}$ or $\omega = -\omega_{mn}$, $c_m$ will contain a term increasing linearly with $t$ and the probability that a transition from the state $E_n$ to the state $E_m$ will be found to have occurred will increase as time elapses. In this case, if $E_m > E_n$ and a transition takes place, the system must absorb energy $E_m - E_n$ from the agent responsible for the perturbing oscillation; but, by equation (6.6.6),

$$E_m - E_n = \omega_{mn}\hbar = \omega\hbar, \tag{6.9.7}$$

taking $\omega$ to be positive (without loss of generality). Now $\omega\hbar = h\nu$, where $\nu$ is the frequency of the perturbation and, by Einstein's equation (4.4.6), this is the magnitude of the quantum of energy we should expect to be associated with an oscillator possessing this frequency. We conclude, therefore, that permanent transitions tend to take place between energy levels requiring the absorption (or emission) of a single quantum of energy from (or to) the source of the perturbation. In particular, if the source is an electromagnetic wave, permanent transitions involve the absorption or emission of a photon. This is in very satisfactory agreement with experiment.

## 6.10. Perturbation of a charged particle by an electromagnetic wave

Consider a particle of mass $m$ and charge $e$ which moves under the action of a force determined by a potential energy function $V$. Suppose that it is perturbed by an electromagnetic field derived from a vector potential

$\mathfrak{A}$ and a scalar potential $\phi$. Then, neglecting spin, it follows from equation (5.6.5) that the Hamiltonian is modified by an additional term

$$\epsilon\hat{K} = -\frac{e}{2mc}\frac{\hbar}{\iota}(\mathfrak{A}.\nabla+\nabla.\mathfrak{A})+e\phi+\frac{e^2}{2mc^2}\mathfrak{A}^2. \tag{6.10.1}$$

We shall take $\mathfrak{A}$ to be of order $\epsilon$ and hence the final term will be neglected.

In the special case of a plane polarised electromagnetic wave having angular frequency $\omega$ and wave number $k$ travelling in the direction of the $x$-axis with its electric component in the $y$-direction and its magnetic component in the $z$-direction, we can take

$$\left.\begin{aligned} A_x = 0, \quad A_y = \frac{\epsilon a}{k}\sin(kx-\omega t+\alpha), \quad A_z = 0,\\ \phi = 0; \end{aligned}\right\} \tag{6.10.2}$$

substitution in equations (5.6.2), (5.6.3), then yields $\mathfrak{B}$ and $\mathfrak{E}$ in the required directions with equal amplitudes $\epsilon a$. With this form for $\mathfrak{A}$,

$$\hat{K} = -\frac{ea\hbar}{mck\iota}\sin(kx-\omega t+\alpha)\frac{\partial}{\partial y}\cdot \tag{6.10.3}$$

Equation (6.9.1) reduces to this form if we take

$$\hat{F} = \frac{ea\hbar}{2mck}\exp[\iota(kx+\alpha)]\frac{\partial}{\partial y}, \quad \hat{G} = -\frac{ea\hbar}{2mck}\exp[-\iota(kx+\alpha)]\frac{\partial}{\partial y}\cdot \tag{6.10.4}$$

Thus, since $ck=\omega$,

$$F_{mn} = \frac{ea\hbar}{2m\omega}\int\psi^{m*}\exp[\iota(kx+\alpha)]\frac{\partial\psi^n}{\partial y}\,d\tau, \tag{6.10.5}$$

$$G_{mn} = -\frac{ea\hbar}{2m\omega}\int\psi^{m*}\exp[-\iota(kx+\alpha)]\frac{\partial\psi^n}{\partial y}\,d\tau, \tag{6.10.6}$$

where $\psi^1$, $\psi^2$, ... are the wave amplitudes for the unperturbed energy eigenstates. It is left as an exercise for the reader to prove that the identities (6.9.4) are satisfied in this case. (Hint: Integrate by parts.) Equation (6.9.5) now yields the elements of the perturbation matrix.

If the wave is plane polarised with the electric intensity vector parallel to the $z$-axis, we take

$$A_x = A_y = 0, \quad A_z = \frac{\epsilon b}{k}\sin(kx-\omega t+\beta), \tag{6.10.7}$$

so that the amplitudes of $\mathfrak{E}$ and $\mathfrak{B}$ are both $\epsilon b$. Then

$$F_{mn} = \frac{eb\hbar}{2mck}\int \psi^{m*}\exp[\iota(kx+\beta)]\frac{\partial\psi^n}{\partial z}\,d\tau. \tag{6.10.8}$$

If either of these waves is incident upon the particle, as explained in the previous section, the probability of a permanent transition from an initial state $E_n$ to a new state $E_m$ being observed is always small unless $\omega$ satisfies equation (6.9.7). However, in practice, the incident wave will normally comprise components spread over a continuous spectrum of frequencies and we shall therefore proceed to calculate the probability of a transition when this spectrum includes the critical frequency.

Suppose, first, that the wave is polarised with $\mathfrak{E}$ in the $y$-drection. Then the vector potential for the infinitesimal component corresponding to the angular frequency range $(\omega, \omega+d\omega)$ is given by equations (6.10.2) with $a$ replaced by $a d\omega$; both $a$ and $\alpha$ will now be functions of $\omega$ and, of course, $k=\omega/c$. The appropriate $F_{mn}$ for this component now follows from equation (6.10.5) and then the corresponding expression for $c_m$ is given by equation (6.9.6). By integrating over the spectrum of values of $\omega$, we then calculate the form of $c_m$ for the whole wave, viz.

$$\begin{aligned} c_m = {} & \frac{\epsilon e}{2m}\int \frac{a(\omega)\exp(\iota\alpha)\{1-\exp[\iota(\omega_{mn}-\omega)t]\}}{\omega(\omega_{mn}-\omega)}\,d\omega \\ & \times \int \psi^{m*}\exp(\iota kx)\frac{\partial\psi^n}{\partial y}\,d\tau \\ & + \frac{\epsilon e}{2m}\int \frac{a(\omega)\exp(-\iota\alpha)\{1-\exp[\iota(\omega_{mn}+\omega)t]\}}{\omega(\omega_{mn}+\omega)}\,d\omega \\ & \times \int \psi^{n}\exp(-\iota kx)\frac{\partial\psi^{m*}}{\partial y}\,d\tau. \end{aligned} \tag{6.10.9}$$

There is a similar expression for $c_m$ for the case of a wave polarised with $\mathfrak{E}$ in the $z$-direction. Then, since any plane wave being propagated in the $x$-direction can be separated into two polarised components, for such a wave $c_m$ will be expressible as the sum of two expressions like (6.10.9).

Now suppose that the light wave is nominally monochromatic, i.e. $a(\omega)$ is negligible except over a narrow band of frequencies centred on a value $\omega=\omega_0$. Any monochromatic beam generated in the laboratory will be of this type, since its ultimate source will be a very large number of similar atoms whose random motions will introduce Doppler frequency shifts of their characteristic emission frequency $\omega_0$. Also, suppose $\omega_{mn}>0$ and $\omega_0=\omega_{mn}$ (the critical frequency). Then, in the right-hand member of equation (6.10.9), the second term can be calculated approximately by putting $\omega=\omega_{mn}$ throughout the integrand except for the factor $a(\omega)\exp(-\iota\alpha)$; the integration with respect to $\omega$ is then elementary and yields the result

$$\frac{\epsilon e A}{4m\omega_{mn}^2}[1-\exp(2\iota\omega_{mn}t)]\int \psi^n \exp(-\iota k_0 x)\frac{\partial\psi^m}{\partial y}d\tau, \tag{6.10.10}$$

where $k_0=\omega_0/c$ and

$$A = \int a(\omega)\exp(-\iota\alpha)\,d\omega. \tag{6.10.11}$$

This clearly constitutes a small oscillatory component of $c_m$ which cannot be responsible for permanent transitions. We shall accordingly neglect this term. However, the integrand of the first term contains a factor

$$\frac{1-\exp[\iota(\omega_{mn}-\omega)t]}{\omega_{mn}-\omega} \tag{6.10.12}$$

and, for values of $\omega$ in the neighbourhood of $\omega_{mn}$, this is approximately $-\iota t$ and, thus, is not oscillatory in $t$. This term will, accordingly, correspond to a steady trend of the system from the eigenstate $E_n$ to the eigenstate $E_m$ and will be retained.

Hence, for an unpolarised wave, $c_m$ can be written in the form

$$c_m = \int P\exp(\iota\alpha)\,d\omega+\int Q\exp(\iota\beta)\,d\omega, \tag{6.10.13}$$

where $P=P(\omega,t)$, $Q=Q(\omega,t)$ and both these quantities are zero except in a neighbourhood of $\omega=\omega_{mn}$.

But, if

$$S = a_1+a_2+\ldots+a_N, \tag{6.10.14}$$

where the $a_i$ are complex quantities whose phases are distributed in a random manner over the interval $(0,2\pi)$, then

$$SS^* = a_1a_1^*+a_2a_2^*+\ldots+a_Na_N^*+\sum_{i\neq j} a_i a_j^* \tag{6.10.15}$$

and $\sum a_i a_j^*$ can be expected to be zero if $N$ is very large. Thus

$$SS^* = \sum_{i=1}^{N} a_i a_i^*. \tag{6.10.16}$$

Now the phases $\alpha(\omega)$, $\beta(\omega)$ can be expected to be distributed in a random manner with respect to variation of $\omega$, since the different frequency components of the light wave will arise from atoms which are radiating independently. Hence, the analogous result to (6.10.16) for (6.10.13) is that

$$c_m c_m^* = \int PP^*\,d\omega + \int QQ^*\,d\omega. \tag{6.10.17}$$

Now

$$PP^* = \frac{\epsilon^2 e^2}{m^2 \omega^2} a^2(\omega)\,|\alpha_{mn}|^2 \left[\frac{\sin \frac{1}{2}(\omega_{mn}-\omega)\,t}{\omega_{mn}-\omega}\right]^2, \tag{6.10.18}$$

where

$$\alpha_{mn} = \int \psi^{m*} \exp(\iota kx) \frac{\partial \psi^n}{\partial y}\,d\tau. \tag{6.10.19}$$

There is a similar formula for $QQ^*$. Assuming that the factor $|\alpha_{mn}|/\omega$ does not vary appreciably in the relevant neighbourhood of $\omega_{mn}$, we can replace it by its value at $\omega = \omega_{mn}$ and then

$$\int PP^*\,d\omega = \left(\frac{\epsilon e |\alpha_{mn}|}{m\omega_{mn}}\right)^2 \int a^2 \left[\frac{\sin \frac{1}{2}(\omega_{mn}-\omega)\,t}{\omega_{mn}-\omega}\right]^2 d\omega, \tag{6.10.20}$$

where $\alpha_{mn}$ is to be calculated with $k = \omega_{mn}/c$.

Consider, now, the integral in formula (6.10.20). Since $a(\omega)$ is being assumed zero outside a neighbourhood of $\omega_{mn}$, we can take the range of integration to be $(-\infty, \infty)$. The factor $\sin^2 \frac{1}{2}(\omega_{mn}-\omega)t/(\omega_{mn}-\omega)^2$ is large, for large values of $t$, only in a neighbourhood of $\omega = \omega_{mn}$; taking this neighbourhood to extend between the two zeros $\omega_{mn} - \omega = \pm 2\pi/t$, we observe that its length approaches zero as $t \to \infty$. Assuming $a(\omega)$ to be continuous near $\omega = \omega_{mn}$, we can accordingly approximate the integral, for large values of $t$, by writing

$$\begin{aligned}\int_{-\infty}^{\infty} a^2 \left[\frac{\sin \frac{1}{2}(\omega_{mn}-\omega)\,t}{\omega_{mn}-\omega}\right]^2 d\omega &= a^2(\omega_{mn}) \int_{-\infty}^{\infty} \left[\frac{\sin \frac{1}{2}(\omega_{mn}-\omega)\,t}{\omega_{mn}-\omega}\right]^2 d\omega,\\ &= \tfrac{1}{2} t a^2(\omega_{mn}) \int_{-\infty}^{\infty} \frac{\sin^2\theta}{\theta^2}\,d\theta,\\ &= \tfrac{1}{2}\pi t a^2(\omega_{mn}),\end{aligned} \tag{6.10.21}$$

having put $\theta = \frac{1}{2}(\omega_{mn}-\omega)t$.

Thus

$$\int PP^*\,d\omega = \tfrac{1}{2}\pi t \left(\frac{\epsilon e |\alpha_{mn}|\,a}{m\omega_{mn}}\right)^2. \tag{6.10.22}$$

Similarly

$$\int QQ^*\,d\omega = \tfrac{1}{2}\pi t \left(\frac{\epsilon e |\beta_{mn}|\,b}{m\omega_{mn}}\right)^2, \tag{6.10.23}$$

where

$$\beta_{mn} = \int \psi^{m*} \exp(\iota kx) \frac{\partial \psi^n}{\partial z} d\tau, \tag{6.10.24}$$

and hence

$$c_m c_m^* = \tfrac{1}{2}\pi t \left(\frac{\epsilon e}{m\omega_{mn}}\right)^2 [|\alpha_{mn}|^2 a^2 + |\beta_{mn}|^2 b^2]. \tag{6.10.25}$$

We note that the rate of transition between the states $E_n$ and $E_m$, for a large number of identical systems, is proportional to $t$; this result will, of course, only be true while $c_m$ is small.

If we assume $\omega_{mn} < 0$, i.e. we are contemplating a transition from a higher to a lower energy state with the *absorption* of a photon from the light beam, then we suppose that $\omega_0 = -\omega_{mn}$ and it is the first integral in the right-hand member of equation (6.10.9) which we neglect. We take $k = \omega_0/c = -\omega_{mn}/c$, so that the sign of $k$ is reversed. The argument proceeds as before to the result (6.10.25) with, however, $\alpha_{mn}$ being replaced by

$$\int \psi^n \exp(\iota kx) \frac{\partial \psi^{m*}}{\partial y} d\tau \tag{6.10.26}$$

(here, $k$ is supposed to take its former value of $+\omega_{mn}/c$). Integrating by parts, we prove that

$$\int_{-\infty}^{\infty} \psi^n \frac{\partial \psi^{m*}}{\partial y} dy = -\int_{-\infty}^{\infty} \psi^{m*} \frac{\partial \psi^n}{\partial y} dy, \tag{6.10.27}$$

provided $\psi^n$, $\psi^m$ are assumed to vanish at infinity. It then follows that

$$\int \psi^n \exp(\iota kx) \frac{\partial \psi^{m*}}{\partial y} d\tau = -\int \psi^{m*} \exp(\iota kx) \frac{\partial \psi^n}{\partial y} d\tau = -\alpha_{mn}. \tag{6.10.28}$$

Hence, equation (6.10.25) is valid for absorption as well as emission.

It is a well-known result from electromagnetic theory that Poynting's vector $\mathfrak{S}$, defined by

$$\mathfrak{S} = \frac{c}{4\pi} \mathfrak{E} \times \mathfrak{B}, \tag{6.10.29}$$

is the energy current density vector for an electromagnetic wave, i.e. gives the rate of transport of energy across a unit area placed normally to the direction of propagation. Consider the plane polarised wave having angular frequency $\omega$ determined by equations (6.10.2). The electric and magnetic intensity vectors are given by

$$\left.\begin{aligned}\mathfrak{E} &= [0, \epsilon a \cos(kx - \omega t + \alpha), 0], \\ \mathfrak{B} &= [0, 0, \epsilon a \cos(kx - \omega t + \alpha)].\end{aligned}\right\} \tag{6.10.30}$$

Thus,

$$\mathfrak{S} = \frac{c}{4\pi}[\epsilon^2 a^2 \cos^2(kx - \omega t + \alpha), 0, 0]. \tag{6.10.31}$$

The mean value of $\cos^2(kx - \omega t + \alpha)$ over a complete cycle is $\frac{1}{2}$ and hence the mean rate of energy flow across unit area is given by

$$I = \frac{c\epsilon^2 a^2}{8\pi}. \tag{6.10.32}$$

$I$ is termed the *intensity* of the radiation.

Hence, if $I_y$, $I_z$ are the intensities of the components of the nominally monochromatic light beam polarised with $\mathfrak{E}$ along the $y$- and $z$-axes respectively, it follows from equations (6.10.25), (6.10.32) that

$$c_m c_m^* = \frac{1}{c}\left(\frac{2\pi e}{m\omega_{mn}}\right)^2 [|\alpha_{mn}|^2 I_y + |\beta_{mn}|^2 I_z]\, t. \tag{6.10.33}$$

It will be noted that the probability of the absorption or emission of a photon from or to the incident beam, is proportional to the beam's intensity. In circumstances where the system undergoes a transition to a lower energy state with the emission of a photon into the beam, the phenomenon is referred to as that of *induced emission of radiation*. However, even in the absence of the perturbing beam, the particle system may jump into a lower energy state, with the radiation of a photon. Such a transition is referred to as *spontaneous emission* and the theory we have developed does not apply to this phenomenon; to provide an adequate explanation, it is necessary to take account of the quantum mechanical properties of the space into which the photon is radiated and thus to regard the phenomenon as an interaction between two coupled quantum mechanical systems, in which energy is transferred from one to the other. This leads to the idea of quantisation of the electromagnetic field, a development which we shall not pursue in this book.

## 6.11. The electric dipole approximation

If the particle of the last section whose state was perturbed by a plane electromagnetic wave is an orbital electron in an atom, then the eigenfunctions $\psi^n$ of its unperturbed stationary states will take non-zero values only over a small region having dimensions of the order of the atomic radius. Let $(x_0, y_0, z_0)$ be the coordinates of the centre of the atom. Then equation (6.10.19) can be written in the form

$$\alpha_{mn} = \exp(\iota k x_0) \int \psi^{m*} \exp[\iota k(x-x_0)] \frac{\partial \psi^n}{\partial y}\, d\tau. \qquad (6.11.1)$$

Since the wavelength of any electromagnetic radiation normally encountered in the laboratory is large by comparison with the atomic radius, the quantity $k(x-x_0)$ will be small throughout the region over which $\psi^{m*}$ and $\partial\psi^n/\partial y$ take non-zero values and equation (6.11.1) can be approximated by expanding the exponential factor thus:

$$\alpha_{mn} = \exp(\iota k x_0) \int \psi^{m*} [1+\iota k(x-x_0) - \tfrac{1}{2}k^2(x-x_0)^2 + \ldots] \frac{\partial \psi^n}{\partial y}\, d\tau. \qquad (6.11.2)$$

Neglect of every term in the expansion except that of zero order, leads to what is termed the *electric dipole approximation*. According to this approximation.

$$\alpha_{mn} = \exp(\iota k x_0) \int \psi^{m*} \frac{\partial \psi^n}{\partial y}\, d\tau. \qquad (6.11.3)$$

Now $\psi^n$, $\psi^m$ satisfy the equations

$$\hat{H}\psi^n = E_n \psi^n, \quad \hat{H}\psi^m = E_m \psi^m. \qquad (6.11.4)$$

It follows that

$$\int [\psi^{m*} y \hat{H} \psi^n - \psi^n y (\hat{H}\psi^m)^*]\, d\tau = (E_n - E_m) \int \psi^{m*}\, y \psi^n\, d\tau. \quad (6.11.5)$$

But, since $\hat{H}$ is Hermitian,

$$\int \psi^n y (\hat{H}\psi^m)^*\, d\tau = \int \psi^{m*} \hat{H}(y\psi^n)\, d\tau. \qquad (6.11.6)$$

Hence, the left-hand member of equation (6.11.5) can be written in the form

$$\int \psi^{m*} [y, \hat{H}] \psi^n\, d\tau. \qquad (6.11.7)$$

Now

$$[y, \hat{H}] = -\frac{\hbar^2}{2m}\left[y, \frac{\partial^2}{\partial y^2}\right] = -\frac{\hbar^2}{m}\frac{\partial}{\partial y}. \qquad (6.11.8)$$

Thus, equation (6.11.5) is equivalent to the result

$$-\frac{\hbar^2}{m}\int \psi^{m*}\frac{\partial \psi^n}{\partial y}\,d\tau = (E_n - E_m)\int \psi^{m*}\,y\psi^n\,d\tau, \tag{6.11.9}$$

or

$$\int \psi^{m*}\frac{\partial \psi^n}{\partial y}\,d\tau = \frac{m}{\hbar}\omega_{mn}\int \psi^{m*}\,y\psi^n\,d\tau. \tag{6.11.10}$$

We conclude that

$$\alpha_{mn} = \frac{m}{\hbar}\omega_{mn}\exp(\iota k x_0)\,y_{mn}, \tag{6.11.11}$$

$y_{mn}$ being a matrix element for $y$ in the representation employing the $\psi^i$ as a basis.

Similarly

$$\beta_{mn} = \frac{m}{\hbar}\omega_{mn}\exp(\iota k x_0)\,z_{mn}. \tag{6.11.12}$$

Substituting for $\alpha_{mn}$, $\beta_{mn}$ in equation (6.10.33), we obtain the result

$$c_m c_m^* = \frac{1}{c}\left(\frac{2\pi e}{\hbar}\right)^2 [|y_{mn}|^2 I_y + |z_{mn}|^2 I_z]t. \tag{6.11.13}$$

Clearly, if $y_{mn}=0$, to the order of approximation to which we are at present working, an atom in the state $E_n$ cannot absorb or emit a photon into a beam which is polarised in the $y$-direction, in order to effect a transition to the energy state $E_m$. Further, if $E_n > E_m$, it may be proved that, in this case, even if the perturbing radiation is absent, the atom cannot jump from the state $E_n$ to the state $E_m$ with the emission of $y$-polarised radiation. Such a transition is said to be *forbidden*. Also, if $x_{mn}=0$ and $z_{mn}=0$, the transition cannot be effected by the emission of $x$-polarised or $z$-polarised radiation either. Such forbidden transitions correspond to the absence of certain groups of lines from the spectrum of the atom, which otherwise might have been expected to be present. The rules by which we identify the possible transitions are called *selection rules*. However, it must be appreciated that such selection rules usually identify transitions which are highly improbable rather than ones which are absolutely impossible, since they have been calculated by a method involving approximation; thus, although the zero order term in the expansion (6.11.2) may vanish, the higher order terms may not and then the transition will be found to be possible but, on account of the presence of the small factor $k(x-x_0)$, the probability of its occurrence will prove to be small. The intensity of the corresponding line in the atom's spectrum will then be weak, but not zero.

## 6.12. Selection rules for the harmonic oscillator

Consider the isotropic harmonic oscillator for which the P.E. function is $\frac{1}{2}m\omega^2 r^2$ (Ex. 27, Chap. 3). The wave amplitude for a steady state can be written

$$\psi_{ijk} = \left(\frac{2m\omega}{h}\right)^{\frac{3}{4}} \left(\frac{1}{2^n\, i!\, j!\, k!}\right)^{\frac{1}{2}} \exp\left(-\tfrac{1}{2}\rho^2\right) H_i(\xi)\, H_j(\eta)\, H_k(\zeta), \quad (6.12.1)$$

where

$$\xi = \sqrt{\left(\frac{m\omega}{\hbar}\right)}\, x, \text{ etc}, \quad (6.12.2)$$

$$\rho^2 = \xi^2+\eta^2+\zeta^2, \quad (6.12.3)$$

$i$, $j$, $k$ are positive integers and

$$n = i+j+k. \quad (6.12.4)$$

The energy in the state is given by

$$E = (n+\tfrac{3}{2})\,\omega\hbar. \quad (6.12.5)$$

The matrix element for $x$ corresponding to a transition from the state $(i,j,k)$ to the state $(i',j',k')$ is given by

$$\begin{aligned} x_{ijk,\, i'j'k'} &= \int \psi_{ijk}^* x \psi_{i'j'k'}\, d\tau, \\ &= \left(\frac{\hbar}{m\omega}\right)^2 \iiint\limits_{-\infty}^{\infty} \psi_{ijk}^*\, \xi \psi_{i'j'k'}\, d\xi\, d\eta\, d\zeta. \end{aligned} \quad (6.12.6)$$

It has been proved in section 3.11 that

$$\left.\begin{aligned} \int_{-\infty}^{\infty} \xi \exp(-\xi^2)\, H_i(\xi)\, H_j(\xi)\, d\xi &= 2^i(i+1)!\,\pi^{\frac{1}{2}}, \quad j = i+1, \\ &= 2^j(j+1)!\,\pi^{\frac{1}{2}}, \quad i = j+1, \\ &= 0, \text{ otherwise.} \end{aligned}\right\} \quad (6.12.7)$$

Also, from equation (C.21) in Appendix C, we have

$$\int_{-\infty}^{\infty} \exp(-\xi^2)\, H_i(\xi)\, H_j(\xi)\, d\xi = 2^i\, i!\, \pi^{\frac{1}{2}}\, \delta_{ij}. \quad (6.12.8)$$

It now follows immediately that

$$\left.\begin{aligned} x_{ijk,\,i'j'k'} &= \sqrt{\left[\frac{\hbar}{2m\omega}(i+1)\right]}, \quad \text{if } i' = i+1, j' = j, k' = k, \\ &= \sqrt{\left[\frac{\hbar}{2m\omega}(i'+1)\right]}, \quad \text{if } i = i'+1, j = j', k = k', \\ &= 0, \quad \text{otherwise.} \end{aligned}\right\} \tag{6.12.9}$$

Similar results can be obtained for the matrix elements of $y$ and $z$.

We can now conclude that, for the harmonic oscillator, all transitions are forbidden except those for which *one* of the quantum numbers $i, j$ or $k$ increases or decreases by unity. For such transitions, $E$ changes by amount $\omega\hbar$ and hence the spectrum of the oscillator has only one bright line.

## 6.13. Selection rules for a particle moving in a spherically symmetric field

The energy eigenfunctions for a particle moving in a spherically symmetric field have been calculated as equation (5.3.9). We have

$$\psi_{lkn} = R_{ln}\,Y_{lk}. \tag{6.13.1}$$

The matrix element for $z$ corresponding to a transition from the state $(l, k, n)$ to the state $(l', k', n')$ is given by

$$\begin{aligned} z_{lkn,\,l'k'n'} &= \int \psi^*_{lkn}\, z\psi_{l'k'n'}\, d\tau, \\ &= \iiint R_{nl}R_{n'l'}P_l^k P_{l'}^{k'} \exp[\iota(k'-k)\phi]\, r^3 \cos\theta \sin\theta \, dr\, d\theta\, d\phi, \end{aligned} \tag{6.13.2}$$

where we have put $z = r\cos\theta$. The integrations can be separated; that with respect to $\phi$ is

$$\left.\begin{aligned} \int_0^{2\pi} \exp[\iota(k'-k)\phi]\, d\phi &= 2\pi, \quad \text{if } k' = k, \\ &= 0, \quad \text{otherwise.} \end{aligned}\right\} \tag{6.13.3}$$

This gives the selection rule

$$k' = k \tag{6.13.4}$$

immediately. Assuming this condition is satisfied, the integration with respect to $\theta$ is

$$\int_0^{\pi} P_l^k(\cos\theta)\, P_{l'}^k(\cos\theta) \cos\theta \sin\theta \, d\theta. \tag{6.13.5}$$

This integral will be found calculated at equation (D.27) of Appendix D, where it is shown to be zero unless

$$l = l'+1, \quad l' = l+1, \tag{6.13.6}$$

giving two further selection rules. The integration with respect to $r$ can only be performed when the form of the central field has been specified; it does not, in general, lead to the formulation of further selection rules.

The matrix elements for $x$ and $y$ can be calculated similarly after putting

$$x = r\sin\theta\cos\phi, \quad y = r\sin\theta\sin\phi. \tag{6.13.7}$$

In the case of $x$, an integration with respect to $\phi$ of the form

$$\int\limits_0^{2\pi} \exp[\iota(k'-k)\phi]\cos\phi\, d\phi \tag{6.13.8}$$

has to be performed. It is easy to show that this integral vanishes unless

$$k = k'+1 \quad \text{or} \quad k' = k+1. \tag{6.13.9}$$

These are new selection rules. If the first equation is satisfied, the integration with respect to $\theta$ is

$$\int\limits_0^{\pi} P_l^k(\cos\theta)P_{l'}^{k-1}(\cos\theta)\sin^2\theta\, d\theta = \int\limits_{-1}^{1} \sqrt{(1-\mu^2)}P_l^k(\mu)P_{l'}^{k-1}(\mu)\, d\mu. \tag{6.13.10}$$

Employing the identity (D.11) from Appendix D, we write this integral in the form

$$\frac{1}{2l'+1}\int\limits_{-1}^{1} P_l^k(P_{l'+1}^k - P_{l'-1}^k)\, d\mu \tag{6.13.11}$$

and it now follows from the equation (D.23) that the integral vanishes unless

$$l = l'+1, \quad \text{or} \quad l' = l+1. \tag{6.13.12}$$

Thus, this integration does not provide any fresh rules.

The matrix elements for $y$ can be dealt with similarly and no further rules arise.

To summarise, the selection rules for a particle moving in a spherically symmetric field (spin being disregarded) are

$$\left.\begin{aligned} \Delta l &= \pm 1, \\ \Delta k &= 0, \pm 1, \end{aligned}\right\} \tag{6.13.13}$$

where $\Delta l = l'-l$, $\Delta k = k'-k$.

CHAPTER 7

# Dirac's relativistic equation

## 7.1. Dirac's Wave Equation

Consider a particle having rest mass $m$, electric charge $e$ and moving in an electromagnetic field derived from a vector potential $\mathfrak{A}$ and a scalar potential $\Phi$. If $Oxyz$ is a rectangular cartesian inertial frame and $t$ is time measured by clocks stationary in the frame, the state of the particle at any instant has been specified by means of a wave function $\psi(x,y,z,t)$ satisfying the equation of motion (4.1.7), with Hamiltonian given by equation (5.6.5). This Hamiltonian was derived from the corresponding classical expression, which leads to non-relativistic equations of motion for the particle and therefore it is found that the wave equation for $\psi$ (equation (5.6.6)) is not covariant with respect to Lorentz transformation between inertial frames. This wave equation is accordingly inconsistent with the special principle of relativity and cannot be accepted as a satisfactory basis for the quantum mechanics of a particle. To obtain an equation having the desired transformation properties, it is necessary to make use of the Hamiltonian which leads to the equations of motion for the particle suggested by the special theory of relativity. This we shall now do.

It is demonstrated in Appendix F that the relativistic Hamiltonian is given by

$$H = c\left[\left(\mathfrak{p}-\frac{e}{c}\mathfrak{A}\right)^2+m^2c^2\right]^{\frac{1}{2}}+e\Phi. \tag{7.1.1}$$

Introducing the usual Minkowski coordinates $x_i$ ($i=1,2,3,4$), where

$$x_1 = x, \quad x_2 = y, \quad x_3 = z, \quad x_4 = \iota ct, \tag{7.1.2}$$

we know† that the $x_i$-space is pseudo-Euclidean and that a 4-vector momentum $\mathfrak{P}$ and potential $\mathfrak{V}$ can be defined in the space by equations

$$P_1 = p_x, \quad P_2 = p_y, \quad P_3 = p_z, \quad P_4 = \frac{\iota}{c}H, \tag{7.1.3}$$

$$V_1 = A_x, \quad V_2 = A_y, \quad V_3 = A_z, \quad V_4 = \iota\Phi. \tag{7.1.4}$$

† See, e.g., *An Introduction to Tensor Calculus and Relativity* by D. F. Lawden, Methuen, 2nd Edn., 1967.

Equation (7.1.1) can then be written in the Lorentz-covariant form

$$\left\{\sum_{i=1}^{4}\left(P_i-\frac{e}{c}V_i\right)^2\right\}^{\frac{1}{2}} = \iota mc. \tag{7.1.5}$$

This last equation implies that the observable defined by the left-hand member is not only invariant with respect to Lorentz transformations, but possesses a single eigenvalue, namely $\iota mc$. The characteristic equation for the observable is accordingly

$$\left\{\sum_{i=1}^{4}\left(\hat{P}_i-\frac{e}{c}\hat{V}_i\right)^2\right\}^{\frac{1}{2}}\psi = \iota mc\psi, \tag{7.1.6}$$

and this must be satisfied by every wave function $\psi$. Employing the $xyz$-representation, the operators representing the $P_i$ $(i=1,2,3)$ have been taken to be $-\iota\hbar\partial/\partial x_i$. Also, since equation (4.1.7) is valid for all $\psi$, we shall have

$$\hat{H} = \iota\hbar\frac{\partial}{\partial t}\cdot \tag{7.1.7}$$

This leads to the equation

$$\hat{P}_4 = -\iota\hbar\frac{\partial}{\partial x_4}\cdot \tag{7.1.8}$$

Also, in this representation, $\hat{V}_i = V_i$. Hence, equation (7.1.6) will tentatively be written in the form

$$\left\{\sum_{i=1}^{4}\left(\iota\hbar\frac{\partial}{\partial x_i}+\frac{e}{c}V_i\right)^2\right\}^{\frac{1}{2}}\psi = \iota mc\psi. \tag{7.1.9}$$

In this form it is clearly Lorentz covariant, but there is difficulty in giving a meaning to the operator occurring in the left-hand member.

By operating on both sides with the square root operator, it is possible to remove the square root and so obtain an unambiguous equation, viz.

$$\left\{\sum_{i=1}^{4}\left(\iota\hbar\frac{\partial}{\partial x_i}+\frac{e}{c}V_i\right)^2\right\}\psi = -m^2c^2\psi. \tag{7.1.10}$$

This is called the *Klein-Gordon equation*. It is unsatisfactory since it involves a second derivative of $\psi$ with respect to $t$ and this would imply that the subsequent motion of the particle was not fully determined by its state $\psi$ at an initial instant ($\partial\psi/\partial t$ would have to be known as well).

Now, in the case of a particle with spin, we know that $\psi$ is a column matrix. Thus, we follow Dirac and replace $\psi$ by a column matrix, so that the square root operator can be written as

$$\left\{\sum_{i=1}^{4}\left(\hat{P}_i-\frac{e}{c}V_i\right)^2\right\}^{\frac{1}{2}}\mathbf{I}, \tag{7.1.11}$$

where $\mathbf{I}$ is a unit matrix possessing the same number of rows and columns as $\boldsymbol{\psi}$ has rows. Then, in the case of zero field (i.e. $V_i=0$), it will be shown that this operator can be expressed linearly in terms of the operators $\hat{P}_i$ thus:

$$\left[\sum_{i=1}^{4}\hat{P}_i^2\right]^{\frac{1}{2}}\mathbf{I} = \sum_{i=1}^{4}\boldsymbol{\gamma}_i\hat{P}_i, \tag{7.1.12}$$

where the $\boldsymbol{\gamma}_i$ are constant matrices yet to be calculated. Squaring both sides of this last equation and noting that the matrices $\boldsymbol{\gamma}_i$ and the operators $\hat{P}_i$ commute, we obtain

$$\sum_{i=1}^{4}\hat{P}_i^2\mathbf{I} = \sum_{i=1}^{4}\sum_{j=1}^{4}\boldsymbol{\gamma}_i\boldsymbol{\gamma}_j\hat{P}_i\hat{P}_j. \tag{7.1.13}$$

Since the $\hat{P}_i$ commute, this will be an identity if the $\boldsymbol{\gamma}_i$ are chosen so that

$$\tfrac{1}{2}(\boldsymbol{\gamma}_i\boldsymbol{\gamma}_j+\boldsymbol{\gamma}_j\boldsymbol{\gamma}_i) = \boldsymbol{\Delta}_{ij}, \tag{7.1.14}$$

where $\boldsymbol{\Delta}_{ij}=\mathbf{I}$ if $i=j$, and is the zero matrix otherwise.

Assuming that a set of matrices $\boldsymbol{\gamma}_i$ can be found to satisfy the relations (7.1.14), for the case of zero field, equation (7.1.6) is replaced by the equation

$$\sum_{i=1}^{4}\boldsymbol{\gamma}_i\hat{P}_i\boldsymbol{\psi} = \iota mc\boldsymbol{\psi}, \tag{7.1.15}$$

where $\boldsymbol{\psi}$ is a column matrix. It is then assumed that, in the presence of a field, this equation becomes

$$\sum_{i=1}^{4}\boldsymbol{\gamma}_i\left(\hat{P}_i-\frac{e}{c}V_i\right)\boldsymbol{\psi}= \iota mc\boldsymbol{\psi}. \tag{7.1.16}$$

Multiplying this last equation through by $\iota\boldsymbol{\gamma}_4^{-1}=\iota\boldsymbol{\gamma}_4$ (since $\boldsymbol{\gamma}_4^2=\mathbf{I}$ by equation (7.1.14)) and substituting for the $\hat{P}_i$ and $V_i$ from equations (7.1.3), (7.1.4), we obtain the equation in the form

$$\left[mc^2\boldsymbol{\beta}+c\sum_r\boldsymbol{\alpha}_r\left(\hat{p}_r-\frac{e}{c}A_r\right)+e\Phi\right]\boldsymbol{\psi} = \iota\hbar\frac{\partial\boldsymbol{\psi}}{\partial t}, \tag{7.1.17}$$

where $r$ ranges over the values $x$, $y$, $z$ and the matrices $\boldsymbol{\alpha}_r$, $\boldsymbol{\beta}$ are given by

$$\boldsymbol{\alpha}_x = \iota\gamma_4\gamma_1, \quad \boldsymbol{\alpha}_y = \iota\gamma_4\gamma_2, \quad \boldsymbol{\alpha}_z = \iota\gamma_4\gamma_3, \quad \boldsymbol{\beta} = \gamma_4. \tag{7.1.18}$$

Equation (7.1.17) is the *Dirac Wave Equation*. We conclude that the relativistic Hamiltonian for a particle in an electromagnetic field is the matrix operator

$$\hat{\mathbf{H}} = mc^2\boldsymbol{\beta}+c\sum_r \boldsymbol{\alpha}_r\left(\hat{p}_r-\frac{e}{c}A_r\right)+e\Phi\mathbf{I}. \tag{7.1.19}$$

It was demonstrated in section 4.1 that the Hamiltonian operator must be Hermitian. It follows from this that the matrices $\boldsymbol{\alpha}_r$, $\boldsymbol{\beta}$ must be Hermitian. Further, from equations (7.1.14), (7.1.18), we calculate that

$$\begin{aligned}\boldsymbol{\alpha}_x\boldsymbol{\alpha}_y+\boldsymbol{\alpha}_y\boldsymbol{\alpha}_x &= -\gamma_4\gamma_1\gamma_4\gamma_2-\gamma_4\gamma_2\gamma_4\gamma_1,\\ &= \gamma_4^2\gamma_1\gamma_2+\gamma_4^2\gamma_2\gamma_1,\\ &= \gamma_1\gamma_2+\gamma_2\gamma_1,\\ &= \mathbf{0}.\end{aligned} \tag{7.1.20}$$

Other counterparts of the identities (7.1.14) for the matrices $\boldsymbol{\alpha}_r$, $\boldsymbol{\beta}$ may be proved similarly; this is left as an exercise for the reader. The complete set of identities equivalent to equations (7.1.14) is as follows:

$$\left.\begin{aligned}\tfrac{1}{2}(\boldsymbol{\alpha}_r\boldsymbol{\alpha}_s+\boldsymbol{\alpha}_s\boldsymbol{\alpha}_r) = \boldsymbol{\Delta}_{rs}, \quad \boldsymbol{\alpha}_r\boldsymbol{\beta}+\boldsymbol{\beta}\boldsymbol{\alpha}_r = \mathbf{0},\\ \boldsymbol{\beta}^2 = \mathbf{I}.\end{aligned}\right\} \tag{7.1.21}$$

Dirac showed that the matrices $\boldsymbol{\alpha}_r$, $\boldsymbol{\beta}$ must be of at least the fourth order to satisfy the identities (7.1.21) and be Hermitian. However, if a set of Hermitian matrices satisfying the identities has been found, it is clear that the matrices $\mathbf{U}\boldsymbol{\alpha}_r\mathbf{U}^\dagger$, $\mathbf{U}\boldsymbol{\beta}\mathbf{U}^\dagger$ obtained from these by a unitary transformation of matrix $\mathbf{U}$, also satisfy the identities and are Hermitian. Taking the matrices to be of the fourth order, it may be proved that any set can be reduced by a unitary transformation to the canonical forms

$$\left.\begin{aligned}\boldsymbol{\alpha}_x = \begin{pmatrix}\mathbf{0} & \boldsymbol{\sigma}_x\\ \boldsymbol{\sigma}_x & \mathbf{0}\end{pmatrix}, \quad \boldsymbol{\alpha}_y = \begin{pmatrix}\mathbf{0} & \boldsymbol{\sigma}_y\\ \boldsymbol{\sigma}_y & \mathbf{0}\end{pmatrix},\\ \boldsymbol{\alpha}_z = \begin{pmatrix}\mathbf{0} & \boldsymbol{\sigma}_z\\ \boldsymbol{\sigma}_z & \mathbf{0}\end{pmatrix}, \quad \boldsymbol{\beta} = \begin{pmatrix}\mathbf{I}_2 & \mathbf{0}\\ \mathbf{0} & -\mathbf{I}_2\end{pmatrix},\end{aligned}\right\} \tag{7.1.22}$$

$\boldsymbol{\sigma}_x$, $\boldsymbol{\sigma}_y$, $\boldsymbol{\sigma}_z$ being the Pauli matrices and $\mathbf{I}_2$ the $2\times 2$ unit matrix. The reader need only verify that the identities (7.1.21) are satisfied by these matrices.

Suppose that the particle is in an energy eigenstate. Then $\boldsymbol{\psi}$ satisfies the characteristic equation

$$\hat{\mathbf{H}}\boldsymbol{\psi} = E\boldsymbol{\psi}. \tag{7.1.23}$$

Substituting for $\hat{\mathbf{H}}$ from equation (7.1.19) and employing the representation (7.1.22), it will be found that this is equivalent to the set of four equations

$$\left.\begin{aligned}(mc^2+e\Phi-E)\,\psi_1+c\hat{\pi}_z\psi_3+c(\hat{\pi}_x-\iota\hat{\pi}_y)\,\psi_4 &= 0,\\ (mc^2+e\Phi-E)\,\psi_2+c(\hat{\pi}_x+\iota\hat{\pi}_y)\,\psi_3-c\hat{\pi}_z\psi_4 &= 0,\\ c\hat{\pi}_z\psi_1+c(\hat{\pi}_x-\iota\hat{\pi}_y)\,\psi_2-(mc^2-e\Phi+E)\,\psi_3 &= 0,\\ c(\hat{\pi}_x+\iota\hat{\pi}_y)\,\psi_1-c\hat{\pi}_z\psi_2-(mc^2-e\Phi+E)\,\psi_4 &= 0,\end{aligned}\right\} \tag{7.1.24}$$

where

$$\hat{\pi}_x = \hat{p}_x-\frac{e}{c}A_x, \text{ etc.} \tag{7.1.25}$$

An approximation to these equations can be found as follows: First, it is clear from equation (7.1.1) that $E-e\Phi$ will differ from $mc^2$ by a small amount unless the particle velocity is large or the electromagnetic field is intense. Hence, the last pair of equations (7.1.24) can be approximated by

$$\left.\begin{aligned}2mc\psi_3 &= \hat{\pi}_z\psi_1+(\hat{\pi}_x-\iota\hat{\pi}_y)\,\psi_2,\\ 2mc\psi_4 &= (\hat{\pi}_x+\iota\hat{\pi}_y)\,\psi_1-\hat{\pi}_z\psi_2.\end{aligned}\right\} \tag{7.1.26}$$

Substituting from these equations for $\psi_3$, $\psi_4$ into the first pair of equations (7.1.24), it will be found that these both reduce to the form

$$\left[\frac{1}{2m}(\hat{\pi}_x^2+\hat{\pi}_y^2+\hat{\pi}_z^2)+e\Phi\right]\psi_i = E'\,\psi_i, \tag{7.1.27}$$

$i=1, 2$, where

$$E' = E-mc^2 \tag{7.1.28}$$

is the counterpart of the classical energy. Equation (7.1.27) is the characteristic equation for the energy when the classical Hamiltonian (5.6.4) is employed.

We have proved, therefore, that for particle velocities which are small by comparison with the speed of light and relatively weak fields, the components $\psi_1$, $\psi_2$ of $\boldsymbol{\psi}$ will be approximately eigenfunctions of the classical Hamiltonian. It will further be proved, in section 7.4, that $\psi_1$, $\psi_2$ are approximately the two components of the spinor specifying the spin state of the particle in the $\sigma_z$-representation.

Equations (7.1.26) indicate that the components $\psi_3$, $\psi_4$ of $\boldsymbol{\psi}$ are of the same order of smallness as $v/c$, where $v$ is the particle speed. Thus, in a non-relativistic theory ($c = \infty$), they will both be approximated by zero and will accordingly never appear.

## 7.2. Conservation of probability

Multiplying both members of equation (7.1.17) on the left by the row matrix $\psi^\dagger$, we obtain the equation

$$mc^2\psi^\dagger\beta\psi + \sum_r (c\psi^\dagger\alpha_r\hat{p}_r\psi - eA_r\psi^\dagger\alpha_r\psi) + e\Phi\psi^\dagger\psi = \iota\hbar\psi^\dagger\frac{\partial\psi}{\partial t}. \tag{7.2.1}$$

Also, taking the conjugate transpose of both members of equation (7.1.17) and then multiplying on the right by $\psi$, we obtain the equation

$$mc^2(\beta\psi)^\dagger\psi + \sum_r [c(\alpha_r\hat{p}_r\psi)^\dagger\psi - eA_r(\alpha_r\psi)^\dagger\psi] + e\Phi\psi^\dagger\psi = -\iota\hbar\frac{\partial\psi^\dagger}{\partial t}\psi. \tag{7.2.2}$$

Subtracting equation (7.2.2) from equation (7.2.1) and remembering that the matrices $\alpha_r$, $\beta$ are Hermitian, we find that

$$c\sum_r [\psi^\dagger\alpha_r\hat{p}_r\psi - (\alpha_r\hat{p}_r\psi)^\dagger\psi] = \iota\hbar\frac{\partial}{\partial t}(\psi^\dagger\psi). \tag{7.2.3}$$

But

$$(\alpha_r\hat{p}_r\psi)^\dagger = (\hat{p}_r\psi)^\dagger\alpha_r^\dagger = -\frac{\hbar}{\iota}\frac{\partial\psi^\dagger}{\partial x_r}\alpha_r, \tag{7.2.4}$$

since $\alpha_r$ is Hermitian. Hence, equation (7.2.3) can be written

$$\frac{\partial}{\partial t}(\psi^\dagger\psi) + c\sum_r \frac{\partial}{\partial x_r}(\psi^\dagger\alpha_r\psi) = 0. \tag{7.2.5}$$

This is an equation of continuity of the type (4.6.5) with

$$\rho = \psi^\dagger\psi, \tag{7.2.6}$$

$$j_r = c\psi^\dagger\alpha_r\psi. \tag{7.2.7}$$

Assuming that $\psi$ has been normalised so that

$$\int\psi^\dagger\psi\,d\tau = 1, \tag{7.2.8}$$

it is now reasonable to interpret $\psi^\dagger\psi\,d\tau$ as the probability of finding the particle in an element $d\tau$ of $xyz$-space at the instant $t$. $j_r$ ($r = x, y, z$) are then the components of the probability current density, specifying the mode of probability flow.

## 7.3. Magnetic moment of a particle

An interesting feature of the Dirac wave equation is that any particle whose motion is governed by the equation behaves as though it possesses intrinsic magnetic moment and spin. In this section, we shall demonstrate the presence of the magnetic moment and, in the next section, the existence of spin.

The magnetic moment is revealed in a non-relativistic approximation to the Dirac Hamiltonian (7.1.19). It is evident from equation (7.1.1) that, for a stationary particle in a zero field, $H=mc^2$. Hence $H_0=H-mc^2$ must correspond to the non-relativistic Hamiltonian (5.6.4). We therefore define

$$\hat{\mathbf{H}}_0 = \hat{\mathbf{H}}-mc^2\mathbf{I} \tag{7.3.1}$$

and will then approximate to $\hat{\mathbf{H}}_0$ by neglecting terms which become small as $c\to\infty$.

Substituting for $\hat{\mathbf{H}}$ from equation (7.1.19) into equation (7.3.1) and rearranging, we obtain

$$\frac{1}{c}\hat{\mathbf{H}}_0+\left(mc-\frac{e\Phi}{c}\right)\mathbf{I} = \boldsymbol{\beta}mc+\sum_r \boldsymbol{\alpha}_r\left(\hat{p}_r-\frac{e}{c}A_r\right). \tag{7.3.2}$$

Squaring both sides and neglecting terms $O(c^{-2})$, it follows that

$$2m\hat{\mathbf{H}}_0+(m^2c^2-2me\Phi)\mathbf{I}$$

$$= m^2c^2\mathbf{I}+\sum_{r,s}\boldsymbol{\alpha}_r\boldsymbol{\alpha}_s\left(\hat{p}_r-\frac{e}{c}A_r\right)\left(\hat{p}_s-\frac{e}{c}A_s\right), \tag{7.3.3}$$

where we have employed the identities (7.1.21). Further,

$$\boldsymbol{\alpha}_x\boldsymbol{\alpha}_y\left(\hat{p}_x-\frac{e}{c}A_x\right)\left(\hat{p}_y-\frac{e}{c}A_y\right)+\boldsymbol{\alpha}_y\boldsymbol{\alpha}_x\left(\hat{p}_y-\frac{e}{c}A_y\right)\left(\hat{p}_x-\frac{e}{c}A_x\right)$$

$$= \boldsymbol{\alpha}_x\boldsymbol{\alpha}_y\left[\hat{p}_x-\frac{e}{c}A_x,\hat{p}_y-\frac{e}{c}A_y\right],$$

$$= \frac{e}{c}\boldsymbol{\alpha}_x\boldsymbol{\alpha}_y\{[\hat{p}_y,A_x]-[\hat{p}_x,A_y]\},$$

$$= \frac{e\hbar}{\iota c}\boldsymbol{\alpha}_x\boldsymbol{\alpha}_y\left(\frac{\partial A_x}{\partial y}-\frac{\partial A_y}{\partial x}\right). \tag{7.3.4}$$

But, employing the representation (7.1.22), we find that

$$\boldsymbol{\alpha}_x\boldsymbol{\alpha}_y = \iota\begin{pmatrix}\boldsymbol{\sigma}_z & \mathbf{0}\\ \mathbf{0} & \boldsymbol{\sigma}_z\end{pmatrix} \tag{7.3.5}$$

(see equation (2.2.27)). There are two similar results for the products $\boldsymbol{\alpha}_y \boldsymbol{\alpha}_z$ and $\boldsymbol{\alpha}_z \boldsymbol{\alpha}_x$. We accordingly put

$$\boldsymbol{\alpha}_y \boldsymbol{\alpha}_z = \iota\boldsymbol{\Sigma}_x, \quad \boldsymbol{\alpha}_z \boldsymbol{\alpha}_x = \iota\boldsymbol{\Sigma}_y, \quad \boldsymbol{\alpha}_x \boldsymbol{\alpha}_y = \iota\boldsymbol{\Sigma}_z, \tag{7.3.6}$$

where

$$\boldsymbol{\Sigma}_x = \begin{pmatrix} \boldsymbol{\sigma}_x & \mathbf{0} \\ \mathbf{0} & \boldsymbol{\sigma}_x \end{pmatrix}, \text{ etc.} \tag{7.3.7}$$

Then, since the magnetic field is given by equation (5.6.3), the right-hand member of equation (7.3.4) can be written in the form

$$-\frac{e\hbar}{c}\boldsymbol{\Sigma}_z B_z. \tag{7.3.8}$$

Equation (7.3.3) now yields the result

$$\hat{\mathbf{H}}_0 = \frac{1}{2m}\sum_r \left(\hat{p}_r - \frac{e}{c}A_r\right)^2 \mathbf{I} + e\Phi\mathbf{I} - \frac{e\hbar}{2mc}\sum_r (\boldsymbol{\Sigma}_r B_r). \tag{7.3.9}$$

Suppose we partition the $4\times 1$ matrix $\boldsymbol{\psi}$ into two $2\times 1$ matrices, thus:

$$\boldsymbol{\psi} = \begin{pmatrix} \boldsymbol{\psi}_1 \\ \boldsymbol{\psi}_2 \end{pmatrix}. \tag{7.3.10}$$

Then the characteristic equation for the energy, viz. $\hat{\mathbf{H}}_0\boldsymbol{\psi} = E\boldsymbol{\psi}$, is equivalent to the pair of equations

$$\hat{\mathbf{H}}'\boldsymbol{\psi}_i = E\boldsymbol{\psi}_i, \quad i = 1, 2, \tag{7.3.11}$$

where

$$\hat{\mathbf{H}}' = \frac{1}{2m}\sum_r \left(\hat{p}_r - \frac{e}{c}A_r\right)^2 \mathbf{I}_2 + e\Phi\mathbf{I}_2 - \frac{e\hbar}{2mc}\sum_r \boldsymbol{\sigma}_r B_r. \tag{7.3.12}$$

But this is exactly the form already obtained in section 5.7 for the Hamiltonian in the case of a particle moving in an electromagnetic field, when the particle possesses an intrinsic magnetic moment

$$\mu_0 \mathfrak{z} = \frac{e\hbar}{2mc}\mathfrak{z}. \tag{7.3.13}$$

We conclude, therefore, that for velocities which are small by comparison with $c$, the Dirac particle will possess an intrinsic magnetic moment given by equation (7.3.13).

## 7.4. Spin

Consider a Dirac particle moving in an electrostatic field which is everywhere directed towards the origin $O$ of the inertial frame. Then it was shown in section 5.3 that the non-relativistic theory predicts that any component of the particle's angular momentum about $O$ will be conserved, in the sense that the probability distribution of this observable is independent of $t$. This result follows from the fact that $\hat{\mathbf{H}}$ and $\hat{L}_z$ commute. However, if $\hat{\mathbf{H}}$ is taken to be the Dirac Hamiltonian, it no longer commutes with $\hat{L}_z$ and the orbital angular momentum is not conserved.

To prove this, we calculate the commutator of $\hat{\mathbf{H}}$ and $\hat{L}_z$. Now

$$\hat{\mathbf{H}} = mc^2\boldsymbol{\beta}+c\sum_r \boldsymbol{\alpha}_r\hat{p}_r+e\Phi\mathbf{I}, \tag{7.4.1}$$

where $\Phi$ depends on the spherical polar coordinate $r$ alone. Also

$$\hat{L}_z = x\hat{p}_y-y\hat{p}_x = \frac{\hbar}{\iota}\frac{\partial}{\partial\phi}, \tag{7.4.2}$$

where $\phi$ is the usual spherical polar angle. It is now evident that $\hat{L}_z$ commutes with the first and last terms in the expression for $\hat{\mathbf{H}}$ and hence

$$[\hat{\mathbf{H}}, \hat{L}_z] = c\sum_r \boldsymbol{\alpha}_r[\hat{p}_r, x\hat{p}_y-y\hat{p}_x]. \tag{7.4.3}$$

Now

$$\begin{aligned}[\hat{p}_x, x\hat{p}_y-y\hat{p}_x] &= [\hat{p}_x, x]\hat{p}_y,\\ &= \frac{\hbar}{\iota}\hat{p}_y.\end{aligned} \tag{7.4.4}$$

Similarly,

$$[\hat{p}_y, x\hat{p}_y-y\hat{p}_x] = -\frac{\hbar}{\iota}\hat{p}_x. \tag{7.4.5}$$

Further, $\hat{p}_z$ commutes with $x\hat{p}_y-y\hat{p}_x$. It follows that

$$[\hat{\mathbf{H}}, \hat{L}_z] = \iota c\hbar(\boldsymbol{\alpha}_y\hat{p}_x-\boldsymbol{\alpha}_x\hat{p}_y). \tag{7.4.6}$$

After the analysis of the previous section, it is easy to guess why the orbital angular momentum is not conserved. We know that the particle behaves like a magnetic dipole and its dipole moment will be affected by interaction with the electrostatic field. Assuming that this moment is induced by the spin of the particle, we expect that this spin will also be acted upon by the field and will not, therefore, be conserved. To ensure that the net angular momentum of the system is conserved, it is accordingly necessary for the orbital angular momentum to vary.

Arising from the analysis of the previous section it seems probable that, in the Dirac theory, a component of the spin axis (say the $z$-component) is represented by the $4\times 4$ matrix $\boldsymbol{\Sigma}_z$. Let us calculate the commutator of $\hat{\mathbf{H}}$ and $\boldsymbol{\Sigma}_z$. We have

$$[\hat{\mathbf{H}}, \boldsymbol{\Sigma}_z] = mc^2[\boldsymbol{\beta}, \boldsymbol{\Sigma}_z]+c\sum_r [\boldsymbol{\alpha}_r, \boldsymbol{\Sigma}_z]\hat{p}_r. \tag{7.4.7}$$

The reader is left to verify that

$$\left.\begin{aligned}&[\boldsymbol{\beta},\boldsymbol{\Sigma}_z] = \mathbf{0},\\ &[\boldsymbol{\alpha}_x,\boldsymbol{\Sigma}_z] = -2\iota\boldsymbol{\alpha}_y, \quad [\boldsymbol{\alpha}_y,\boldsymbol{\Sigma}_z] = 2\iota\boldsymbol{\alpha}_x,\\ &[\boldsymbol{\alpha}_z,\boldsymbol{\Sigma}_z] = \mathbf{0}.\end{aligned}\right\} \tag{7.4.8}$$

Thus

$$[\hat{\mathbf{H}},\boldsymbol{\Sigma}_z] = 2\iota c(\boldsymbol{\alpha}_x\hat{p}_y - \boldsymbol{\alpha}_y\hat{p}_x). \tag{7.4.9}$$

Comparing equations (7.4.6), (7.4.9), it is clear that

$$[\hat{\mathbf{H}},\hat{L}_z\mathbf{I}+\tfrac{1}{2}\hbar\boldsymbol{\Sigma}_z] = \mathbf{0}. \tag{7.4.10}$$

Thus, the Dirac theory requires that the spin and spin axis should be related by the equation (2.2.1), if the net angular momentum of a particle is to be conserved in a central force field. This prediction of the correct relationship between the intrinsic magnetic moment and angular momentum of a particle, is one of the most remarkable features of the Dirac theory.

Since the $z$-component of the spin axis is represented by the matrix $\boldsymbol{\Sigma}_z$, it follows that the $\sigma_z$-eigenstates are given by the characteristic equation

$$\boldsymbol{\Sigma}_z\boldsymbol{\psi} = \lambda\boldsymbol{\psi}. \tag{7.4.11}$$

This is equivalent to the set of equations

$$\psi_1 = \lambda\psi_1, \quad \psi_2 = -\lambda\psi_2, \quad \psi_3 = \lambda\psi_3, \quad \psi_4 = -\lambda\psi_4. \tag{7.4.12}$$

For a non-zero solution, $\lambda$ must take one of the eigenvalues $\pm 1$. If $\lambda = +1$, then $\psi_1$, $\psi_3$ are indeterminate and $\psi_2 = \psi_4 = 0$. If $\lambda = -1$, then $\psi_2$, $\psi_4$ are indeterminate and $\psi_1 = \psi_3 = 0$.

Employing the non-relativistic approximation, we take $\psi_3 = \psi_4 = 0$. Then, in the two eigenstates $\sigma_z = +1, -1$,

$$\boldsymbol{\psi} = \begin{pmatrix}\alpha_+\\0\\0\\0\end{pmatrix}, \quad \boldsymbol{\psi} = \begin{pmatrix}0\\\alpha_-\\0\\0\end{pmatrix}, \tag{7.4.12}$$

respectively.

Suppose the particle is in the state $\boldsymbol{\psi} = \{\psi_1,\psi_2,0,0\}$. Consider the states

$$\boldsymbol{\psi} = \boldsymbol{\psi}_+ = a_+\begin{pmatrix}\psi_1\\0\\0\\0\end{pmatrix}, \quad \boldsymbol{\psi} = \boldsymbol{\psi}_- = a_-\begin{pmatrix}0\\\psi_2\\0\\0\end{pmatrix}, \tag{7.4.13}$$

where $a_+$, $a_-$ are chosen to normalise $\psi_+$, $\psi_-$ respectively; i.e.

$$a_+ = \left[\iint |\psi_1|^2\,d\tau\right]^{-\frac{1}{2}}, \quad a_- = \left[\iint |\psi_2|^2\,d\tau\right]^{-\frac{1}{2}}. \tag{7.4.14}$$

Then $\psi_+$, $\psi_-$ represent $\sigma_z$-eigenstates. Associate with $\sigma_z$ a sufficient number of additional observables to create a maximal set and suppose these observables to be selected so that $\psi_+$, $\psi_-$ are two eigenfunctions for the set (this can be done since $\psi_+$, $\psi_-$ are orthogonal). Then $\psi$ can be expanded in terms of the eigenfunctions of the set thus,

$$\psi = a_+^{-1}\psi_+ + a_-^{-1}\psi_- \tag{7.4.15}$$

and it follows immediately from this expansion that the probabilities of finding the particle in the eigenstates $\psi_+$, $\psi_-$ are

$$a_+^{-2} = \int |\psi_1|^2\,d\tau, \quad a_-^{-2} = \int |\psi_2|^2\,d\tau \tag{7.4.16}$$

respectively. The probability of finding the particle in any other eigenstate is zero. Thus, the probability of measuring $\sigma_z = +1$ is $a_+^{-2}$ and of measuring $\sigma_z = -1$ is $a_-^{-2}$. This implies that, in the non-relativistic approximation, $\{\psi_1, \psi_2\}$ can be regarded as a spinor in the $\sigma_z$-representation.

## 7.5. Free motion of a Dirac particle. The positron

In the absence of an electromagnetic field, we have

$$\hat{H} = mc^2\beta + c\sum_r \alpha_r \hat{p}_r. \tag{7.5.1}$$

For a steady state of energy $E$, the $4 \times 1$ matrix representing the particle's state has the form

$$\psi \exp(-\iota Et/\hbar), \tag{7.5.2}$$

where $\psi$ is independent of $t$. $\psi$ satisfies the energy characteristic equation

$$\hat{H}\psi = E\psi \tag{7.5.3}$$

which, written out at length, is equivalent to the four equations

$$\left.\begin{aligned}
(mc^2 - E)\psi_1 + c\hat{p}_z\psi_3 + c(\hat{p}_x - \iota\hat{p}_y)\psi_4 &= 0,\\
(mc^2 - E)\psi_2 + c(\hat{p}_x + \iota\hat{p}_y)\psi_3 - c\hat{p}_z\psi_4 &= 0,\\
c\hat{p}_z\psi_1 + c(\hat{p}_x - \iota\hat{p}_y)\psi_2 - (mc^2 + E)\psi_3 &= 0,\\
c(\hat{p}_x + \iota\hat{p}_y)\psi_1 - c\hat{p}_z\psi_2 - (mc^2 + E)\psi_4 &= 0.
\end{aligned}\right\} \tag{7.5.4}$$

Since $\hat{\mathbf{H}}$ commutes with $\hat{p}_x$, $\hat{p}_y$ and $\hat{p}_z$, states exist in which the linear momentum is sharp for all $t$. In such a state

$$\psi = \mathbf{A}\exp[\iota(p_x x+p_y y+p_z z)/\hbar] \tag{7.5.5}$$

and hence the elements of $\mathbf{A}$ satisfy the equations

$$\left.\begin{aligned}(mc^2-E)A_1+cp_z A_3+c(p_x-\iota p_y)A_4 &= 0,\\ (mc^2-E)A_2+c(p_x+\iota p_y)A_3-cp_z A_4 &= 0,\\ cp_z A_1+c(p_x-\iota p_y)A_2-(mc^2+E)A_3 &= 0,\\ c(p_x+\iota p_y)A_1-cp_z A_2-(mc^2+E)A_4 &= 0.\end{aligned}\right\} \tag{7.5.6}$$

These equations possess a non-zero solution in the $A_i$ if, and only if, the determinant of their coefficients vanishes. Expanding the determinant, this condition can be written

$$[c^2(p_x^2+p_y^2+p_z^2+m^2c^2)-E^2]^2 = 0. \tag{7.5.7}$$

Thus $E$ has two eigenvalues given by the equation

$$E = \pm c(p^2+m^2c^2)^{\frac{1}{2}}, \tag{7.5.8}$$

where $p^2=p_x^2+p_y^2+p_z^2$.

The positive eigenvalue is precisely the relativistic formula for the energy of a particle whose rest mass is $m$ and whose momentum is $p$. The negative eigenvalue cannot be assumed by the energy in special relativity theory since the positive and negative energy values given by equation (7.5.8) are separated by an interval $(-mc^2, mc^2)$ of values which can never be attained by $E$; it follows that $E$ can never change its sign by a continuous variation. However, in quantum mechanics, discontinuous variations are acceptable and the negative eigenvalues cannot be disregarded.

The fact that electrons in negative energy states are not encountered in nature, led Dirac to postulate that normally all such states are filled and hence, by the exclusion principle, to conclude that no electron can be observed undergoing a transition into such a state. However, by supplying energy in excess of the minimum $2mc^2$, it should be possible on this theory to raise an electron from one of the negative energy states to a positive state, in which it immediately becomes observable (by undergoing transitions to lower positive energy states with the emission of radiation). If such a transition from a negative to a positive energy state takes place, a vacant negative state or 'hole' is created in the continuum of negative states. Dirac showed that such a 'hole' would behave like a *positron* (a particle having the same mass as but opposite charge to the electron). Thus the overall effect would be that of the creation of an electron-positron pair from a photon supplying energy in excess of $2mc^2$. Such pair creations have now been observed. The reverse process of the mutual annihilation of an electron and a positron with the liberation of energy, has also been observed and, in Dirac's theory, this can be explained as an encounter between an electron and a 'hole', in which the electron falls into the vacant negative energy state with the liberation of the energy it loses thereby.

Substituting for $E$ from equation (7.5.8) into equations (7.5.6), it will be found that the rank of the determinant of the coefficients is 2. Thus, two of the $A_i$ can be assigned arbitrary values and the other two are then determined. Giving arbitrary values to $A_1$, $A_2$, it will be found that

$$A_3 = \frac{p_z A_1 + (p_x - \iota p_y) A_2}{mc + E/c}, \quad A_4 = \frac{(p_x + \iota p_y) A_1 - p_z A_2}{mc + E/c}. \quad (7.5.9)$$

Thus, if the positive energy eigenvalue is being considered, $E \geqslant mc^2$ and hence $A_3$, $A_4$ will be small quantities; the small magnitudes of the components $\psi_3$, $\psi_4$ of $\boldsymbol{\psi}$ has already been noted. However, if $E$ is negative and $p$ is not too large, $mc + E/c$ will be small and hence $A_3$, $A_4$ will be large; $\psi_3$, $\psi_4$ will then be the principal components of $\psi$ and $\psi_1$, $\psi_2$ will be negligible.

## 7.6. A complete set of commuting operators for a Dirac particle in a central field

As a preliminary to the solution employing Dirac's theory of the problem of an electron moving under the attraction of an atomic nucleus, we will demonstrate that, for a particle moving in a central force field, there exists a certain set of four operators which commute in pairs. This gives rise to a complete set of simultaneous eigenfunctions from which any state can be derived by superposition. The operators are (i) the Hamiltonian $\hat{\mathbf{H}}$, (ii) the $z$-component of the overall angular momentum $\hat{\mathbf{J}}_z$, (iii) the square of the magnitude of the total angular momentum $\hat{\mathbf{J}}^2$ and (iv) an operator $\hat{\mathbf{K}}$. Denoting by $V(r)$ the P.E. of the particle in the central force field, these operators are defined by the equations

$$\hat{\mathbf{H}} = mc^2 \boldsymbol{\beta} + c\mathfrak{A} . \hat{\mathfrak{p}} + V(r)\mathbf{I}, \quad (7.6.1)$$

$$\hat{\mathbf{J}}_z = \hat{L}_z \mathbf{I} + \tfrac{1}{2}\hbar \boldsymbol{\Sigma}_z, \quad (7.6.2)$$

$$\hat{\mathfrak{J}}^2 = \mathfrak{J}_x^2+\mathfrak{J}_y^2+\mathfrak{J}_z^2,$$
$$= (\hat{L}^2+\tfrac{3}{4}\hbar^2)\mathbf{I}+\hbar\mathfrak{S}.\hat{\mathfrak{L}}, \tag{7.6.3}$$
$$\hat{\mathfrak{K}} = \boldsymbol{\beta}\mathfrak{S}.\hat{\mathfrak{L}}+\hbar\boldsymbol{\beta}, \tag{7.6.4}$$

where $\mathfrak{A}$, $\hat{\mathfrak{p}}$, $\hat{\mathfrak{L}}$, $\mathfrak{S}$ are vector quantities with components $(\boldsymbol{\alpha}_x, \boldsymbol{\alpha}_y, \boldsymbol{\alpha}_z)$, $(\hat{p}_x, \hat{p}_y, \hat{p}_z)$, $(\hat{L}_x, \hat{L}_y, \hat{L}_z)$, $(\boldsymbol{\Sigma}_x, \boldsymbol{\Sigma}_y, \boldsymbol{\Sigma}_z)$, respectively. In establishing the formula (7.6.3), we have made use of the matrix identity

$$\boldsymbol{\Sigma}_x^2+\boldsymbol{\Sigma}_y^2+\boldsymbol{\Sigma}_z^2 = 3\mathbf{I}. \tag{7.6.5}$$

This follows from equations (2.2.5), (2.2.6).

We first remark that all matrices commute with all differential operators and simple multipliers. Further, $\boldsymbol{\beta}$ commutes with all components of $\mathfrak{S}$ (equations (7.4.8)). Also, equations (7.4.6), (7.4.9), are equivalent to

$$[\hat{\mathfrak{H}}, \hat{\mathfrak{L}}] = -\iota c\hbar\mathfrak{A}\times\hat{\mathfrak{p}}, \tag{7.6.6}$$

$$[\hat{\mathfrak{H}}, \mathfrak{S}] = 2\iota c\mathfrak{A}\times\hat{\mathfrak{p}}. \tag{7.6.7}$$

The reader should also verify that

$$\mathfrak{S}\times\mathfrak{A} = 2\iota\mathfrak{A}. \tag{7.6.8}$$

It has already been proved that $\hat{\mathfrak{H}}$ and $\hat{\mathfrak{J}}_z$ commute (equation (7.4.10)). Now

$$[\hat{\mathfrak{H}}, \hat{\mathfrak{J}}^2] = [\hat{\mathfrak{H}}, \hat{L}^2]+\hbar[\hat{\mathfrak{H}}, \mathfrak{S}.\hat{\mathfrak{L}}] \tag{7.6.9}$$

Also

$$[\hat{\mathfrak{H}}, \hat{L}^2] = c[\mathfrak{A}.\hat{\mathfrak{p}}, \hat{L}^2],$$
$$= c\mathfrak{A}.[\hat{\mathfrak{p}}, \hat{L}^2]$$
$$= \iota c\hbar\mathfrak{A}.(\hat{\mathfrak{L}}\times\hat{\mathfrak{p}}-\hat{\mathfrak{p}}\times\hat{\mathfrak{L}}), \tag{7.6.10}$$

by the result quoted in Ex. 3(d) of Chap. 5. Further, employing the identity from Ex. 9(c) of Chap. 1, we calculate that

$$[\hat{\mathfrak{H}}, \mathfrak{S}.\hat{\mathfrak{L}}] = [\hat{\mathfrak{H}}, \mathfrak{S}].\hat{\mathfrak{L}}+\mathfrak{S}.[\hat{\mathfrak{H}}, \hat{\mathfrak{L}}],$$
$$= 2\iota c\mathfrak{A}\times\hat{\mathfrak{p}}.\hat{\mathfrak{L}}-\iota c\hbar\mathfrak{S}.\mathfrak{A}\times\hat{\mathfrak{p}},$$
$$= 2\iota c\mathfrak{A}.\hat{\mathfrak{p}}\times\hat{\mathfrak{L}}-\iota c\hbar\mathfrak{S}\times\mathfrak{A}.\hat{\mathfrak{p}},$$
$$= 2\iota c\mathfrak{A}.\hat{\mathfrak{p}}\times\hat{\mathfrak{L}}+2c\hbar\mathfrak{A}.\hat{\mathfrak{p}}, \tag{7.6.11}$$

using equations (7.6.6)–(7.6.8). Substituting into equation (7.6.9) from equations (7.6.10), (7.6.11), we obtain

$$[\hat{\mathfrak{H}}, \hat{\mathfrak{J}}^2] = \iota c\hbar\mathfrak{A}.(\hat{\mathfrak{L}}\times\hat{\mathfrak{p}}+\hat{\mathfrak{p}}\times\hat{\mathfrak{L}})+2c\hbar^2\mathfrak{A}.\hat{\mathfrak{p}},$$
$$= 0, \tag{7.6.12}$$

by identity Ex. 3(e) of Chap. 5. Thus $\hat{\mathfrak{H}}$ and $\hat{\mathfrak{J}}^2$ commute.

Now consider

$$[\hat{\mathfrak{H}}, \hat{\mathfrak{K}}] = [\hat{\mathfrak{H}}, \boldsymbol{\beta}\mathfrak{S}.\hat{\mathfrak{L}}]+\hbar[\hat{\mathfrak{H}}, \boldsymbol{\beta}]. \tag{7.6.13}$$

We have that

$$[\hat{\mathfrak{H}}, \boldsymbol{\beta}\mathfrak{S}.\hat{\mathfrak{L}}] = [\hat{\mathfrak{H}}, \boldsymbol{\beta}]\,\mathfrak{S}.\hat{\mathfrak{L}}+\boldsymbol{\beta}[\hat{\mathfrak{H}}, \mathfrak{S}.\hat{\mathfrak{L}}]. \tag{7.6.14}$$

Also

$$\begin{aligned}[\hat{\mathfrak{H}}, \boldsymbol{\beta}] &= c[\mathfrak{A}.\hat{\mathfrak{p}}, \boldsymbol{\beta}],\\ &= -2c\boldsymbol{\beta}\mathfrak{A}.\hat{\mathfrak{p}},\end{aligned} \tag{7.6.15}$$

since, as the reader can verify

$$[\boldsymbol{\alpha}_r, \boldsymbol{\beta}] = -2\boldsymbol{\beta}\boldsymbol{\alpha}_r. \tag{7.6.16}$$

It follows, using equation (7.6.11), that

$$[\hat{\mathfrak{H}}, \boldsymbol{\beta}\mathfrak{S}.\hat{\mathfrak{L}}] = 2c\boldsymbol{\beta}[-\mathfrak{A}.\hat{\mathfrak{p}}\mathfrak{S}.\hat{\mathfrak{L}}+\iota\mathfrak{A}.\hat{\mathfrak{p}}\times\hat{\mathfrak{L}}+\hbar\mathfrak{A}.\hat{\mathfrak{p}}]. \tag{7.6.17}$$

Employing the repeated subscript summation convention, it may be verified that

$$\boldsymbol{\alpha}_r\boldsymbol{\Sigma}_s = \iota\epsilon_{rst}\boldsymbol{\alpha}_t, \tag{7.6.18}$$

where $\epsilon_{rst}$ is the usual permutation symbol. Hence

$$\begin{aligned}\mathfrak{A}.\hat{\mathfrak{p}}\mathfrak{S}.\hat{\mathfrak{L}} &= \boldsymbol{\alpha}_r\hat{p}_r\boldsymbol{\Sigma}_s\hat{L}_s,\\ &= \iota\epsilon_{rst}\boldsymbol{\alpha}_t\hat{p}_r\hat{L}_s,\\ &= \iota\mathfrak{A}.\hat{\mathfrak{p}}\times\hat{\mathfrak{L}}.\end{aligned} \tag{7.6.19}$$

Equation (7.6.17) now reduces to

$$\begin{aligned}[\hat{\mathfrak{H}}, \boldsymbol{\beta}\mathfrak{S}.\hat{\mathfrak{L}}] &= 2c\hbar\boldsymbol{\beta}\mathfrak{A}.\hat{\mathfrak{p}},\\ &= -\hbar[\hat{\mathfrak{H}}, \boldsymbol{\beta}].\end{aligned} \tag{7.6.20}$$

Hence

$$[\hat{\mathfrak{H}}, \hat{\mathfrak{K}}] = 0 \tag{7.6.21}$$

and $\hat{\mathfrak{H}}$ commutes with $\hat{\mathfrak{K}}$.

Next, we have

$$\begin{aligned}[\hat{\mathfrak{J}}_z, \hat{\mathfrak{J}}^2] &= \hbar(\hat{L}_z, \mathfrak{S}.\hat{\mathfrak{L}}]+\tfrac{1}{2}\hbar^2[\boldsymbol{\Sigma}_z, \mathfrak{S}.\hat{\mathfrak{L}}],\\ &= \hbar\mathfrak{S}.(\hat{L}_z, \hat{\mathfrak{L}}]+\tfrac{1}{2}\hbar^2[\boldsymbol{\Sigma}_z, \mathfrak{S}].\hat{\mathfrak{L}},\\ &= \iota\hbar^2(\boldsymbol{\Sigma}_x\hat{L}_y-\boldsymbol{\Sigma}_y\hat{L}_x)\\ &\qquad+\tfrac{1}{2}\hbar^2.2\iota(\boldsymbol{\Sigma}_y\hat{L}_x-\boldsymbol{\Sigma}_x\hat{L}_y),\\ &= 0.\end{aligned} \tag{7.6.22}$$

$\hat{\mathfrak{J}}_z$ and $\hat{\mathfrak{J}}^2$ accordingly commute.

Next

$$
\begin{aligned}
[\hat{\mathbf{J}}_z, \hat{\mathbf{K}}] &= [\hat{L}_z, \boldsymbol{\beta}\mathfrak{S}.\hat{\mathfrak{L}}] + \tfrac{1}{2}\hbar[\boldsymbol{\Sigma}_z, \boldsymbol{\beta}\mathfrak{S}.\hat{\mathfrak{L}}],\\
&= \boldsymbol{\beta}\mathfrak{S}.[\hat{L}_z, \hat{\mathfrak{L}}] + \tfrac{1}{2}\hbar\boldsymbol{\beta}[\boldsymbol{\Sigma}_z, \mathfrak{S}].\hat{\mathfrak{L}},\\
&= \boldsymbol{\beta}\iota\hbar(\boldsymbol{\Sigma}_x\hat{L}_y - \boldsymbol{\Sigma}_y\hat{L}_x)\\
&\qquad + \tfrac{1}{2}\hbar\boldsymbol{\beta}.2\iota(\boldsymbol{\Sigma}_y\hat{L}_x - \boldsymbol{\Sigma}_x\hat{L}_y),\\
&= 0.
\end{aligned}
\tag{7.6.23}
$$

It follows that $\hat{\mathbf{J}}_z$ and $\hat{\mathbf{K}}$ commute.

Finally, we have to consider

$$
\begin{aligned}
[\hat{\mathbf{J}}^2, \hat{\mathbf{K}}] &= \hbar[\mathfrak{S}.\hat{\mathfrak{L}}, \boldsymbol{\beta}\mathfrak{S}.\hat{\mathfrak{L}}] + \hbar^2[\mathfrak{S}.\hat{\mathfrak{L}}, \boldsymbol{\beta}],\\
&= \hbar[\mathfrak{S}.\hat{\mathfrak{L}}, \boldsymbol{\beta}]\,\mathfrak{S}.\hat{\mathfrak{L}} + \hbar^2[\mathfrak{S}.\hat{\mathfrak{L}}, \boldsymbol{\beta}],\\
&= \hbar[\mathfrak{S}.\hat{\mathfrak{L}}, \boldsymbol{\beta}]\,(\mathfrak{S}.\hat{\mathfrak{L}} + \hbar\mathbf{I}).
\end{aligned}
\tag{7.6.24}
$$

But

$$
[\mathfrak{S}.\hat{\mathfrak{L}}, \boldsymbol{\beta}] = [\mathfrak{S}, \boldsymbol{\beta}].\hat{\mathfrak{L}} = 0, \tag{7.6.25}
$$

since $\boldsymbol{\beta}$ commutes with all components of $\mathfrak{S}$. Hence, $\hat{\mathbf{J}}^2$ commutes with $\hat{\mathbf{K}}$.

Having proved that $\hat{\mathbf{J}}_z$, $\hat{\mathbf{J}}^2$ and $\hat{\mathbf{K}}$ commute with $\hat{\mathbf{H}}$, we conclude that the observables represented by these operators are constants of the motion, i.e. their probability distributions do not change with the time and, in particular, if they are sharp at an initial instant, they remain sharp and constant. Thus, the simultaneous eigenfunctions of this set of operators will represent steady states of the particle in which the quantities $J_z$, $J^2$, $K$ and $H$ are sharp and constant.

## 7.7. Simultaneous angular momentum eigenfunctions for a Dirac particle in a central field

In this section, we shall calculate a set of simultaneous eigenfunctions for the operators $\hat{\mathbf{J}}_z$, $\hat{\mathbf{K}}$ and $\hat{\mathbf{J}}^2$.

First, suppose that $\boldsymbol{\psi}$ is an eigenfunction for $\hat{\mathbf{J}}_z$ corresponding to the eigenvalue $q\hbar$. Then

$$
\hat{\mathbf{J}}_z\boldsymbol{\psi} = q\hbar\boldsymbol{\psi}. \tag{7.7.1}
$$

Substituting for $\hat{\mathbf{J}}_z$ from equation (7.6.2), this is seen to be equivalent to the set of four equations

$$
\left.
\begin{aligned}
\hat{L}_z\psi_i &= (q - \tfrac{1}{2})\,\hbar\psi_i, \quad i = 1, 3,\\
\hat{L}_z\psi_i &= (q + \tfrac{1}{2})\,\hbar\psi_i, \quad i = 2, 4.
\end{aligned}
\right\}
\tag{7.7.2}
$$

Employing spherical polar coordinates $(r,\theta,\phi)$ and proceeding as in section 5.2, we now deduce that $q=s+\frac{1}{2}$, where $s$ is an integer and that

$$\left.\begin{aligned} \psi_i &= A_i \exp(\iota s\phi), \quad i = 1,3, \\ \psi_i &= A_i \exp[\iota(s+1)\phi], \quad i = 2,4, \end{aligned}\right\} \tag{7.7.3}$$

where the $A_i$ remain to be calculated as functions of $r$ and $\theta$. Thus, as in section 5.5, we have shown that the eigenvalues of $J_z$ are

$$\ldots -\tfrac{5}{2}\hbar, -\tfrac{3}{2}\hbar, -\tfrac{1}{2}\hbar, \tfrac{1}{2}\hbar, \ldots . \tag{7.7.4}$$

Next, suppose that $\boldsymbol{\psi}$ is an eigenfunction of $K$ corresponding to an eigenvalue $k\hbar$. Then

$$\hat{\mathbf{K}}\boldsymbol{\psi} = k\hbar\boldsymbol{\psi} \tag{7.7.5}$$

and this is equivalent to the set of equations

$$\left.\begin{aligned} [\hat{L}_z-(k-1)\hbar]\psi_1+\hat{L}_-\psi_2 &= 0, \\ \hat{L}_+\psi_1-[\hat{L}_z+(k-1)\hbar]\psi_2 &= 0, \\ [\hat{L}_z+(k+1)\hbar]\psi_3+\hat{L}_-\psi_4 &= 0, \\ \hat{L}_+\psi_3-[\hat{L}_z-(k+1)\hbar]\psi_4 &= 0, \end{aligned}\right\} \tag{7.7.6}$$

$$\hat{L}_+ = \hat{L}_x+\iota\hat{L}_y, \quad \hat{L}_- = \hat{L}_x-\iota\hat{L}_y. \tag{7.7.7}$$

Operating on the first of equations (7.7.6) with $[\hat{L}_z+(k-1)\hbar]$ and on the second with $\hat{L}_-$ and adding, we obtain

$$[\hat{L}^2-(k-1)^2\hbar^2]\psi_1-\hbar(\hat{L}_z\psi_1+\hat{L}_-\psi_2) = 0, \tag{7.7.8}$$

having employed the identities

$$\hat{L}_-\hat{L}_+ = \hat{L}_x^2+\hat{L}_y^2-\hbar\hat{L}_z, \tag{7.7.9}$$

$$[\hat{L}_z,\hat{L}_-] = -\hbar\hat{L}_-. \tag{7.7.10}$$

But, rearranging the first of equations (7.7.6) into the form

$$\hat{L}_z\psi_1+\hat{L}_-\psi_2 = (k-1)\hbar\psi_1, \tag{7.7.11}$$

it is clear that equation (7.7.8) reduces to

$$\hat{L}^2\psi_1 = k(k-1)\hbar^2\psi_1. \tag{7.7.12}$$

$\psi_2$ satisfies the same equation.

$\psi_3$ satisfies the equation

$$\hat{L}^2\psi_3 = k(k+1)\hbar^2\psi_3 \tag{7.7.13}$$

and $\psi_4$ satisfies the same equation.

It now follows, as in section 5.2, that $k$ must be a positive or negative integer or zero, if the $\psi_i$ are not all to vanish. Taking $l$ to be a positive integer or zero, we find that if $k=l$, then

$$\left.\begin{aligned} \psi_1 &= B_1\, Y_{l-1,\,s}(\theta,\phi), \\ \psi_2 &= B_2\, Y_{l-1,\,s+1}(\theta,\phi), \\ \psi_3 &= B_3\, Y_{l,\,s}(\theta,\phi), \\ \psi_4 &= B_4\, Y_{l,\,s+1}(\theta,\phi). \end{aligned}\right\} \tag{7.7.14}$$

If $k=-l$, then

$$\left.\begin{aligned} \psi_1 &= B_1\, Y_{l,\,s}(\theta,\phi), \\ \psi_2 &= B_2\, Y_{l,\,s+1}(\theta,\phi), \\ \psi_3 &= B_3\, Y_{l-1,\,s}(\theta,\phi), \\ \psi_4 &= B_4\, Y_{l-1,\,s+1}(\theta,\phi). \end{aligned}\right\} \tag{7.7.15}$$

Finally, we require that $\boldsymbol{\psi}$ should be an eigenfunction of $\hat{\mathbf{J}}^2$. Taking the eigenvalue to be $j(j+1)\hbar^2$ $(j>0)$, this leads to the set of equations

$$\left.\begin{aligned} (\hat{L}^2+\hbar\hat{L}_z+\tfrac{3}{4}\hbar^2)\,\psi_1+\hbar\hat{L}_-\,\psi_2 &= j(j+1)\,\hbar^2\,\psi_1, \\ \hbar\hat{L}_+\,\psi_1+(\hat{L}^2-\hbar\hat{L}_z+\tfrac{3}{4}\hbar^2)\,\psi_2 &= j(j+1)\,\hbar^2\,\psi_2, \end{aligned}\right\} \tag{7.7.16}$$

together with an identical pair of equations for $\psi_3$, $\psi_4$.

In the case $k=l$, substituting from equations (7.7.14), it will be found that $B_1$, $B_2$ must satisfy the equations

$$\left.\begin{aligned} [l(l-1)-j(j+1)+s+\tfrac{3}{4}]\,B_1-\sqrt{[(l+s)\,(l-s-1)]}\,B_2 &= 0, \\ \sqrt{[l+s)\,(l-s-1)]}\,B_1+[l(l-1)-j(j+1)+s+\tfrac{3}{4}]\,B_2 &= 0, \end{aligned}\right\} \tag{7.7.17}$$

with a similar pair of equations for $B_3$ and $B_4$ (the identities given in Ex. 4, Chap. 5, should be used). These equations should be compared with equations (5.5.31), (5.5.32); they possess a non-zero solution in $\{B_1, B_2, B_3, B_4\}$ if, and only if,

$$j = l-\tfrac{3}{2},\ l-\tfrac{1}{2} \text{ or } l+\tfrac{1}{2}. \tag{7.7.18}$$

If $j=l-\frac{3}{2}$, then $B_3=B_4=0$. If $j=l+\frac{1}{2}$, then $B_1=B_2=0$. If $j=l-\frac{1}{2}$, then

$$\left.\begin{aligned} B_1 &= \sqrt{(l+s)}\,F, \\ B_2 &= -\sqrt{(l-s-1)}\,F, \\ B_3 &= \sqrt{(l-s)}\,G, \\ B_4 &= \sqrt{(l+s+1)}\,G, \end{aligned}\right\} \tag{7.7.19}$$

where $F$, $G$ are functions of $r$ alone.

In the case $k = -l$, it will similarly be found that the permissible values of $j$ are as given at (7.7.18) and that the expressions for $B_1$, $B_2$ are interchanged with the expressions for $B_3$, $B_4$.

## 7.8. Eigenfunctions of the Dirac Hamiltonian

We shall now further require that $\psi$ shall be an eigenfunction of the Dirac Hamiltonian $\hat{\mathfrak{H}}$ given by equation (7.6.1). By substituting directly from equations (7.7.14) or (7.7.15) into the characteristic equation

$$\hat{\mathfrak{H}}\psi = E\psi, \tag{7.8.1}$$

it is possible to derive two first order differential equations for the unknown functions $F(r)$, $G(r)$, introduced in the last section. However, the working is laborious and it is easier first to express $\hat{\mathfrak{H}}$ in an alternative form before making use of equation (7.8.1).

To do this, we first note that

$$\mathfrak{S}.\mathfrak{r} = \begin{pmatrix} \mathbf{X} & \mathbf{0} \\ \mathbf{0} & \mathbf{X} \end{pmatrix}, \tag{7.8.2}$$

where

$$\mathbf{X} = \begin{pmatrix} z & x - \iota y \\ x + \iota y & -z \end{pmatrix}. \tag{7.8.3}$$

Similarly

$$\mathfrak{S}.\hat{\mathfrak{p}} = \begin{pmatrix} \hat{\mathbf{Y}} & \mathbf{0} \\ \mathbf{0} & \hat{\mathbf{Y}} \end{pmatrix}, \tag{7.8.4}$$

where

$$\hat{\mathbf{Y}} = \begin{pmatrix} \hat{p}_z & \hat{p}_x - \iota \hat{p}_y \\ \hat{p}_x + \iota \hat{p}_y & -\hat{p}_z \end{pmatrix}. \tag{7.8.5}$$

Multiplying these matrices, it is easy to verify that

$$(\mathfrak{S}.\mathfrak{r})(\mathfrak{S}.\hat{\mathfrak{p}}) = \mathfrak{r}.\hat{\mathfrak{p}}\mathbf{I} + \iota\mathfrak{S}.\hat{\mathfrak{L}}. \tag{7.8.6}$$

Now

$$\mathbf{X}^2 = r^2\mathbf{I} \tag{7.8.7}$$

and hence

$$(\mathfrak{S}.\mathfrak{r})^2 = r^2\mathbf{I}. \tag{7.8.8}$$

Multiplying both members of the identity (7.8.6) on the left by $\mathfrak{S}.\mathfrak{r}$, we therefore obtain

$$r^2(\mathfrak{S}.\hat{\mathfrak{p}}) = (\mathfrak{S}.\mathfrak{r})\{(\mathfrak{r}.\hat{\mathfrak{p}})\mathbf{I} + \iota\mathfrak{S}.\hat{\mathfrak{L}}\}. \tag{7.8.9}$$

Defining a $4 \times 4$ matrix $\boldsymbol{\rho}$ by the equation

$$\boldsymbol{\rho} = \begin{pmatrix} \mathbf{0} & \mathbf{I} \\ \mathbf{I} & \mathbf{0} \end{pmatrix}, \tag{7.8.10}$$

it is obvious that

$$\boldsymbol{\rho}\boldsymbol{\Sigma}_x = \boldsymbol{\alpha}_x, \text{ etc.} \tag{7.8.11}$$

Multiplication of the identity (7.8.9) on the left by $\boldsymbol{\rho}$ accordingly transforms it to the form

$$r^2(\mathfrak{A}.\hat{\mathfrak{p}}) = (\mathfrak{A}.\mathfrak{r})\{(\mathfrak{r}.\hat{\mathfrak{p}})\mathbf{I} + \iota\mathfrak{S}.\hat{\mathfrak{L}}\}. \tag{7.8.12}$$

But, from equation (7.6.4) by multiplication on the left by $\boldsymbol{\beta}$, we have

$$\mathfrak{S}.\hat{\mathfrak{L}} = \boldsymbol{\beta}\hat{\mathbf{K}} - \hbar\mathbf{I}. \tag{7.8.13}$$

Hence

$$\mathfrak{A}.\hat{\mathfrak{p}} = \frac{1}{r^2}(\mathfrak{A}.\mathfrak{r})\{(\mathfrak{r}.\hat{\mathfrak{p}} - \iota\hbar)\mathbf{I} + \iota\boldsymbol{\beta}\hat{\mathbf{K}}\}. \tag{7.8.14}$$

$\hat{\mathbf{H}}$ can now be expressed in the form

$$\hat{\mathbf{H}} = mc^2\boldsymbol{\beta} + V\mathbf{I} + \frac{c}{r^2}(\mathfrak{A}.\mathfrak{r})\{(\mathfrak{r}.\hat{\mathfrak{p}} - \iota\hbar)\mathbf{I} + \iota\boldsymbol{\beta}\hat{\mathbf{K}}\}. \tag{7.8.15}$$

Now

$$\mathfrak{A}.\mathfrak{r} = \boldsymbol{\rho}\mathfrak{S}.\mathfrak{r} = \begin{pmatrix} \mathbf{0} & \mathbf{X} \\ \mathbf{X} & \mathbf{0} \end{pmatrix} = r\begin{pmatrix} \mathbf{0} & \boldsymbol{\tau} \\ \boldsymbol{\tau} & \mathbf{0} \end{pmatrix}, \tag{7.8.16}$$

where, in spherical polar coordinates,

$$\boldsymbol{\tau} = \begin{pmatrix} \cos\theta & \exp(-\iota\phi)\sin\theta \\ \exp(\iota\phi)\sin\theta & -\cos\theta \end{pmatrix}. \tag{7.8.17}$$

Also

$$\begin{aligned} \mathfrak{r}.\hat{\mathfrak{p}} &= \frac{\hbar}{\iota}\left(x\frac{\partial}{\partial x} + y\frac{\partial}{\partial y} + z\frac{\partial}{\partial z}\right), \\ &= \frac{\hbar}{\iota} r \frac{\partial}{\partial r}. \end{aligned} \tag{7.8.18}$$

Thus, if we partition $\boldsymbol{\psi}$ into two $2 \times 1$ matrices thus,

$$\boldsymbol{\psi} = \begin{pmatrix} \boldsymbol{\xi} \\ \boldsymbol{\eta} \end{pmatrix}, \tag{7.8.19}$$

where

$$\boldsymbol{\xi} = \begin{pmatrix} \psi_1 \\ \psi_2 \end{pmatrix}, \quad \boldsymbol{\eta} = \begin{pmatrix} \psi_3 \\ \psi_4 \end{pmatrix}, \tag{7.8.20}$$

and recall that $\boldsymbol{\psi}$ is an eigenfunction of $\hat{\mathbf{K}}$ so that

$$\hat{\mathbf{K}}\boldsymbol{\psi} = k\hbar\boldsymbol{\psi}, \tag{7.8.21}$$

it follows that equation (7.8.1) is equivalent to the pair of equations

$$\left.\begin{aligned}(V+mc^2-E)\boldsymbol{\xi}-\frac{\iota c\hbar}{r}\boldsymbol{\tau}\left(r\frac{\partial}{\partial r}+k+1\right)\boldsymbol{\eta} &= 0,\\ -\frac{\iota c\hbar}{r}\boldsymbol{\tau}\left(r\frac{\partial}{\partial r}-k+1\right)\boldsymbol{\xi}+(V-mc^2-E)\boldsymbol{\eta} &= 0.\end{aligned}\right\} \tag{7.8.22}$$

Supposing $k$ to be positive, substituting for $\psi_3$, $\psi_4$ from equations (7.7.14) and employing the identities (D.17), (D.18) from Appendix D, it will be found that

$$\boldsymbol{\tau}\begin{pmatrix}\psi_3\\ \psi_4\end{pmatrix} = \sqrt{\left(\frac{2l+1}{2l-1}\right)}\,G\begin{pmatrix}\sqrt{(l+s)}\,Y_{l-1,\,s}\\ -\sqrt{(l-s-1)}\,Y_{l-1,\,s+1}\end{pmatrix}. \tag{7.8.23}$$

Thus, the first of equations (7.8.22) is equivalent to the equation

$$(V+mc^2-E)F-\frac{c\hbar}{r}\left\{r\frac{dH}{dr}+(k+1)H\right\} = 0, \tag{7.8.24}$$

where

$$H = \iota\sqrt{\left(\frac{2l+1}{2l-1}\right)}\,G. \tag{7.8.25}$$

Similarly, it can be shown that the second of equations (7.8.22) is equivalent to the equation

$$\frac{c\hbar}{r}\left\{r\frac{dF}{dr}-(k-1)F\right\}+(V-mc^2-E)H = 0. \tag{7.8.26}$$

If $k$ is negative, we make use of the equations (7.7.15) and are led by a similar argument to the same equations (7.8.24), (7.8.26), for $F$ and $H$.

Equations (7.8.24), (7.8.26), imply immediately that, if either of the functions $F$ or $H$ is identically zero, then so is the other. Hence, it is impossible for only one of the pairs $(\psi_1, \psi_2)$, $(\psi_3, \psi_4)$ to vanish. This has the effect of eliminating the possibilities $j = l-\frac{3}{2}$, $l+\frac{1}{2}$ at (7.7.18), leaving only the one possibility

$$j = l-\tfrac{1}{2} = |k|-\tfrac{1}{2}. \tag{7.8.27}$$

We now put

$$F = \frac{u}{r}, \quad H = \frac{v}{r}, \tag{7.8.28}$$

$$\frac{mc^2-E}{\hbar c} = a_1, \quad \frac{mc^2+E}{\hbar c} = a_2. \tag{7.8.29}$$

This reduces equations (7.8.24), (7.8.26), to the forms

$$v' - \left(\frac{V}{\hbar c} + a_1\right)u + \frac{k}{r}v = 0, \tag{7.8.30}$$

$$u' - \frac{k}{r}u + \left(\frac{V}{\hbar c} - a_2\right)v = 0, \tag{7.8.31}$$

where primes denote differentiations with respect to $r$.

## 7.9. Energy levels of the hydrogen atom

We will now consider the solution of equations (7.8.30), (7.8.31), in the case when the electron is moving in a Coulomb field for which

$$V = -\frac{Ze^2}{r}. \tag{7.9.1}$$

For a bound state, $E < mc^2$ and $a_1$, $a_2$ are both positive constants. Defining the fine-structure constant $\alpha$ as at equation (6.4.8), equations (7.8.30), (7.8.31) assume the form

$$u' - \frac{k}{r}u - \left(\frac{Z\alpha}{r} + a_2\right)v = 0, \tag{7.9.2}$$

$$v' + \left(\frac{Z\alpha}{r} - a_1\right)u + \frac{k}{r}v = 0. \tag{7.9.3}$$

For large $r$, we have approximately

$$u' - a_2 v = 0, \quad v' - a_1 u = 0, \tag{7.9.4}$$

showing that at great distances from the centre of attraction, $u$ and $v$ can be expected to behave like $\exp(-ar)$, where

$$a^2 = a_1 a_2. \tag{7.9.5}$$

Thus, we shall put

$$u = \exp(-ar)f, \quad v = \exp(-ar)g, \tag{7.9.6}$$

transforming equations (7.9.2), (7.9.3), to the forms

$$f' = \left(a + \frac{k}{r}\right)f + \left(a_2 + \frac{Z\alpha}{r}\right)g, \tag{7.9.7}$$

$$g' = \left(a_1 - \frac{Z\alpha}{r}\right)f + \left(a - \frac{k}{r}\right)g. \tag{7.9.8}$$

We shall now apply the method of Frobenius to these equations: Substituting

$$f = \sum_{p=0}^{\infty} f_p r^{p+\rho}, \quad g = \sum_{p=0}^{\infty} g_p r^{p+\rho}, \tag{7.9.9}$$

and equating coefficients of $r^{\rho-1}$, we obtain the indicial equations

$$\left.\begin{aligned}(\rho-k)f_0 - Z\alpha g_0 &= 0,\\ Z\alpha f_0 + (\rho+k)g_0 &= 0.\end{aligned}\right\} \tag{7.9.10}$$

Also, equating coefficients of $r^{\rho+p}$, we obtain the recurrence relationships

$$\left.\begin{aligned}(\rho+p-k+1)f_{p+1} - Z\alpha g_{p+1} &= af_p + a_2 g_p,\\ Z\alpha f_{p+1} + (\rho+p+k+1)g_{p+1} &= a_1 f_p + a g_p,\end{aligned}\right\} \tag{7.9.11}$$

where $p=0, 1, 2, \ldots$. It is clear from this last pair of equations that, if $f_0$, $g_0$ both vanish, then $f_p$, $g_p$ vanish for all $p$; this leads to the trivial solution $u=v=0$. However, if $f_0$, $g_0$ are not both zero, it follows from the indicial equations (7.9.10) that

$$\begin{vmatrix} \rho-k & -Z\alpha \\ Z\alpha & \rho+k \end{vmatrix} = 0. \tag{7.9.12}$$

Thus,

$$\rho = \pm(k^2 - Z^2\alpha^2)^{\frac{1}{2}}. \tag{7.9.13}$$

To permit normalisation of $\psi$, it is necessary that $r\psi_i \to 0$ as $r \to 0$. Hence, $u$, $v \to 0$ as $r \to 0$ and this requires that $f$, $g \to 0$. It follows that $\rho \geqslant 0$ and that only the positive sign in equation (7.9.13) is acceptable.

We now multiply the first of equations (7.9.11) by $a_1$ and the second by $a$ and subtract. This yields the result

$$[a_1(\rho+p-k+1) - aZ\alpha]f_{p+1} = [a(\rho+p+k+1) + a_1 Z\alpha]g_{p+1}. \tag{7.9.14}$$

We conclude that, if one of the series terminates, so does the other and at the same power of $r$. Assuming that the series do not terminate, it follows from the last equation that

$$\frac{f_p}{g_p} \to \frac{a}{a_1} = \sqrt{\left(\frac{a_2}{a_1}\right)}, \tag{7.9.15}$$

as $p \to \infty$. But, from the first of equations (7.9.11), we find that

$$\frac{f_{p+1}}{f_p} = \frac{a + a_2 \dfrac{g_p}{f_p}}{\rho+p-k+1-Z\alpha \dfrac{g_{p+1}}{f_{p+1}}}. \tag{7.9.16}$$

Hence

$$\frac{f_{p+1}}{f_p}\bigg/\frac{2a}{p} \to 1 \tag{7.9.17}$$

as $p \to \infty$. Thus, the coefficients $f_p$ are ultimately all positive or all negative and, by the theorem proved in Appendix B, we conclude that either

$$f \geqslant K\exp(\beta r) + P(r), \tag{7.9.18}$$

or

$$f \leqslant -K\exp(\beta r) - P(r), \tag{7.9.19}$$

where $K > 0$, $P(r)$ is a polynomial and $\beta = 2a - \eta$, $\eta$ being positive and arbitrarily small. It now follows from equations (7.9.6) that, in these circumstances, $u$ must tend to infinity exponentially as $r \to \infty$. Such behaviour is unacceptable for reasons which will now be familiar. We conclude, therefore, that the series for $f$ and $g$ must both terminate.

Suppose the series terminate with terms in $r^{\rho+N}$. Putting $p = N$ in equations (7.9.11), we thus obtain

$$\left.\begin{aligned} af_N + a_2 g_N &= 0, \\ a_1 f_N + a g_N &= 0. \end{aligned}\right\} \tag{7.9.20}$$

Also, from equation (7.9.14) after putting $p = N-1$, we get

$$[a_1(\rho+N-k) - aZ\alpha]f_N = [a(\rho+N+k) + a_1 Z\alpha]g_N. \tag{7.9.21}$$

Since neither of $f_N$, $g_N$, vanishes, this last equation can only be made consistent with equations (7.9.20) by assuming that

$$a_1[a(\rho+N+k) + a_1 Z\alpha] + a[a_1(\rho+N-k) - aZ\alpha] = 0, \tag{7.9.22}$$

i.e. that

$$2a(\rho+N) = (a_2 - a_1)Z\alpha. \tag{7.9.23}$$

Substituting for $a_1$, $a_2$ from equations (7.8.29) and for $\rho$ from equations (7.9.13), this yields an eigenvalue for $E$ in the form

$$E = mc^2\left[1 + \left(\frac{Z\alpha}{N + \sqrt{(k^2 - Z^2\alpha^2)}}\right)^2\right]^{-\frac{1}{2}}. \tag{7.9.24}$$

It will be observed that the energy eigenvalues are dependent upon two quantum numbers $N$ and $k$. Since $\alpha$ is small, it is permissible to expand the expression (7.9.24) in ascending powers of $\alpha$ thus,

$$\begin{aligned} E &= mc^2\left[1 - \frac{Z^2\alpha^2}{2(N+|k|)^2} + O(\alpha^4)\right], \\ &= mc^2 - \frac{mZ^2e^4}{2\hbar^2}\cdot\frac{1}{(N+|k|)^2} + O(\alpha^4). \end{aligned} \tag{7.9.25}$$

It is now evident that, as a first approximation, the energy levels depend only upon the single quantum number $(N+|k|)$. This, of course, is the approximation already derived as equation (5.4.14), where the integer $N+|k|$ has been denoted by $n$. Thus, we put

$$n = N+|k| = N+j+\tfrac{1}{2} \tag{7.9.26}$$

and then the energy levels are given by the formula

$$E = mc^2\left[1+\left(\frac{Z\alpha}{n-j-\frac{1}{2}+\sqrt{\{(j+\frac{1}{2})^2-Z^2\alpha^2\}}}\right)^2\right]^{-\frac{1}{2}}. \tag{7.9.27}$$

The coarse structure of these levels is determined by the number $n$; only the fine structure is dependent upon the number $j$.

APPENDIX A

# Solution of a tensor equation

It is given that $\chi_{ij}$ is a second rank tensor with respect to all orthogonal transformations

$$x'_i = a_{ij}x_j \tag{A.1}$$

and has the same elements in every frame. (We shall assume the $x_i$-space has $N$ dimensions and shall employ the summation convention.) Thus

$$\chi_{ij} = a_{ir}a_{js}\chi_{rs}. \tag{A.2}$$

Multiplying equation (A.2) by $a_{ik}$ and summing over $i$, we find that

$$\begin{aligned} a_{ik}\chi_{ij} &= a_{ik}a_{ir}a_{js}\chi_{rs}, \\ &= \delta_{kr}a_{js}\chi_{rs}, \\ &= a_{js}\chi_{ks}, \end{aligned} \tag{A.3}$$

since

$$a_{ik}a_{ir} = \delta_{kr} \tag{A.4}$$

if the transformation is orthogonal.

Consider the orthogonal transformation which only changes the sense of the $x_1$-axis, viz.

$$x'_1 = -x_1, \quad x'_2 = x_2, \quad \ldots, \quad x'_N = x_N. \tag{A.5}$$

Then, $a_{11} = -1$, $a_{22} = a_{33} = \ldots = a_{NN} = 1$ and the remaining $a_{ij}$ vanish. Thus, taking $k=1, j \neq 1$, in equation (A.3), we obtain

$$-\chi_{1j} = \chi_{1j}, \tag{A.6}$$

implying that $\chi_{1j}$ vanishes. Similarly, it may be proved that $\chi_{ij}$ vanishes whenever $i \neq j$.

Now consider the orthogonal transformation which exchanges the $x_1$- and $x_2$-axes, viz.

$$x'_1 = x_2, \quad x'_2 = x_1, \quad x'_3 = x_3, \quad \ldots, \quad x'_N = x_N. \tag{A.7}$$

Then, $a_{12} = a_{21} = a_{33} = \ldots = a_{NN} = 1$ and the remaining $a_{ij}$ are zero. Taking $k=1, j=2$, in equation (A.3), we find that

$$\chi_{22} = \chi_{11}. \tag{A.8}$$

Similarly it may be proved that all the elements $\chi_{ij}$ for which $i=j$ are equal.

It follows that

$$\chi_{ij} = \gamma\delta_{ij}, \tag{A.9}$$

where $\gamma$ is an invariant.

APPENDIX B

# Theorem on series

We state the theorem as follows:

*If* (i) $\sum_0^\infty u_n x^n, \sum_0^\infty v_n x^n$ *are power series, convergent for* $|x| < \rho$,

(ii) $u_n > 0, v_n > 0$ *for* $n \geqslant n_0$,

(iii) $\frac{u_n}{u_{n-1}} \geqslant \frac{v_n}{v_{n-1}}$ *for* $n > n_0$,

*then, for* $0 < x < \rho$,

$$\sum_0^\infty u_n x^n \geqslant k \sum_0^\infty v_n x^n + p(x),$$

*where* $k > 0$ *is a constant and* $p(x)$ *is a polynomial of degree less than* $n_0$.

If $n \geqslant n_0$,

$$\begin{aligned} u_n &= \frac{u_n}{u_{n-1}} \cdot \frac{u_{n-1}}{u_{n-2}} \cdots \frac{u_{n_0+1}}{u_{n_0}} u_{n_0}, \\ &\geqslant \frac{v_n}{v_{n-1}} \cdot \frac{v_{n-1}}{v_{n-2}} \cdots \frac{v_{n_0+1}}{v_{n_0}} u_{n_0}, \\ &= kv_n, \end{aligned}$$

where $k = u_{n_0}/v_{n_0} > 0$.

It now follows that, if $0 < x < \rho$, then

$$\sum_{n_0}^\infty u_n x^n \geqslant k \sum_{n_0}^\infty v_n x^n.$$

Taking

$$p(x) = \sum_0^{n_0-1} u_n x^n - k \sum_0^{n_0-1} v_n x^n,$$

the inequality to be established becomes identical with the inequality just proved.

Condition (iii) may be replaced by the alternative condition

$$\text{(iii)} \quad \frac{u_n}{u_{n-1}} \Big/ \frac{v_n}{v_{n-1}} \to L > 1, \textit{ as } n \to \infty,$$

and the theorem still remains true, for this new condition implies that the original condition is valid for a sufficiently large value of $n_0$.

APPENDIX C

# Hermite polynomials

In this appendix we shall study the polynomial solution possessed by equation (3.10.8) when the parameter $\lambda = 2n+1$, is an odd integer. Thus, the equation to be considered may be written

$$\frac{d^2 y}{dx^2} - 2x\frac{dy}{dx} + 2ny = 0. \tag{C.1}$$

Putting

$$y = \exp(x^2)\, u, \tag{C.2}$$

the equation becomes

$$\frac{d^2 u}{dx^2} + 2x\frac{du}{dx} + 2(n+1)\, u = 0. \tag{C.3}$$

Changing the dependent variable again by

$$u = \frac{d^n v}{dx^n} = v^{(n)}, \tag{C.4}$$

the equation takes the form

$$v^{(n+2)} + 2xv^{(n+1)} + 2(n+1)\, v^{(n)} = 0. \tag{C.5}$$

By Leibniz' theorem, this can be written

$$\frac{d^{n+1}}{dx^{n+1}}(v' + 2xv) = 0. \tag{C.6}$$

One solution of this last equation is clearly

$$v' + 2xv = 0, \tag{C.7}$$

from which we deduce that

$$v = A\exp(-x^2), \tag{C.8}$$

where $A$ is an arbitrary constant. It now follows that

$$y = A\exp(x^2)\frac{d^n}{dx^n}[\exp(-x^2)] \tag{C.9}$$

is a solution of equation (C.1). But this solution is evidently a polynomial and it was shown in section 3.10 that the equation (C.1) possesses only one such solution (ignoring the arbitrary constant multiplier); equation (C.9) is therefore this particular solution.

Choosing $A$ so that the coefficient of $x^n$ is $2^n$, equation (C.9) defines the Hermite polynomial of degree $n$. A little consideration reveals that $A=(-1)^n$. Thus

$$H_n(x) = (-1)^n \exp(x^2) \frac{d^n}{dx^n}[\exp(-x^2)]. \tag{C.10}$$

This result leads immediately to a generating function for the polynomials. For taking $f(t)=\exp(-t^2)$ and expanding $f(x-t)$ in a series of powers of $t$ by Taylor's theorem, we obtain

$$\begin{aligned}\exp[-(x-t)^2] &= f(x)-tf'(x)+\frac{t^2}{2!}f''(x)-\ldots+\frac{(-t)^n}{n!}f^{(n)}(x)+\ldots,\\ &= \exp(-x^2)\left\{H_0(x)+tH_1(x)+\ldots+\frac{t^n}{n!}H_n(x)+\ldots\right\}.\end{aligned} \tag{C.11}$$

Thus

$$\exp(-t^2+2xt) = \sum_0^\infty H_n(x)\frac{t^n}{n!}. \tag{C.12}$$

Also, by differentiating equation (C.10), we find

$$\begin{aligned}H'_n &= (-1)^n\left\{\exp(x^2)\frac{d^{n+1}}{dx^{n+1}}[\exp(-x^2)]+2x\exp(x^2)\frac{d^n}{dx^n}[\exp(-x^2)]\right\},\\ &= -H_{n+1}+2xH_n,\end{aligned}$$

i.e.

$$H_{n+1} = 2xH_n-H'_n. \tag{C.13}$$

The polynomials are easily calculated in succession employing this recurrence relationship, commencing with $H_0=1$.

$H_n$ must satisfy equation (C.1) and hence

$$H''_n-2xH'_n+2nH_n = 0. \tag{C.14}$$

Now, by differentiating equation (C.13), we obtain

$$H''_n = 2xH'_n+2H_n-H'_{n+1}. \tag{C.15}$$

Substituting for $H_n''$ in equation (C.14), we arrive at a further recurrence relationship, viz.

$$H_{n+1}' = 2(n+1)\,H_n.$$

This is equivalent to

$$H_n' = 2nH_{n-1}. \tag{C.16}$$

Consider the integral

$$\int_{-\infty}^{\infty} \exp(-x^2)\,H_m(x)\,H_n(x)\,dx = (-1)^n \int_{-\infty}^{\infty} H_m(x)\frac{d^n}{dx^n}[\exp(-x^2)]\,dx. \tag{C.17}$$

Then performing $n$ integrations by parts upon the right-hand integral, the 'integrated-out-part' will, at each step, be zero, since it takes the form of a polynomial multiplying $\exp(-x^2)$ to be calculated at $x = \pm\infty$. The final result we are left with will accordingly be

$$\int_{-\infty}^{\infty} \frac{d^n H_m}{dx^n}\exp(-x^2)\,dx. \tag{C.18}$$

If, now, $m \neq n$, we can suppose, without loss of generality, that $m < n$. In this case, $d^n H_m/dx^n = 0$, since $H_m$ is a polynomial of degree less than $n$. It follows that the integral vanishes. This is expressed by saying that the Hermite polynomials satisfy an *orthogonality condition.* If, however, $m = n$, then

$$\begin{aligned}\frac{d^n H_m}{dx^n} &= \frac{d^n}{dx^n}(2^n x^n + \text{terms of lower degree}),\\ &= 2^n\,n!,\end{aligned} \tag{C.19}$$

and the integral reduces to

$$2^n\,n! \int_{-\infty}^{\infty} \exp(-x^2)\,dx = 2^n\,n!\,\pi^{\frac{1}{2}}. \tag{C.20}$$

We have proved, therefore, that

$$\int_{-\infty}^{\infty} \exp(-x^2)\,H_m\,H_n\,dx = 2^n\,n!\,\pi^{\frac{1}{2}}\,\delta_{mn}. \tag{C.21}$$

It is now easy to normalise $\psi^n$ as given by equation (3.10.33). The normalisation condition is that

$$1 = \int_{-\infty}^{\infty} \psi^n \psi^{n*}\, dx = \sqrt{\left(\frac{\hbar}{m\omega}\right)} \int_{-\infty}^{\infty} A_n^2 \exp(-\xi^2)\, H_n(\xi)\, H_n(\xi)\, d\xi, \tag{C.22}$$

having employed equation (3.10.4). Thus, by equation (C.21),

$$1 = A_n^2 \sqrt{\left(\frac{\hbar}{m\omega}\right)}\, 2^n\, n!\, \pi^{\frac{1}{2}}, \tag{C.23}$$

leading immediately to the result (3.10.34).

APPENDIX D

# Legendre functions

The Legendre polynomial $P_l(x)$ of degree $l$ has been defined in section 5.2 as the polynomial solution of the equation

$$(1-x^2)\frac{d^2y}{dx^2}-2x\frac{dy}{dx}+l(l+1)y = 0 \tag{D.1}$$

for which the coefficient of $x^l$ is $(2l)!/2^l(l!)^2$. Writing $\xi = 1+x$, we deduce from equation (5.2.20) that

$$\begin{aligned}
P_l(x) &= (-1)^l \sum_{r=0}^{l} (-1)^r \frac{(l+r)!}{(r!)^2(l-r)!}\left(\frac{\xi}{2}\right)^r, \\
&= (-1)^l \frac{2^l}{l!} \sum_{r=0}^{l} (-1)^r \frac{l!}{r!(l-r)!}\frac{d^l}{d\xi^l}\left(\frac{\xi}{2}\right)^{r+l}, \\
&= \frac{(-1)^l}{l!}\frac{d^l}{d\xi^l}\left[\xi^l \sum_{r=0}^{l} \frac{l!}{r!(l-r)!}\left(-\frac{\xi}{2}\right)^r\right], \\
&= \frac{(-1)^l}{l!}\frac{d^l}{d\xi^l}[\xi^l(1-\tfrac{1}{2}\xi)^l], \\
&= \frac{1}{2^l l!}\frac{d^l}{d\xi^l}[\xi^l(\xi-2)^l], \\
&= \frac{1}{2^l l!}\frac{d^l}{dx^l}(x^2-1)^l.
\end{aligned} \tag{D.2}$$

This is *Rodrigues' formula.*

Various recurrence relationships are readily obtainable from Rodrigues' formula thus:

$$\begin{aligned}
\frac{dP_l}{dx} &= \frac{1}{2^l l!}\frac{d^{l+1}}{dx^{l+1}}(x^2-1)^l, \\
&= \frac{1}{2^l l!}\frac{d^l}{dx^l}[2lx(x^2-1)^{l-1}],
\end{aligned}$$

$$= \frac{1}{2^{l-1}(l-1)!}\left[x\frac{d^l}{dx^l}(x^2-1)^{l-1}+l\frac{d^{l-1}}{dx^{l-1}}(x^2-1)^{l-1}\right],$$

$$= x\frac{dP_{l-1}}{dx}+lP_{l-1}, \tag{D.3}$$

where Leibniz' formula has been used in the penultimate step. Again

$$P_l = \frac{1}{2^l l!}\frac{d^l}{dx^l}[(x^2-1)(x^2-1)^{l-1}],$$

$$= \frac{1}{2^l l!}\frac{d^l}{dx^l}[x^2(x^2-1)^{l-1}]-\frac{1}{2^l l!}\frac{d^l}{dx^l}(x^2-1)^{l-1},$$

$$= \frac{1}{2l.2^l l!}\frac{d^l}{dx^l}\left[x\frac{d}{dx}(x^2-1)^l\right]-\frac{1}{2l}\frac{dP_{l-1}}{dx},$$

$$= \frac{1}{2l.2^l l!}\left[x\frac{d^{l+1}}{dx^{l+1}}(x^2-1)^l+l\frac{d^l}{dx^l}(x^2-1)^l\right]-\frac{1}{2l}\frac{dP_{l-1}}{dx},$$

$$= \frac{x}{2l}\frac{dP_l}{dx}+\tfrac{1}{2}P_l-\frac{1}{2l}\frac{dP_{l-1}}{dx}.$$

Whence, we obtain

$$\frac{dP_{l-1}}{dx} = x\frac{dP_l}{dx}-lP_l. \tag{D.4}$$

Eliminating $dP_{l-1}/dx$ between (D.3) and (D.4), we now find that

$$(x^2-1)\frac{dP_l}{dx} = lxP_l-lP_{l-1}. \tag{D.5}$$

Eliminating $dP_l/dx$ between the same equations, we also obtain

$$(x^2-1)\frac{dP_{l-1}}{dx} = -lxP_{l-1}+lP_l. \tag{D.6}$$

Replacing $l$ by $l+1$ in equation (D.3) and subtracting equation (D.4), we get

$$\frac{dP_{l+1}}{dx}-\frac{dP_{l-1}}{dx} = (2l+1)P_l. \tag{D.7}$$

Replacing $l$ by $l+1$ in equation (D.4) and subtracting it from equation (D.3), we find

$$0 = x\left(\frac{dP_{l-1}}{dx}-\frac{dP_{l+1}}{dx}\right)+lP_{l-1}+(l+1)P_{l+1}.$$

Using equation (D.7), this reduces to

$$(l+1)P_{l+1}-(2l+1)xP_l+lP_{l-1} = 0. \tag{D.8}$$

The associated Legendre functions of the $k$th order $P_l^k(x)$ are defined by the equation

$$P_l^k(x) = (1-x^2)^{\frac{1}{2}k}\frac{d^kP_l}{dx^k}, \tag{D.9}$$

where we assume $k$ to be a positive integer. Clearly, unless $l \geqslant k$, $P_l^k$ vanishes. (For negative $k$, we take $P_l^k = P_l^{-k}$.)

Differentiating equation (D.3) $k$ times, we obtain

$$\frac{d^{k+1}P_l}{dx^{k+1}} = x\frac{d^{k+1}P_{l-1}}{dx^{k+1}}+(k+l)\frac{d^kP_{l-1}}{dx^k}.$$

After multiplying through by $(1-x^2)^{\frac{1}{2}(k+1)}$, this equation can be written

$$P_l^{k+1} = xP_{l-1}^{k+1}+(l+k)(1-x^2)^{\frac{1}{2}}P_{l-1}^k. \tag{D.10}$$

Differentiation of equation (D.7) $k$ times and multiplication through by $(1-x^2)^{\frac{1}{2}(k+1)}$ yields the relationship

$$P_{l+1}^{k+1}-P_{l-1}^{k+1} = (2l+1)(1-x^2)^{\frac{1}{2}}P_l^k. \tag{D.11}$$

Replacing $l$ by $l+1$ in the identity (D.10) and then eliminating $P_l^k$ between this equation and equation (D.11), we obtain (after replacing $k+1$ by $k$) the identity

$$(l+k)P_{l-1}^k+(l-k+1)P_{l+1}^k = (2l+1)xP_l^k. \tag{D.12}$$

Differentiating equation (D.6) $k$ times and multiplying through by $(1-x^2)^{\frac{1}{2}k}$, it will be found that the result can be written in the form

$$(1-x^2)^{\frac{1}{2}}[P_{l-1}^{k+1}-k(l+k-1)P_{l-1}^{k-1}] = (2k+l)xP_{l-1}^k-lP_l^k. \tag{D.13}$$

Replacing $k$ by $k-1$ in equation (D.10) and then substituting for $P_{l-1}^{k-1}$ from this equation into equation (D.13), this last equation reduces to

$$(1-x^2)^{\frac{1}{2}}P_{l-1}^{k+1} = (l+k)xP_{l-1}^k-(l-k)P_l^k. \tag{D.14}$$

Now

$$\begin{aligned}(1-x^2)\frac{dP_{l-1}^k}{dx} &= (1-x^2)\frac{d}{dx}\left[(1-x^2)^{\frac{1}{2}k}\frac{d^kP_{l-1}}{dx^k}\right],\\ &= (1-x^2)^{\frac{1}{2}k+1}\frac{d^{k+1}P_{l-1}}{dx^{k+1}}-kx(1-x^2)^{\frac{1}{2}k}\frac{d^kP_{l-1}}{dx^k},\\ &= (1-x^2)^{\frac{1}{2}}P_{l-1}^{k+1}-kxP_{l-1}^k. \end{aligned}\tag{D.15}$$

Substituting for $(1-x^2)^{\frac{1}{2}} P_{l-1}^{k+1}$ in equation (D.14) from equation (D.15), we find that

$$(1-x^2)\frac{dP_{l-1}^k}{dx} = lxP_{l-1}^k-(l-k)P_l^k. \tag{D.16}$$

Replacing $l$ by $l+1$ in the identity (D.14) and then eliminating $P_{l+1}^k$ between this equation and equation (D.12), a new identity

$$(1-x^2)^{\frac{1}{2}} P_l^{k+1} = (l+k)P_{l-1}^k-(l-k)xP_l^k \tag{D.17}$$

results. Eliminating $P_{l-1}^k$ between this identity and identity (D.14), we also obtain

$$xP_l^{k+1} = P_{l-1}^{k+1}+(l-k)(1-x^2)^{\frac{1}{2}} P_l^k. \tag{D.18}$$

Consider next the integral

$$\int_{-1}^{1} P_l^k P_m^k\,dx = \int_{-1}^{1} (1-x^2)^k \frac{d^k P_l}{dx^k}\frac{d^k P_m}{dx^k}\,dx,$$

$$= \frac{1}{2^{l+m}\,l!\,m!}\int_{-1}^{1} (1-x^2)^k \frac{d^{k+l}}{dx^{k+l}}(x^2-1)^l \frac{d^{k+m}}{dx^{k+m}}(x^2-1)^m\,dx. \tag{D.19}$$

We shall assume, without loss of generality, that $l \geqslant m$. Integrating by parts $(k+l)$ times, the last expression is transformed into

$$\frac{(-1)^{k+l}}{2^{l+m}\,l!\,m!}\int_{-1}^{1} (x^2-1)^l \frac{d^{k+l}}{dx^{k+l}}\left[(1-x^2)^k \frac{d^{k+m}}{dx^{k+m}}(x^2-1)^m\right]dx, \tag{D.20}$$

the 'integrated-out-part' vanishing on each occasion since, if $Q(x)$ is any polynomial,

$$\frac{d^n}{dx^n}[(1-x^2)^p\,Q(x)] = 0 \tag{D.21}$$

at $x = \pm 1$, provided $p > n$.

Now, by consideration of the term of highest degree in $x$, we see that

$$\left.\begin{aligned} \frac{d^{k+l}}{dx^{k+l}}\left[(1-x^2)^k \frac{d^{k+m}}{dx^{k+m}}(x^2-1)^m\right] &= 0, \quad \text{if } l > m, \\ &= (-1)^k \frac{(2l)!\,(l+k)!}{(l-k)!}, \quad \text{if } l = m. \end{aligned}\right\} \tag{D.22}$$

We deduce immediately that the integral (D.19) vanishes unless $l=m$.

In the case $l=m$, we find that

$$\int_{-1}^{1} (P_l^k)^2\,dx = \frac{(2l)!\,(l+k)!}{2^{2l}(l!)^2\,(l-k)!}\int_{-1}^{1}(1-x^2)^l\,dx. \tag{D.23}$$

But

$$\int_{-1}^{1}(1-x^2)^l\,dx = 2\int_{0}^{\frac{1}{2}\pi}\cos^{2l+1}\theta\,d\theta = \frac{2^{2l+1}(l!)^2}{(2l+1)!}. \tag{D.24}$$

Hence

$$\int_{-1}^{1} P_l^k P_m^k\,dx = \frac{2}{2l+1}\cdot\frac{(l+k)!}{(l-k)!}\,\delta_{lm}. \tag{D.25}$$

In the particular case $k=0$, we have that

$$\int_{-1}^{1} P_l P_m\,dx = \frac{2}{2l+1}\,\delta_{lm}. \tag{D.26}$$

Another integral result which is useful may be established thus: Employing the identity (D.12), it is clear that

$$\left.\begin{aligned}\int_{-1}^{1} xP_l^k P_{l'}^k\,dx &= \frac{1}{2l+1}\int_{-1}^{1}[(l+k)P_{l-1}^k+(l-k+1)P_{l+1}^k]P_{l'}^k\,dx,\\ &= \frac{2(l+k)!}{(4l^2-1)(l-k-1)!},\quad \text{if } l=l'+1,\\ &= \frac{2(l'+k)!}{(4l'^2-1)(l'-k-1)!},\quad \text{if } l'=l+1,\\ &= 0,\quad \text{otherwise.}\end{aligned}\right\} \tag{D.27}$$

APPENDIX E

# Laguerre polynomials

The equation

$$xy'' + (k+1-x)\,y' + ny = 0, \tag{E.1}$$

where $n$, $k$ are positive integers or zero, was encountered in section 5.4 and it was shown that the equation possesses a polynomial solution which is unique except for an arbitrary constant multiplying factor. This solution is of degree $n$. If the multiplying factor is chosen so that the coefficient of the term in $x^n$ is $(-1)^{n+k}(n+k)!/n!$, then the polynomial is denoted by $L_n^k(x)$ and is called the generalised Laguerre polynomial of degree $n$ and order $k$.

The ordinary Laguerre polynomials are obtained by taking $k=0$, that of degree $n$ being denoted by $L_n(x)$. $L_{n+k}$ satisfies the equation

$$xy'' + (1-x)\,y' + (n+k)\,y = 0. \tag{E.2}$$

Differentiating this last equation $k$ times, we obtain

$$xy^{(k+2)} + (k+1-x)\,y^{(k+1)} + ny^{(k)} = 0. \tag{E.3}$$

Comparison of equations (E.1) and (E.3) indicates that if $L_{n+k}$ is the polynomial solution of equation (E.2), then $L_{n+k}^{(k)}$ is a polynomial solution of equation (E.1). Thus

$$L_n^k(x) = CL_{n+k}^{(k)}(x), \tag{E.4}$$

where $C$ is a constant. Since the coefficient of $x^{n+k}$ in $L_{n+k}$ is $(-1)^{n+k}$, the coefficient of $x^n$ in $L_{n+k}^{(k)}$ is $(-1)^{n+k}(n+k)!/n!$. Thus $C=1$ and

$$L_n^k(x) = \frac{d^k}{dx^k}L_{n+k}(x). \tag{E.5}$$

We now introduce a new dependent variable $z$ into the equation (E.1) by putting

$$y = \exp(x)\,x^{-k}z. \tag{E.6}$$

It will be found that

$$xz'' + (x-k+1)\,z' + (n+1)\,z = 0. \tag{E.7}$$

Then, setting

$$z = \frac{d^n w}{dx^n}, \tag{E.8}$$

equation (E.7) becomes

$$xw^{(n+2)} + (x-k+1)\,w^{(n+1)} + (n+1)\,w^{(n)} = 0, \tag{E.9}$$

which is equivalent to

$$\frac{d^{n+1}}{dx^{n+1}}\{xw' + (x-n-k)\,w\} = 0. \tag{E.10}$$

But the equation

$$xw' + (x-n-k)\,w = 0, \tag{E.11}$$

has the solution

$$w = A\exp(-x)\,x^{n+k}. \tag{E.12}$$

It follows that

$$y = A\exp(x)\,x^{-k}\frac{d^n}{dx^n}[\exp(-x)\,x^{n+k}] \tag{E.13}$$

is a solution of equation (E.1) and, by application of Leibniz' theorem, it is clear that this solution is a polynomial. The coefficient of $x^n$ in this polynomial is $(-1)^n A$; thus, taking $A = (-1)^k(n+k)!/n!$, the polynomial must become identical with $L_n^k$. We have proved, therefore, that

$$L_n^k(x) = (-1)^k\frac{(n+k)!}{n!}\exp(x)\,x^{-k}\frac{d^n}{dx^n}[\exp(-x)\,x^{n+k}]. \tag{E.14}$$

In particular, taking $k=0$, we have

$$L_n(x) = \exp(x)\frac{d^n}{dx^n}[\exp(-x)\,x^n]. \tag{E.15}$$

We will now derive some properties of the ordinary Laguerre polynomials. Firstly, applying Leibniz' theorem to the formula (E.15), we obtain explicitly

$$L_n(x) = n!\sum_{r=0}^{n}\binom{n}{r}\frac{(-x)^r}{r!}, \tag{E.16}$$

where $\binom{n}{r}$ is a binomial coefficient. The first five polynomials are

$$\left.\begin{aligned} &L_0 = 1,\quad L_1 = 1-x,\quad L_2 = 2-4x+x^2,\\ &L_3 = 6-18x+9x^2-x^3,\quad L_4 = 24-96x+72x^2-16x^3+x^4.\end{aligned}\right\} \tag{E.17}$$

Generalised polynomials can now be computed from these by differentiation according to equation (E.5).

A recurrence relationship between the polynomials $L_n$ can also be derived from the formula (E.15). Thus, writing this formula in the form

$$L_n = \exp(x)\frac{d^n}{dx^n}(\exp(-x)\,x^{n-1}.x) \tag{E.18}$$

and applying Leibniz' theorem, we deduce that

$$\begin{aligned} L_n &= \exp(x)\left\{x\frac{d^n}{dx^n}[\exp(-x)\,x^{n-1}]+n\frac{d^{n-1}}{dx^{n-1}}[\exp(-x)\,x^{n-1}]\right\}, \\ &= x\exp(x)\frac{d}{dx}[\exp(-x)\,L_{n-1}]+nL_{n-1}, \\ &= (n-x)\,L_{n-1}+xL'_{n-1}. \end{aligned} \tag{E.19}$$

Since $L_{n-1}$ satisfies equation (E.1) with $k=0$ and $n$ replaced by $(n-1)$,

$$xL''_{n-1}+(1-x)\,L'_{n-1}+(n-1)\,L_{n-1} = 0. \tag{E.20}$$

Differentiating equation (E.19), we obtain

$$L'_n = xL''_{n-1}+(n+1-x)\,L'_{n-1}-L_{n-1}. \tag{E.21}$$

Elimination of $L''_{n-1}$ between equations (E.20), (E.21), now yields the result

$$L'_n = n(L'_{n-1}-L_{n-1}). \tag{E.22}$$

From equations (E.19), (E.22), it now follows that

$$xL'_n = nL_n-n^2\,L_{n-1}. \tag{E.23}$$

Also, writing equation (E.23) with $n$ replaced by $(n-1)$ and then substituting from this equation and (E.23) for $L'_{n-1}$, $L'_n$ respectively into equation (E.22), the following recurrence relationship is found:

$$L_{n+1}-(2n+1-x)\,L_n+n^2\,L_{n-1} = 0. \tag{E.24}$$

Various recurrence formulae involving the generalised polynomials can now be derived by differentiating the formulae found above and employing equation (E.5).

The generalised polynomials satisfy an orthogonality condition. For, consider the integral

$$\int_0^\infty \exp(-x)\,x^k\,L_n^k\,L_m^k\,dx, \tag{E.25}$$

where, without loss of generality, we can assume $n \geqslant m$. Substituting for $L_n^k$ from equation (E.14), the integral takes the form

$$(-1)^k \frac{(n+k)!}{n!} \int_0^\infty \frac{d^n}{dx^n} [\exp(-x)\, x^{n+k}]\, L_m^k\, dx. \tag{E.26}$$

Performing $n$ integrations by parts and noting that the 'integrated-out-part' invariably possesses a factor $x \exp(-x)$ and so vanishes at both limits, we arrive at the result

$$(-1)^{n+k} \frac{(n+k)!}{n!} \int_0^\infty \exp(-x)\, x^{n+k} \frac{d^n}{dx^n} L_m^k\, dx. \tag{E.27}$$

If $n > m$, the $n$th derivative of the $m$th degree polynomial $L_m^k$ is identically zero and the integral vanishes therefore. If $n = m$, however,

$$\frac{d^n}{dx^n} L_m^k = (-1)^{n+k} (n+k)! \tag{E.28}$$

and, since

$$\int_0^\infty \exp(-x)\, x^{n+k}\, dx = (n+k)!, \tag{E.29}$$

it follows that

$$\int_0^\infty \exp(-x)\, x^k\, L_n^k\, L_m^k\, dx = \frac{\{(n+k)!\}^3}{n!}\, \delta_{mn}. \tag{E.30}$$

However, a number of other integrals of a similar type have arisen in the calculations given in the main text and these may be calculated by the above method. Thus

$$\begin{aligned} \int_0^\infty \exp(-x)\, x^{k+1} \{L_n^k\}^2\, dx &= (-1)^k \frac{(n+k)!}{n!} \int_0^\infty \frac{d^n}{dx^n} [\exp(-x)\, x^{n+k}]\, x L_n^k\, dx, \\ &= (-1)^{n+k} \frac{(n+k)!}{n!} \\ &\quad \times \int_0^\infty \exp(-x)\, x^{n+k} \frac{d^n}{dx^n} (x L_n^k)\, dx, \end{aligned} \tag{E.31}$$

after $n$ integrations by parts as before. From the formula (E.14), by use of Leibniz' theorem we find that

$$L_n^k = (-1)^{n+k}\frac{(n+k)!}{n!}$$
$$\times[x^n - n(n+k)x^{n-1} + \tfrac{1}{2}n(n-1)(n+k)(n+k-1)x^{n-2} + \ldots]. \qquad \text{(E.32)}$$

Hence,

$$\frac{d^n}{dx^n}(xL_n^k) = (-1)^{n+k}(n+k)!\,[(n+1)x - n(n+k)]. \qquad \text{(E.33)}$$

Thus, employing equation (E.29), equation (E.31) yields

$$\int_0^\infty \exp(-x)\,x^{k+1}\{L_n^k\}^2\,dx = \frac{(2n+k+1)\{(n+k)!\}^3}{n!}. \qquad \text{(E.34)}$$

Next, supposing $k \geqslant 1$, we have

$$\int_0^\infty \exp(-x)\,x^{k-1}\{L_n^k\}^2\,dx = (-1)^k\frac{(n+k)!}{n!}$$
$$\times \int_0^\infty x^{-1}L_n^k\frac{d^n}{dx^n}[\exp(-x)\,x^{n+k}]\,dx,$$
$$= (-1)^{n+k}\frac{(n+k)!}{n!}$$
$$\times \int_0^\infty \exp(-x)\,x^{n+k}\frac{d^n}{dx^n}(x-1\,L_n^k)\,dx. \qquad \text{(E.35)}$$

From equation (E.14), it follows that

$$L_n^k = (-1)^k\frac{\{(n+k)!\}^2}{n!\,k!}\left[1 - \frac{n}{k+1}x + \ldots\right] \qquad \text{(E.36)}$$

and hence

$$\frac{d^n}{dx^n}(x^{-1}\,L_n^k) = (-1)^{n+k}\frac{\{(n+k)!\}^2}{k!}\cdot\frac{1}{x^{n+1}}. \qquad \text{(E.37)}$$

Thus,

$$\int_0^\infty \exp(-x)\,x^{k-1}\{L_n^k\}^2\,dx = \frac{\{(n+k)!\}^3}{n!\,k}. \qquad \text{(E.38)}$$

Similarly, if $k \geqslant 2$, we prove that

$$\int_0^\infty \exp(-x)\, x^{k-2}\{L_n^k\}^2\, dx = \frac{\{(n+k)!\}^3\,(2n+k+1)}{n!\, k(k^2-1)}. \qquad \text{(E.39)}$$

Finally, we proceed to calculate, for $k \geqslant 2$,

$$\int_0^\infty \exp(-x)\, x^{k+1}\, L_n^k\, L_{n+1}^{k-2}\, dx$$

$$= (-1)^k \frac{(n+k-1)!}{(n+1)!} \int_0^\infty x^3\, L_n^k \frac{d^{n+1}}{dx^{n+1}}[\exp(-x)\, x^{n+k-1}]\, dx,$$

$$= (-1)^{n+k+1} \frac{(n+k-1)!}{(n+1)!} \int_0^\infty \exp(-x)\, x^{n+k-1} \frac{d^{n+1}}{dx^{n+1}}(x^3\, L_n^k)\, dx. \qquad \text{(E.40)}$$

Substituting for $L_n^k$ from equations (E.32), we now find that

$$\int_0^\infty \exp(-x)\, x^{k+1}\, L_n^k\, L_{n+1}^{k-2}\, dx = -\frac{3(2n+k+1)\{(n+k)!\}^3}{n!(n+k)}. \qquad \text{(E.41)}$$

APPENDIX F

# Motion of a charge in an electromagnetic field

Consider a charge $e$ having mass $m$ moving in a field specified by a vector potential $\mathfrak{A}$ and a scalar potential $\Phi$. We will verify that the classical Hamiltonian is

$$H = \frac{1}{2m}\left(\mathfrak{p}-\frac{e}{c}\mathfrak{A}\right)^2+e\Phi. \tag{F.1}$$

With this Hamiltonian, Hamilton's equations are

$$\dot{p}_x = -\frac{\partial H}{\partial x} = \frac{e}{mc}\left(\mathfrak{p}-\frac{e}{c}\mathfrak{A}\right).\frac{\partial \mathfrak{A}}{\partial x}-e\frac{\partial \Phi}{\partial x}, \tag{F.2}$$

$$\dot{x} = \frac{\partial H}{\partial p_x} = \frac{1}{m}\left(p_x-\frac{e}{c}A_x\right), \tag{F.3}$$

with similar equations for the $y$- and $z$-components. From equation (F.3) and two similar equations, we deduce that

$$\mathfrak{p} = m\mathfrak{v}+\frac{e}{c}\mathfrak{A}, \tag{F.4}$$

where $\mathfrak{v}$ is the velocity of the charge. Eliminating $\mathfrak{p}$ from equation (F.2) and two similar equations, we now find that

$$m\dot{\mathfrak{v}}+\frac{e}{c}\dot{\mathfrak{A}} = \frac{e}{c}\nabla(\mathfrak{v}\,.\,\mathfrak{A})-e\nabla\Phi. \tag{F.5}$$

But

$$\dot{\mathfrak{A}} = \frac{\partial \mathfrak{A}}{\partial t}+\mathfrak{v}\,.\,\nabla\mathfrak{A} \tag{F.6}$$

and

$$\nabla(\mathfrak{v}\,.\,\mathfrak{A}) = \mathfrak{v}\,.\,\nabla\mathfrak{A}+\mathfrak{v}\times\operatorname{curl}\mathfrak{A}. \tag{F.7}$$

Hence

$$m\dot{\mathfrak{v}} = \frac{e}{c}\mathfrak{v}\times\operatorname{curl}\mathfrak{A}-\frac{e}{c}\frac{\partial\mathfrak{A}}{\partial t}-e\nabla\Phi,$$

$$= \frac{e}{c}\mathfrak{v}\times\mathfrak{B}+e\mathfrak{E}, \tag{F.8}$$

where we have employed equations (5.6.2), (5.6.3). But, the right-hand member of equation (F.8) is the Lorentz force acting upon the charge and it follows that this is the correct equation of motion.

We will next verify that the Hamiltonian which yields the correct relativistic equation of motion is

$$H = c\left[\left(\mathfrak{p}-\frac{e}{c}\mathfrak{A}\right)^2+m^2c^2\right]^{\frac{1}{2}}+e\Phi. \tag{F.9}$$

Hamilton's equations are

$$\dot{p}_x = \frac{e}{S}\left(\mathfrak{p}-\frac{e}{c}\mathfrak{A}\right).\frac{\partial\mathfrak{A}}{\partial x}-e\frac{\partial\Phi}{\partial x}, \tag{F.10}$$

$$\dot{x} = \frac{c}{S}\left(p_x-\frac{e}{c}A_x\right), \tag{F.11}$$

etc., where

$$S = \left[\left(\mathfrak{p}-\frac{e}{c}\mathfrak{A}\right)^2+m^2c^2\right]^{\frac{1}{2}}. \tag{F.12}$$

From equation (F.11), we deduce that

$$S\mathfrak{v} = c\left(\mathfrak{p}-\frac{e}{c}\mathfrak{A}\right), \tag{F.13}$$

from which it follows, after squaring, that

$$S^2 = m^2c^2\Big/\left(1-\frac{v^2}{c^2}\right). \tag{F.14}$$

Eliminating $\mathfrak{p}$ between equations (F.10) and (F.13), we calculate, as in the classical case, that

$$\frac{d}{dt}\left(\frac{S}{c}\mathfrak{v}\right) = \frac{e}{c}\mathfrak{v}\times\mathfrak{B}+e\mathfrak{E}. \tag{F.15}$$

Using equation (F.14), this takes the form of the relativistic equation of motion, viz.

$$\frac{d}{dt}\left\{\frac{m\mathfrak{v}}{\sqrt{(1-v^2/c^2)}}\right\} = \frac{e}{c}\mathfrak{v}\times\mathfrak{B}+e\mathfrak{E}. \tag{F.16}$$

APPENDIX G

# Schlaff's method

The problem is to expand the determinant occurring in equation (6.5.15).

Ignoring for the moment the dependence of the elements $b_i$ upon the quantum numbers $n$, $k$, we shall denote this $(n-k)$-row determinant by $D_{n-k}$. Expanding by elements in the last row, we calculate that

$$D_{r+1} = \alpha D_r - b_r^2 D_{r-1}, \quad r = 1, 2, \ldots, \tag{G.1}$$

where $D_0 = 0$, $D_1 = \alpha$. Then, all the determinants $D_r$ can be found in terms of $\alpha$ and the $b_i$ by using this recurrence relationship. If, now, the elements $b_i$ in the lower diagonal are replaced by $\beta_i b_i$ and those in the upper diagonal by $b_i/\beta_i$, the recurrence relationship is unchanged. We conclude that such a modification of $D_{n-k}$ has no effect upon its value. Taking

$$\beta_i = \sqrt{\left(\frac{i(n+k+i)(2k+2i-1)}{(2k+i)(n-k-i)(2k+2i+1)}\right)}, \tag{G.2}$$

the determinant is thus altered to the form

$$\begin{vmatrix} \alpha & \frac{(2k+1)(n-k-1)}{2k+1} & 0 & \ldots & 0 & 0 \\ \frac{1.(n+k+1)}{2k+3} & \alpha & \frac{(2k+2)(n-k-2)}{2k+3} & \ldots & 0 & 0 \\ 0 & \frac{2.(n+k+2)}{2k+5} & \alpha & \ldots & 0 & 0 \\ \ldots & \ldots & \ldots & \ldots & \ldots & \ldots \\ \ldots & \ldots & \ldots & \ldots & \alpha & \frac{(k+n-1).1}{2n-3} \\ \ldots & \ldots & \ldots & \ldots & \frac{(n-k-1)(2n-1)}{2n-1} & \alpha \end{vmatrix} \tag{G.3}$$

The sum of the elements in the $r$th row on either side of the principal diagonal is

$$\frac{(r-1)(n+k+r-1)}{2k+2r-1} + \frac{(2k+r)(n-k-r)}{2k+2r-1} = n-k-1. \tag{G.4}$$

It follows that, if to each column of the determinant we add every alternate column on its right, the elements in the principal diagonal and above will be unaffected, whereas the elements below the principal diagonal will become $\alpha$ and $n-k-1$ alternately; thus:

$$\begin{vmatrix} \alpha & \dfrac{(2k+1)(n-k-1)}{2k+1} & 0 & \cdots & 0 & 0 \\ n-k-1 & \alpha & \dfrac{(2k+2)(n-k-2)}{2k+3} & \cdots & 0 & 0 \\ \alpha & n-k-1 & \alpha & \cdots & 0 & 0 \\ \cdots & \cdots & \cdots & \cdots & \cdots & \cdots \\ \cdots & \cdots & \cdots & \cdots & \alpha & \dfrac{(k+n-1).1}{2n-3} \\ \cdots & \cdots & \cdots & \cdots & n-k-1 & \alpha \end{vmatrix} \tag{G.5}$$

Working from the bottom row, we now subtract from each row, the row next but one above. Employing equation (G.4) again, this will be found to yield

$$\begin{vmatrix} \alpha & \dfrac{(2k+1)(n-k-1)}{2k+1} & 0 & \cdots & 0 & 0 \\ n-k-1 & \alpha & \dfrac{(2k+2)(n-k-2)}{2k+3} & \cdots & 0 & 0 \\ 0 & 0 & \alpha & \cdots & 0 & 0 \\ \cdots & \cdots & \cdots & \cdots & \cdots & \cdots \\ \cdots & \cdots & \cdots & \cdots & \alpha & \dfrac{(k+n-1).1}{2n-3} \\ \cdots & \cdots & \cdots & \cdots & \dfrac{(n-k-3)(2n-3)}{2n-5} & \alpha \end{vmatrix} \tag{G.6}$$

I.e., the final effect is that the diagonal below the principal diagonal in the determinant (G.3) has been shifted downwards by two steps along its length and the two gaps so formed have been filled by elements $n-k-1$ and 0.

The determinant (G.6) can now be expanded by the minors in its first two columns to yield

$$[\alpha^2-(n-k-1)^2]\begin{vmatrix} \alpha & \dfrac{(2k+3)(n-k-3)}{2k+5} & 0 & \cdots \\ \dfrac{1.(n+k+1)}{2k+3} & \alpha & \dfrac{(2k+4)(n-k-4)}{2k+7} & \cdots \\ 0 & \dfrac{2(n+k+2)}{2k+5} & \alpha & \cdots \\ \cdots & \cdots & \cdots & \cdots \end{vmatrix} \qquad \text{(G.7)}$$

The new determinant so obtained is identical with the original determinant (G.3), after $n$ has been replaced by $n-1$ and $k$ has been replaced by $k+1$, thus reducing the number of rows and columns by 2. Hence this determinant can now be factorised in the same way and will yield a factor $[\alpha^2-(n-k-3)^2]$. Proceeding in this manner, we shall ultimately factorise the determinant (G.3) into the factors

$$[\alpha^2-(n-k-1)^2][\alpha^2-(n-k-3)^2][\alpha^2-(n-k-5)^2]\ldots, \qquad \text{(G.8)}$$

where the final factor is $\alpha^2-1$ or $\alpha$ according as $n-k$ is even or odd respectively.

# Exercises

## Chapter 1

1. If $\mathbf{a}$, $\mathbf{b}$ are Hermitian matrices, prove that $\iota[\mathbf{a},\mathbf{b}]$ is Hermitian.

2. For a certain system, the observables $a$, $b$ can be represented by $2 \times 2$ Hermitian matrices. If a simultaneous eigenstate exists, prove that the matrices commute. (Hint: employ a representation in which the eigenstate is one of the two basic eigenstates.)

3. Employing a certain representation, the observables $\sigma_x$, $\sigma_y$, $\sigma_z$ have matrices

$$\begin{pmatrix} 0 & 1 \\ 1 & 0 \end{pmatrix}, \quad \begin{pmatrix} 0 & -\iota \\ \iota & 0 \end{pmatrix}, \quad \begin{pmatrix} 1 & 0 \\ 0 & -1 \end{pmatrix},$$

respectively. Prove that each observable has eigenvalues $\pm 1$ and calculate the corresponding normalised eigenvectors.

4. Prove that any $2 \times 2$ unitary matrix can be expressed in the form

$$\begin{pmatrix} \exp(\iota\alpha)\cos\theta & \exp(\iota\beta)\sin\theta \\ -\exp(\iota\gamma)\sin\theta & \exp(\iota\delta)\cos\theta \end{pmatrix},$$

where $\alpha+\delta=\beta+\gamma+n\pi$. Subject the matrices of Ex. 3 (Chap. 1) to this transformation (see equation (1.8.16)) and verify that the resulting matrices still have eigenvalues $\pm 1$.

5. Calculate the matrices representing the observables $\sigma_x^2$, $\sigma_y^2$, $\sigma_z^2$, $\sigma_y\sigma_z$, $\sigma_z\sigma_x$, $\sigma_x\sigma_y$, when the matrices representing $\sigma_x$, $\sigma_y$, $\sigma_z$ are as given in Ex. 3 (Chap. 1).

6. $a_{ij}$ are the elements of a Hermitian matrix and

$$a_{ij}\,\alpha_i^*\,\alpha_j = 0$$

for arbitrary vectors $\alpha$. Prove that the $a_{ij}$ all vanish. (Hint: take $\alpha=(1,0,0,\ldots,0)$, $\alpha=(1,1,0,\ldots,0)$, $\alpha=(1,\iota,0,\ldots,0)$.)

If

$$a_{ij}\,\alpha_i^*\,\alpha_j = a,$$

where $a$ is a constant, for arbitrary *normalised* vectors $\alpha$, prove that $a_{ij}=a\delta_{ij}$.

7. If a linear operator $\hat{a}$ is such that

$$\langle \hat{a}\alpha | \alpha \rangle = \langle \alpha | \hat{a}\alpha \rangle$$

for arbitrary vectors $\alpha$, by putting $\alpha = \alpha_1 + c\alpha_2$, where $c$ is an arbitrary complex number and $\alpha_1$, $\alpha_2$ are arbitrary vectors, show that

$$c[\langle \hat{a}\alpha_1 | \alpha_2 \rangle - \langle \alpha_1 | \hat{a}\alpha_2 \rangle] = c^*[\langle \hat{a}\alpha_1 | \alpha_2 \rangle - \langle \alpha_1 | \hat{a}\alpha_2 \rangle]^*.$$

Deduce that $\hat{a}$ is Hermitian.

8. Transforming the matrix $\mathbf{a}$ into the matrix $\mathbf{a}'$ by means of equation (1.8.16), verify that $\mathbf{a}'$ is Hermitian if $\mathbf{a}$ is Hermitian.

9. Verify the identities

(a) $[\mathbf{a}, \mathbf{b}] = -[\mathbf{b}, \mathbf{a}]$,

(b) $[\mathbf{a}, \mathbf{b}+\mathbf{c}] = [\mathbf{a}, \mathbf{b}] + [\mathbf{a}, \mathbf{c}]$,

(c) $[\mathbf{a}, \mathbf{bc}] = [\mathbf{a}, \mathbf{b}]\,\mathbf{c} + \mathbf{b}[\mathbf{a}, \mathbf{c}]$,

(d) $[\mathbf{ab}, \mathbf{c}] = [\mathbf{a}, \mathbf{c}]\,\mathbf{b} + \mathbf{a}[\mathbf{b}, \mathbf{c}]$.

10. $a$, $b$, ... is a maximal set of compatible observables. In the $i$th eigenstate, the observables' eigenvalues are $a_i$, $b_i$, .... The eigenstates are ordered so that the first $r$ of the $a_i$ are identical, the next $s$ are identical and so on. Employing the $a, b, \ldots$ representation, the matrix representing an observable $x$ commutes with that representing $a$. Show that the non-zero elements of $\mathbf{x}$ form square blocks whose principal diagonals coincide with the principal diagonal of $\mathbf{x}$.

11. If $f(\hat{a})$ is defined as the operator which is such that

$$f(\hat{a})\,\alpha = f(a)\,\alpha$$

whenever $\hat{a}\alpha = a\alpha$, prove that $f(\hat{a})$ and $\hat{b}$ commute if $\hat{a}$ and $\hat{b}$ commute. (Hint: expand an arbitrary vector $\psi$ in terms of a complete set of eigenvectors of $\hat{a}$ and operate upon it with $\hat{b}f(\hat{a})$.)

12. Employing the complete orthonormal set $\psi^1$, $\psi^2$, ..., $\psi^N$ as a basis, the observable $a$ is represented by the matrix $(a_{ij})$. Prove that

$$\hat{a}\psi^j = a_{ij}\,\psi^i.$$

13. Transforming the matrix $\mathbf{a}$ into the matrix $\mathbf{a}'$ by means of equation (1.8.16), show that both matrices possess the same eigenvalues.

14. Show that the eigenvalues of a unitary matrix have absolute value unity.

15. Show that if $\hat{p}$, $\hat{q}$ are Hermitian operators and $\alpha$, $\beta$ are arbitrary vectors, then

$$\langle\alpha|\hat{p}\hat{q}\beta\rangle = \langle\hat{q}\hat{p}\alpha|\beta\rangle.$$

Deduce that the operators

$$\tfrac{1}{2}(\hat{p}\hat{q}+\hat{q}\hat{p}), \quad \tfrac{1}{2}\iota(\hat{p}\hat{q}-\hat{q}\hat{p})$$

are Hermitian.

16. $\phi$ is a given normalised vector. If $\alpha$ is any vector, the operator $\hat{P}$ is defined by the equation

$$\hat{P}\alpha = \langle\phi|\alpha\rangle\,\phi.$$

Prove that $\hat{P}$ is Hermitian and that $\hat{P}^2=\hat{P}$. ($\hat{P}$ is the operator projecting the vector $\alpha$ on to the direction of $\phi$.)

17. If a unitary transformation is made to a new reference frame, show that the expected value of an observable $a$ as given by equation (1.8.4) is unchanged.

18. In a certain representation, $\boldsymbol{\alpha}$, $\boldsymbol{\alpha}'$, $\boldsymbol{\alpha}''$, ..., $\boldsymbol{\alpha}^{(n)}$ are column matrices representing a complete orthonormal set of vectors. A square matrix $\mathbf{A}$ is defined by the equation

$$\mathbf{A} = \boldsymbol{\alpha\alpha}^{\dagger}.$$

Prove that (i) $\mathbf{A}$ is Hermitian; (ii) tr$\mathbf{A}=1$; (iii) if $x$ is an observable represented by the matrix $\mathbf{x}$, then $\bar{x}=\mathrm{tr}(\mathbf{Ax})=\mathrm{tr}(\mathbf{xA})$ gives its expected value in the state $\boldsymbol{\alpha}$; (iv) $\mathbf{A}^2=\mathbf{A}$ and hence that $\mathbf{A}$ has eigenvalues 0 and $+1$; (v) $\boldsymbol{\alpha}$ is an eigenvector of $\mathbf{A}$ corresponding to the eigenvalue $+1$; (vi) $\boldsymbol{\alpha}'$, $\boldsymbol{\alpha}''$, ..., $\boldsymbol{\alpha}^{(n)}$ are eigenvectors of $\mathbf{A}$ corresponding to the eigenvalue 0; (vii) if $\hat{A}$ is the operator corresponding to $\mathbf{A}$ and $\beta$ is any vector, then $\hat{A}\beta=\langle\alpha|\beta\rangle\,\alpha$; (viii) if $A$ is the observable represented by $\mathbf{A}$, its expected value in the state $\beta$ is given by $\bar{A}=|\langle\alpha|\beta\rangle|^2$; (ix) if $\mathbf{B}=\boldsymbol{\beta\beta}^{\dagger}$, then $|\langle\alpha|\beta\rangle|^2=\mathrm{tr}(\mathbf{AB})$. (Note: tr$\,\mathbf{A}$ = sum of the elements of $\mathbf{A}$ in the principal diagonal.) [$\mathbf{A}$ is termed the *density matrix* for the state $\alpha$. (vii) shows that $\hat{A}$ is a projection operator (cf. Ex. 16, Chap. 1). (iii) and (ix) show that the theory of quantum mechanics can be developed on the basis that states are represented by density matrices.]

## Chapter 2

1. Employing the $\sigma_z$-representation, calculate the eigenvectors of $\sigma_y$ and show that the probability of a transition between the eigenstates $\sigma_x=+1$, $\sigma_y=+1$ is $\frac{1}{2}$.

2. $\sigma_0$ is the component of spin axis in the direction $(\theta, \phi)$ (see equations (2.3.3)) and $\sigma_0'$ is the component in the direction $(\theta', \phi')$. Employing the result (2.3.12), verify that the probability of a transition between the eigenstates $\sigma_0 = +1$, $\sigma_0' = +1$ is $\cos^2 \frac{1}{2}\psi$, where $\psi$ is the angle between the two directions.

3. A particle's spin state is specified by the spinor (2.2.2). Show that, in this state

$$\begin{aligned} \operatorname{var} \sigma_x &= 1 - 4(\mathscr{R} \alpha_+^* \alpha_-)^2, \\ \operatorname{var} \sigma_y &= 1 - 4(\mathscr{I} \alpha_+^* \alpha_-)^2, \\ \bar{\sigma}_z &= \alpha_+ \alpha_+^* - \alpha_- \alpha_-^*. \end{aligned}$$

Hence verify the uncertainty principle (2.3.20) and prove that the equality sign may be taken if, and only if, at least one of the possibilities

$$\alpha_+ = 0, \quad \alpha_- = 0, \quad \arg \alpha_+ - \arg \alpha_- = \tfrac{1}{2} k\pi,$$

where $k$ is a positive or negative integer, is true.

4. The component of the spin of an electron in a direction $\alpha$ takes the sharp value $\frac{1}{2}\hbar$. $\beta$ is a direction making an angle $\theta$ with $\alpha$. Show that the expected value of the component of spin in the direction $\beta$ is $\frac{1}{2}\hbar \cos \theta$ and that its variance is $\frac{1}{2}\hbar^2 \sin^2 \theta$.

5. The spin state of a particle in the $\sigma_z$-representation is given by the spinor (2.2.2). $P$ has spherical polar coordinates $(r, \theta, \phi)$ relative to $Oxyz$. Show that the probabilities that the component of spin in the direction $OP$ takes the values $\frac{1}{2}\hbar$, $-\frac{1}{2}\hbar$ are

$$\begin{aligned} &\alpha_+ \alpha_+^* \cos^2 \tfrac{1}{2}\theta + \alpha_- \alpha_-^* \sin^2 \tfrac{1}{2}\theta + \mathscr{R} \exp(\iota\phi) \sin\theta\, \alpha_+ \alpha_-^*, \\ &\alpha_+ \alpha_+^* \sin^2 \tfrac{1}{2}\theta + \alpha_- \alpha_-^* \cos^2 \tfrac{1}{2}\theta - \mathscr{R} \exp(\iota\phi) \sin\theta\, \alpha_+ \alpha_-^*, \end{aligned}$$

respectively.

6. The *principal direction of the spin axis* of an electron in any spin state is defined to be the direction in which the component of spin axis takes the value $+1$ with certainty. If $Oxyz$ is a rectangular frame and in the $\sigma_z$-representation the electron's state is specified by the spinor $\{\alpha_+, \alpha_-\}$, show that the principal direction is determined by polar angles $(\theta, \phi)$, where

$$\theta = 2 \tan^{-1} \left| \frac{\alpha_-}{\alpha_+} \right|, \quad \phi = \arg \left( \frac{\alpha_-}{\alpha_+} \right).$$

7. $\{\alpha_+, \alpha_-\}$ is a spinor. $Oxyz$ is a set of rectangular axes. Taking $Oxy$ as an Argand plane, the complex number

$$\zeta = \iota\alpha_+/\alpha_-$$

is represented by the point $P$. $P$ is projected on to the surface of the unit sphere, centre $O$, from a centre of projection at the point $(0,0,-1)$. If $Q$ is the projection of $P$, show that its coordinates are given by

$$x = \iota(\alpha_+\alpha_-^* - \alpha_-\alpha_+^*), \quad y = \alpha_+\alpha_-^* + \alpha_-\alpha_+^*,$$
$$z = \alpha_-\alpha_-^* - \alpha_+\alpha_+^*.$$

All spinors $\{\alpha_+, \alpha_-\}$ now undergo the unitary unimodular transformation

$$\alpha_+' = a\alpha_+ + b\alpha_-,$$
$$\alpha_-' = -b^*\alpha_+ + a^*\alpha_-,$$

where $aa^* + bb^* = 1$. If $P_1$, $P_2$ correspond to two such spinors and $Q_1$, $Q_2$ are the points' projections on the sphere, show that the transformation leaves the angle between $OQ_1$, $OQ_2$ invariant (i.e. the transformation leads to a rotation of the unit sphere).

8. A boson has $s_z$-eigenvalues of $+\hbar, 0, -\hbar$. Employing the $s_z$-representation, calculate the matrix representing $s_0$, the component of spin in a direction determined by the polar angles $(\theta, \phi)$ relative to the axes $Oxyz$. Hence show that $s_0$ possesses the same eigenvalues as $s_z$ and calculate the corresponding eigenspinors. Hence show that, if the component of spin in a certain direction $\alpha$ takes the sharp value $+\hbar$, then the probabilities that the component in a direction $\beta$ inclined at an angle $\theta$ to $\alpha$ will be measured to take values $+\hbar$, $0$, $-\hbar$ are $\cos^4\frac{1}{2}\theta$, $\frac{1}{2}\sin^2\theta$, $\sin^4\frac{1}{2}\theta$ respectively. Show also that, if the spin component in the direction $\alpha$ takes the sharp value 0, the corresponding probabilities are $\frac{1}{2}\sin^2\theta$, $\cos^2\theta$, $\frac{1}{2}\sin^2\theta$.

9. Show that the density matrix $\mathbf{S}$ (Chap. 1, Ex. 18) for the spin state $\sigma_0 = +1$, where $\sigma_0$ is the component of spin axis in the direction having cosines $(l, m, n)$, is given by

$$\mathbf{S} = \tfrac{1}{2}(\mathbf{I} + l\boldsymbol{\sigma}_x + m\boldsymbol{\sigma}_y + n\boldsymbol{\sigma}_z).$$

Verify that $\mathbf{S}^2 = \mathbf{S}$ and that $\operatorname{tr}\mathbf{S} = 1$.

## Chapter 3

1. $\hat{l}_+$, $\hat{l}_-$ are operators defined by the equations

$$\hat{l}_+ = x + \frac{\partial}{\partial x}, \quad \hat{l}_- = x - \frac{\partial}{\partial x}.$$

Show that

$$[\hat{l}_+, \hat{l}_-] = 2.$$

2. If $\psi(x_i)$ is any wave function, show that

(a) $[\psi, \hat{p}_i] = \iota\hbar\partial\psi/\partial x_i$,

(b) $[\psi, \hat{p}_i^2] = \hbar^2\left(\frac{\partial^2\psi}{\partial x_i^2} + 2\frac{\partial\psi}{\partial x_i}\frac{\partial}{\partial x_i}\right)$,

(c) $[\psi, \hat{p}^2] = \hbar^2(\nabla^2\psi + 2\nabla\psi.\nabla)$,

where $\hat{p}^2 = \hat{p}_i\hat{p}_i$.

3. Employing the coordinate representation, show that the operators $\hat{x}_i$, $\hat{p}_i$ are Hermitian. (Hint: use equation (3.4.5) and integrate by parts with respect to $x_i$ for the second operator.)

4. Prove that, if $n$ is a positive integer,

$$[\hat{p}_i^n, \hat{x}_i] = \frac{\hbar}{\iota}n\hat{p}_i^{n-1}, \quad [\hat{p}_i, \hat{x}_i^n] = \frac{\hbar}{\iota}n\hat{x}_i^{n-1}.$$

5. If $\hat{q}$ is any linear Hermitian operator and $\hat{r}$ is another operator such that

$$\hat{q}\hat{r} = \hat{r}\hat{q} = 1,$$

$\hat{r}$ is termed the *inverse* of $\hat{q}$ and we write $\hat{r} = \hat{q}^{-1}$. Prove that the inverse operator is linear and Hermitian. Prove, also, that

$$[\hat{p}, \hat{q}^{-1}] = -\hat{q}^{-1}[\hat{p}, \hat{q}]\hat{q}^{-1}.$$

6. A particle moves in a field of force which is such that there is a steady ground state of energy $E_0$ for which the wave function is $\psi_0$. If $H$ is the Hamiltonian, prove that

$$\int \psi^* \hat{H}\psi \, d\tau$$

is minimised with respect to the class of normalised wave functions $\psi$, when $\psi = \psi_0$, and the minimum value is $E_0$.

In the case of the one-dimensional harmonic oscillator, by assuming $\psi_0 = A\exp(-\alpha x^2)$, prove that $\alpha = m\omega/2\hbar$ and $E_0 = \frac{1}{2}\omega\hbar$.

7. A particle of mass $m$ moves parallel to the $x$-axis in a field of force for which $V = \omega^2|x|$. Assuming that, in the steady state $\psi_0$ can be approximated by $A\exp(-\alpha x^2)$, using the method of Ex. 6 (Chap. 3) show that

$$\alpha^3 = \frac{\omega^4 m^2}{2\pi\hbar^4}, \quad E_0 = \frac{3}{2}\left(\frac{\omega^4\hbar^2}{2\pi m}\right)^{1/3},$$

approximately.

8. A particle moves in a field of force which is such that there is a ground state of energy $E_0$ and a first excited state of energy $E_1$. If $\psi_0$, $\psi_1$ are the normalised wave functions corresponding to these states and $H$ is the Hamiltonian, prove that

$$\int \psi^* \hat{H} \psi \, dx$$

is minimised with respect to the class of normalised wave functions orthogonal to $\psi_0$, when $\psi = \psi_1$ and that the minimum value is $E_1$.

In the case of the one-dimensional harmonic oscillator, by assuming $\psi_1 = Ax \exp(-\alpha x^2)$, prove that $\alpha = m\omega/2\hbar$ and $E_1 = \frac{3}{2}\omega\hbar$. (Note that $\psi_1$ is orthogonal to $\psi_0$ for all $\alpha$.)

9. A particle is moving in one dimension under the influence of a potential energy $V = ax^4$. Use a trial function

$$\psi_0 = \alpha \exp(-\gamma x^2),$$

to estimate the ground state energy of the system.

Why is the trial function

$$\psi_1 = \beta x \exp(-\gamma x^2)$$

a suitable function to use to estimate the energy of the first excited state? Use it to estimate the energy of this state.

10. A particle of mass $m$ moves in one dimension in a potential $V(x)$ which vanishes as $x \to \pm\infty$. The particle is in a bound state of energy $E$ with a wave function

$$\psi(x) = C \operatorname{sech} x,$$

where $C$ is a normalisation constant. Calculate $C$, $E$ and $V(x)$.

11. A particle of mass $m$ moves parallel to the $x$-axis between two perfectly reflecting walls at $x = \pm\alpha(a+b)$, where $\alpha = \hbar/(2mV_0)^{\frac{1}{2}}$. The potential energy of the particle between the walls is given by the equations

$$\begin{aligned} V &= 0, & -\alpha(a+b) < x < -\alpha a, \\ &= V_0(>0), & -\alpha a < x < \alpha a, \\ &= 0, & \alpha a < x < \alpha(a+b). \end{aligned}$$

Show that the total energy $\lambda V_0$ of the particle is quantised such that, for $\lambda < 1$, either

$$(1-\lambda)^{-\frac{1}{2}} \tanh(1-\lambda)^{\frac{1}{2}} a = -\lambda^{-\frac{1}{2}} \tan \lambda^{\frac{1}{2}} b,$$

or

$$(1-\lambda)^{-\frac{1}{2}}\coth(1-\lambda)^{\frac{1}{2}}a = -\lambda^{-\frac{1}{2}}\tan\lambda^{\frac{1}{2}}b,$$

and for $\lambda > 1$, either

$$(\lambda-1)^{-\frac{1}{2}}\tan(\lambda-1)^{\frac{1}{2}}a = -\lambda^{-\frac{1}{2}}\tan\lambda^{\frac{1}{2}}b,$$

or

$$(\lambda-1)^{-\frac{1}{2}}\cot(\lambda-1)^{\frac{1}{2}}a = \lambda^{-\frac{1}{2}}\tan\lambda^{\frac{1}{2}}b.$$

Deduce that, if $b \leqslant \frac{1}{2}\pi$, the total energy must exceed $V_0$.

12. The particle in the harmonic oscillator possesses an electric charge and the oscillator is perturbed by a uniform electric field parallel to its line of motion. Show that the perturbed system behaves like a harmonic oscillator with a displaced centre of oscillation and that the $n$th energy eigenvalue is

$$E_n + V_0,$$

where $E_n$ is the $n$th eigenvalue for the unperturbed oscillator and $V_0$ is the P.E. of the particle in the electric field at a point midway between the old and new centres of oscillation.

13. A particle of mass $m$ moves parallel to a uniform force field, which subjects it to a force $F$ in the direction of the negative $x$-axis. Show that the energy spectrum is continuous and that the wave function corresponding to a total energy $E$($x=0$ being the datum for the P.E.) is given by

$$\psi = C\,\mathrm{Ai}\left[\left(\frac{2m}{\hbar^2 F^2}\right)^{1/3}(Fx-E)\right],$$

where $C$ is a constant normalisation factor and $\mathrm{Ai}(\xi)$ is a solution of the equation

$$\frac{d^2y}{d\xi^2} = \xi y$$

which tends to zero as $\xi \to \pm\infty$. [$\mathrm{Ai}(\xi)$ is *Airy's Function* and may be shown to be given by

$$\mathrm{Ai}(\xi) = \frac{1}{\pi}\int_0^\infty \cos(x\xi + \tfrac{1}{3}x^3)\,dx;$$

e.g. see *Methods of Mathematical Physics* by H. and B. S. Jeffreys, C.U.P., p. 508 *et seq.*]

14. Putting

$$\psi = \exp(\iota S/\hbar)$$

in the one-dimensional Schrödinger equation, show that $S$ satisfies

$$\iota\hbar\frac{d^2S}{dx^2} - \left(\frac{dS}{dx}\right)^2 + 2m(E-V) = 0.$$

Regarding $\hbar$ as a parameter which may be given arbitrarily small values and assuming that $S$ can be expanded in a power series thus

$$S = S_0 + \hbar S_1 + \hbar^2 S_2 + \ldots,$$

show that

$$S_0 = \pm \int p\,dx,$$
$$S_1 = \tfrac{1}{2}\iota \log p,$$
$$S_2 = \frac{p'}{4p^2} - \tfrac{1}{8}\int \frac{p'^2}{p^3}\,dx,$$

where $p = \sqrt{[2m(E-V)]}$ is the classical momentum of the particle in any position.

Deduce an approximate solution to the wave equation in the form

$$\psi = p^{-\frac{1}{2}}\left[A\exp\left\{\frac{\iota}{\hbar}\int p\,dx\right\} + B\exp\left\{-\frac{\iota}{\hbar}\int p\,dx\right\}\right],$$

and show that this approximation is of good quality provided

$$\left|\frac{\hbar p'}{p^2}\right| \ll 1.$$

(This is the WKB-approximation.)

15. For the case of the uniform field of Ex. 13 (Chap. 3), show that the WKB-approximation gives

$$\psi = \xi^{-\frac{1}{4}}[A\exp(\tfrac{2}{3}\xi^{3/2}) + B\exp(-\tfrac{2}{3}\xi^{3/2})],$$

where

$$\xi = \left(\frac{2m}{\hbar^2F^2}\right)^{1/3}(Fx-E),$$

but that the approximation is only valid when $\xi$ is large (positive or negative).

16. A particle moves under no field forces perpendicularly to two parallel perfectly reflecting walls at a distance $a$ apart. Show that the variance of its distance from one wall when in the $n$th excited energy eigenstate is

$$\tfrac{1}{12}a^2\left(1-\frac{6}{n^2\pi^2}\right)$$

and shows that this agrees with the classical result for large $n$.

17. Show that, for the particle of the previous exercise, the momentum probability density for the $n$th excited energy state is

$$\frac{2n^2\pi a}{\hbar}\cdot\frac{1-(-1)^n\cos\dfrac{pa}{\hbar}}{\left(\dfrac{a^2p^2}{\hbar^2}-n^2\pi^2\right)^2}.$$

Sketch the graphs of this density function in the two cases (a) $n$ even and (b) $n$ odd.

18. Show that, by putting $\hat{x}=\alpha\hat{X}$, $\hat{p}=\beta\hat{P}$, where $\alpha$, $\beta$ are suitable numerical factors, the Schrödinger equation of the harmonic oscillator can be put in the form

$$\tfrac{1}{2}(\hat{P}^2+\hat{X}^2)\psi = \epsilon\psi,$$

where $E=\epsilon\hbar\omega$.

Show that $[\hat{X},\hat{P}]=\iota$ and deduce that

$$\hat{P}^2+\hat{X}^2 = (\hat{X}+\iota\hat{P})(\hat{X}-\iota\hat{P})-1.$$

Hence prove that, if $\psi$ is the eigenfunction corresponding to the eigenvalue $\epsilon$, then $(\hat{X}+\iota\hat{P})\psi$ is the eigenfunction corresponding to the eigenvalue $\epsilon-1$.

Deduce that, for some positive integral $n$,

$$(\hat{X}+\iota\hat{P})^n\psi$$

vanishes identically and that, if $\psi=\psi_0$ in the ground state,

$$(\hat{X}+\iota\hat{P})\psi_0 = 0.$$

(Hint: $E$ must be positive.) Derive equations (3.10.32), (3.10.33).

19. The harmonic oscillator is in its $n$th excited energy eigenstate. Show that the variances of $x$ and $p$ are

$$\frac{\hbar}{m\omega}(n+\tfrac{1}{2}),\quad m\hbar\omega(n+\tfrac{1}{2}),$$

respectively.

20. The harmonic oscillator is in its ground state. Show that the probability that the particle will be found outside the zone to which it is restricted according to Newtonian theory is 0·16 ....

21. A particle of mass $m$ is moving under no field forces normally to two parallel perfectly reflecting walls at a distance $a$ apart. At the instant under consideration, its wave function is given by

$$\psi = Ax(a-x),$$

where $x$ is distance measured from one of the walls. Calculate the probability distribution for the particle's energy (i.e. the probabilities of transitions into the various energy eigenstates) and show that its expected value is $5\hbar^2/ma^2$ and that its variance is $5\hbar^4/m^2a^4$. (If $n$ is odd, $\sum n^{-2} = \pi^2/8$, $\sum n^{-4} = \pi^4/96$.)

22. A particle of mass $m$ moving parallel to the $x$-axis is confined in a field-free region between impenetrable walls $x=0$ and $x=a$. It is in its eigenstate of lowest energy when the wall $x=a$ is rapidly displaced outwards to $x=2a$. Assuming its state to be unaffected by the displacement, show that the probability that a subsequent measurement finds the particle energy unaltered is $\frac{1}{2}$.

23. A particle of mass $m$ is moving parallel to the $x$-axis under a force $m\omega^2 x$ directed towards the origin. Its energy is sharp with value $\frac{1}{2}\omega\hbar$. $\omega$ is suddenly reduced to a fifth of its original value and the particle's energy is measured immediately afterwards. Show that the probability that this will be found to be unchanged is $2\sqrt{5}/27$.

24. A particle moves parallel to the $x$-axis between two perfectly reflecting walls at $x=0$, $x=a$. Taking the energy eigenfunctions as basis, show that the matrix representing the $x$-component of momentum has elements $p_{mn}$, where

$$\begin{aligned} p_{mn} &= 0, \quad \text{if } m, n \text{ have the same parity,} \\ &= \frac{\hbar}{\iota a} \cdot \frac{4mn}{m^2-n^2}, \quad \text{if } m, n \text{ have opposite parities.} \end{aligned}$$

By considering the matrix representation of $p^2$, deduce that, in the $n$th eigenstate, the expected value of $p^2$ is $n^2h^2/4a^2$. (Assume that

$$\sum_m \frac{m^2}{(m^2-n^2)^2} = \frac{\pi^2}{16},$$

where $m$ ranges over all odd integers when $n$ is even and over all even integers when $n$ is odd.)

25. For the particle of the previous exercise, employing the same basis, show that the matrix representation of $x$ is $(x_{mn})$ where

$$
\begin{aligned}
x_{mn} &= \tfrac{1}{2}a, && \text{if } m = n, \\
&= 0, && \text{if } m, n \text{ have the same parity and } m \neq n, \\
&= -\frac{8amn}{\pi^2(m^2-n^2)^2}, && \text{if } m, n \text{ have opposite parities.}
\end{aligned}
$$

If the particle is in the state represented by the vector $\frac{1}{2}\sqrt{2}(1,1,0,0,\ldots)$, show that

$$\bar{x} = \tfrac{1}{2}a(1-32/9\pi^2).$$

26. A particle of mass $m$ moves freely within a box having perfectly reflecting walls at $x=0$, $x=a$, $y=0$, $y=b$, $z=0$, $z=c$. Applying the method of separation of variables to the Schrödinger equation expressed in cartesian coordinates $x, y, z$, show that a possible energy eigenstate for the particle has wave amplitude

$$\psi = \sqrt{\left(\frac{8}{abc}\right)} \sin\frac{p\pi x}{a} \sin\frac{q\pi y}{b} \sin\frac{r\pi z}{c},$$

where the energy $E$ of the particle in this state is quantised and is given by

$$E = \frac{h^2}{8m}\left(\frac{p^2}{a^2}+\frac{q^2}{b^2}+\frac{r^2}{c^2}\right),$$

$p, q, r$ being positive integers.

27. The three-dimensional harmonic oscillator comprises a particle of mass $m$, moving within a potential well, centre $O$, for which the P.E. function is $V=\frac{1}{2}m\omega^2r^2$, $r$ being distance measured from $O$. Applying the method of separation of variables as in the previous exercise, show that a possible energy eigenstate for the particle is such that

$$\psi = \left(\frac{2m\omega}{h}\right)^{\frac{3}{4}} \left(\frac{1}{2^n\, i!\, j!\, k!}\right)^{\frac{1}{2}} \exp(-m\omega r^2/2\hbar)\, H_i(\xi)\, H_j(\eta)\, H_k(\zeta),$$

where

$$\xi = \sqrt{\left(\frac{m\omega}{\hbar}\right)}\, x, \text{ etc,}$$

$i, j, k$ are positive integers or zero and $n=i+j+k$.

Show that the particle's energy in this state is given by

$$E = \omega\hbar(n+\tfrac{3}{2}).$$

28. Express the Schrödinger wave equation in terms of spherical polar coordinates $(r, \theta, \phi)$ and hence show that, if $\psi = \chi/r$, where $\chi$ is a function of $r$ alone, then $\chi$ satisfies

$$\frac{d^2\chi}{dr^2} + \frac{2m}{\hbar^2}(E - V)\chi = 0.$$

(Such a solution is spherically symmetric.)

29. Show that the three-dimensional harmonic oscillator can exist in energy eigenstates which are spherically symmetric about the centre of attraction and the wave function for such a state is given by

$$\psi = \left(\frac{2m\omega}{h}\right)^{\frac{3}{4}} \frac{1}{2^{n+1}\sqrt{[(2n+1)!]}} \frac{1}{\xi} \exp\left(-\tfrac{1}{2}\xi^2\right) H_{2n+1}(\xi),$$

where

$$\xi = \sqrt{\left(\frac{m\omega}{\hbar}\right)}\, r,$$

and $n$ is a positive integer or zero.

Prove that the energy in such states is quantised according to the equation

$$E = \omega\hbar(2n + \tfrac{3}{2})$$

and identify the ground state with the ground state of Ex. 27 (Chap. 3).

30. A particle having mass $m$ moves freely with kinetic energy $T$. Assuming its wave function $\psi$ to be spherically symmetric about a point $O$, show that

$$\psi = A\frac{\sin kr}{r},$$

where $k^2 = 2mT/\hbar^2$. Deduce that, in this state, the most probable distances of the particle from $O$ are given by

$$r = (2n+1)\pi/2k,$$

where $n = 0, 1, 2$, etc. (Hint: calculate the probability the particle lies between spheres of radii $r$, $r + dr$ and maximise.)

31. Taking, as a simple model for the hydrogen atom, an electron moving in a spherically symmetric potential 'well' given by

$$V = -V_0, \quad 0 < r < a,$$

$$= 0, \quad r > a,$$

show that the eigenvalues for the energy $-W$ of the spherically symmetric bound states are determined by the equation

$$\sin\theta = \pm k\theta,$$

where

$$\theta = \frac{a}{\hbar}\sqrt{[2m(V_0 - W)]}, \quad k = \frac{\hbar}{a}(2mV_0)^{-\frac{1}{2}},$$

and $\theta$ is taken in the second or fourth quadrants.

Deduce that the minimum depth of the well for there to exist a bound state is given by

$$V_{0,\,\min} = \frac{h^2}{32ma^2}$$

and also that, if the well is very deep, the early members of the series of eigenvalues for $W$ are given by

$$W_n = V_0 - \frac{n^2 h^2}{8ma^2},$$

$n=1, 2, \ldots,$ approximately.

32. A particle of mass $m$ moves in a field which is spherically symmetric about a point $O$ and for which the potential energy function is $-V_0\exp(-r/a)$, $r$ being the distance measured from $O$. Show that the particle can exist in bound, spherically symmetric, energy eigenstates, with energies $E_1, E_2, \ldots, E_n$, where

$$E_s = -\frac{\hbar^2}{8ma^2}\nu_s^2,$$

the $\nu_s$ being the sequence of positive values of $\nu$ for which

$$J_\nu(\lambda) = 0,$$

where $\lambda = (8ma^2 V_0/\hbar^2)^{\frac{1}{2}}$. Deduce that there are no such states if

$$ma^2 V_0 < 0{\cdot}723\hbar^2,$$

and only one such state if

$$0{\cdot}723\hbar^2 < ma^2 V_0 < 3{\cdot}809\hbar^2.$$

Show also that, in the $s$th state,

$$\psi = \frac{A}{r}J_{\nu_s}[\lambda\exp(-r/2a)]$$

and sketch the graph of $\psi$ against $r$.

33. A system, described by one coordinate $x$, is in a state given by $\psi = Ax\exp(-\lambda x)$, $\lambda > 0$ for $x \geqslant 0$ and $\psi = 0$ for $x < 0$. Show that the probability that the momentum shall lie in the range $-\lambda\hbar$ to $\lambda\hbar$ is $(\frac{1}{2} + 1/\pi)$.

34. A particle is moving freely with energy $E$. Write down the equation satisfied by the wave function $\phi$ in the momentum representation. Deduce that $\phi$ vanishes except over the sphere

$$p_i p_i = 2mE$$

in the $p_1 p_2 p_3$-space. If the particle's momentum magnitude $p$ is measured, what will be the result?

35. A particle of mass $m$ is moving along the $x$-axis with energy $E$ under the action of a uniform field of force of magnitude $F$ directed in the negative sense of $Ox$. Write down the equation for the wave function $\phi$ in the momentum representation and deduce that

$$\phi = A\exp\left[\frac{\iota}{F\hbar}\left(\frac{p^3}{6m} - Ep\right)\right].$$

Hence show that all momenta are present with equal probability.

36. The parity operators $\hat{\Pi}_x$, $\hat{\Pi}_y$, $\hat{\Pi}_z$ are defined by the equations

$$\hat{\Pi}_x \psi(x, y, z) = \psi(-x, y, z),$$
$$\hat{\Pi}_y \psi(x, y, z) = \psi(x, -y, z),$$
$$\hat{\Pi}_z \psi(x, y, z) = \psi(x, y, -z).$$

Show that these operators are linear and Hermitian.

Show that $\hat{\Pi}_x^2 = \hat{\Pi}_y^2 = \hat{\Pi}_z^2 = 1$ and deduce that the eigenvalues of each of the parity operators are $\pm 1$. Show that any even function of $x$, $\psi_E$, is an eigenfunction of $\hat{\Pi}_x$ corresponding to the eigenvalue $+1$ and any odd function $\psi_o$ is an eigenfunction corresponding to the eigenvalue $-1$. Show that $\psi_E$, $\psi_o$ are orthogonal and that any function $\psi$ can be expressed linearly in terms of such functions thus

$$\psi = a_E \psi_E + a_o \psi_o,$$

where $a_E$, $a_o$ are normalised. Show that

$$a_E^* a_E + a_o^* a_o = 1.$$

37. If $\Pi_x$ is the observable represented by the operator $\hat{\Pi}_x$ (i.e. the result of an observation procedure having such a nature that the particle's

wave function is forced to become momentarily either even or odd in $x$), prove, in the notation of the previous exercise, that

$$\overline{\Pi_x^n} = 1, \quad n \text{ an even integer,}$$
$$= a_E^* a_E - a_o^* a_o, \quad n \text{ odd.}$$

Deduce that $\Pi_x$ assumes the value $+1$ with probability $a_E^* a_E$ and the value $-1$ with probability $a_o^* a_o$.

38. A particle is in motion parallel to $Ox$ and its wave function at some instant is given by

$$\psi = \phi(x) \exp(\iota p_0 x/\hbar).$$

Show that the expected value of its momentum is $p_0$.

## Chapter 4

1. A particle of mass $m$ has coordinates $x_i$ and components of linear momentum $p_i$ $(i=1,2,3)$. It moves in a conservative field of force with respect to which its P.E. is $V(x_1,x_2,x_3)$. Prove Ehrenfest's Theorem, viz.

$$m\ddot{\bar{x}}_i = \bar{F}_i,$$

where $F_i$ are the components of force acting upon the particle. (Hint: employ equations (4.3.6).)

2. Two particles having masses $m_1$, $m_2$ move freely except for a conservative force of interaction for which the P.E. function is $V(r)$, where $r$ is the distance between the particles. Taking the coordinates of the particles relative to an inertial frame to be $(x_1,y_1,z_1)$, $(x_2,y_2,z_2)$ respectively, write down the Hamiltonian operator for the system in the coordinate representation. Deduce that the expected value of the net linear momentum is independent of the time.

3. The energy of interaction of a pair of distinguishable particles, each having spin $\frac{1}{2}\hbar$, is given by

$$V = k\mathfrak{z}_1 \cdot \mathfrak{z}_2,$$

where $\mathfrak{z}_1$, $\mathfrak{z}_2$ are the spin axis vectors for the particles. Neglecting all other terms of the Hamiltonian $H$, employing the representation introduced in section 2.5, calculate the $4 \times 4$ matrix representing $H$. Deduce that any component of the net spin of the system has an expected value which is independent of the time.

4. Write down the matrix equation of motion for the system described in the previous exercise and deduce that, if the column matrix specifying the state of the system at $t=0$ is $\{\alpha_1, \alpha_2, \alpha_3, \alpha_4\}$, then the matrix determining the state at a later instant $t$ is

$$\begin{pmatrix} \alpha_1 \exp(-\iota kt/\hbar) \\ \frac{1}{2}(\alpha_2+\alpha_3)\exp(-\iota kt/\hbar)+\frac{1}{2}(\alpha_2-\alpha_3)\exp(3\iota kt/\hbar) \\ \frac{1}{2}(\alpha_2+\alpha_3)\exp(-\iota kt/\hbar)-\frac{1}{2}(\alpha_2-\alpha_3)\exp(3\iota kt/\hbar) \\ \alpha_4 \exp(-\iota kt/\hbar) \end{pmatrix}.$$

If, initially, $\sigma_{1z}=+1$, $\sigma_{2z}=-1$, show that these components of the spins are again sharp after a time $\pi\hbar/4k$, but with their eigenvalues interchanged.

5. By applying the method of separation of variables to the one-dimensional wave equation for a particle moving freely parallel to the $x$-axis, obtain the solution

$$\psi = \sum_{n=-\infty}^{\infty} a_n \exp \iota\left(n\omega x - \frac{\hbar n^2 \omega^2}{2m} t\right),$$

where $\omega$ is arbitrary.

6. A particle moves freely parallel to the $x$-axis between two perfectly reflecting walls at $x=0$ and $x=\pi$. Assuming that $\psi$ is identically zero outside the interval $(0, \pi)$ and is continuous at the walls, show that the boundary conditions for the problem can be satisfied by taking $\omega=1$, $a_{-n}=-a_n$ and $a_0=0$ in the series for $\psi$ of the previous exercise. Also, if $\psi$ is known at $t=0$, show that

$$a_n = \frac{1}{\pi\iota}\int_0^{\pi} \psi(x,0)\sin nx\,dx.$$

7. For the particle of the previous exercise, assume that $\psi$ is initially real and in the shape of a rectangular pulse extending from $x=0$ to $x=\delta$. In this case, show that

$$\psi(x,t) = \frac{2}{\pi}\sum_{n=1}^{\infty}\frac{1}{n\delta^{\frac{1}{2}}}(1-\cos n\delta)\exp[-(\iota\hbar n^2/2m)t]\sin nx.$$

8. Taking $\delta=\frac{1}{2}\pi$ in the previous problem and neglecting all but the first two terms of the series, show that the probability density for the particle is given by

$$\rho = \frac{8}{\pi^3}\left(\sin^2 x+\sin^2 2x+2\sin x\sin 2x\cos\frac{3\hbar}{2m}t\right).$$

Plot $\rho$ against $x$ for a succession of values of $t$. (Note that the effect is roughly that of a pulse which oscillates between the reflecting walls.)

9. A particle of mass $m$ is in motion parallel to $Ox$ and

$$\psi(x,t) = A\exp\left[\frac{m\omega}{\hbar}\left\{-\tfrac{1}{2}(x-a\sin\omega t)^2+\iota xa\cos\omega t-\tfrac{1}{4}\iota a^2\sin 2\omega t-\frac{\iota\hbar}{2m}t\right\}\right].$$

Calculate $A$ if $\psi$ is normalised. Calculate the field of force which must be present and describe the manner in which the probability distribution for the particle changes with the time. (A normal error pulse oscillates with S.H.M., without spreading.)

10. A cloud of identical non-interacting particles is in motion in the vicinity of axes $Oxyz$. The wave function for the cloud is

$$\psi = \frac{1}{r}\exp\left[\iota(kr-\omega t)\right],$$

where $r$ is radial distance from $O$. Calculate $\mathbf{j}$ and show that the total mass flux across any sphere centre $O$ is $2kh$.

11. $(r,\theta,\phi)$ are spherical polar coordinates relative to an inertial frame. At a certain instant, the wave function for a particle of mass $m$ moving in the frame is given by

$$\psi = Ar\exp\left(-\tfrac{1}{2}r\right)\sin\theta\exp\left(\iota\phi\right),$$

where $A$ is a constant. Show that the components of $\mathbf{j}$ are given by

$$j_r = j_\theta = 0, \quad j_\phi = \frac{\hbar}{64\pi m}r\exp\left(-r\right)\sin\theta.$$

12. A potential barrier is specified by the equations

$$\begin{aligned} V &= 0, & x &< 0,\\ &= V_0, & 0 &< x < d,\\ &= 2V_0, & x &> d.\end{aligned}$$

A stream of particles, each having total energy $3V_0$, is incident from the left. Show that the transmission coefficient for the barrier has a value lying between $2(2\sqrt{3}-3)$ and $8(7\sqrt{3}-12)$ for all values of $d$.

13. Identical particles, possessing linear momentum $p$, are moving parallel to the $x$-axis in the positive sense and form a steady uniform beam in the region $x<0$. The number of particles per unit length of the

beam is $N$. They are subject to a conservative field of force for which the potential energy function is given by the equations

$$\begin{aligned} V &= 0, & x &< 0, \\ &= V_0, & 0 &< x < a, \\ &= (1+\alpha^2)\,V_0, & a &< x, \end{aligned}$$

where $\alpha > 0$ and $V_0$ is such that, according to classical principles, each particle would just fail to enter the region $x > 0$. Show that, at any instant, the number of particles in the region $x > a$ is

$$\frac{2N\hbar}{\alpha p[\alpha^2+(1+ap\alpha/\hbar)^2]}.$$

14. A steady uniform stream of particles, each of mass $m$ and having momentum $p$, is in motion parallel to the $x$-axis in the positive sense. The stream is incident upon a potential 'well' extending from $x = -h/p$ to $x = h/p$ over which the P.E. of a particle is $-3p^2/2m$. Elsewhere the P.E. is zero. Prove that no particles are reflected by the well. If $\rho_0$ is the density of the particles in the incident stream, show that their density within the well is given by

$$\rho = \tfrac{1}{8}\rho_0\left(5+3\cos\frac{8\pi px}{h}\right).$$

15. Write down, in the momentum representation, the equation of motion (4.1.7) for a particle moving freely. Deduce that the probability distribution for the components of momentum does not change with the time.

16. $\mathbf{r}$ is the position vector of a particle of mass $m$ which is moving freely. If $\sigma^2$ is the variance of $\mathbf{r}$, show that

$$\sigma^2 = \overline{\mathbf{r}^2}-(\bar{\mathbf{r}})^2$$

and deduce that

$$\begin{aligned} \frac{d}{dt}(\sigma^2) &= \frac{2}{m}(\overline{\mathbf{r}.\mathbf{p}}-\bar{\mathbf{r}}.\bar{\mathbf{p}}), \\ \frac{d^2}{dt^2}(\sigma^2) &= \frac{2}{m^2}(\overline{\mathbf{p}^2}-(\bar{\mathbf{p}})^2), \\ \frac{d^3}{dt^3}(\sigma^2) &= 0, \end{aligned}$$

where $\mathbf{p}$ is the linear momentum and $\mathbf{r}.\mathbf{p}$ is written as a symmetrised product.

Prove that the second of these derivatives is positive or zero and deduce that $\sigma^2$ ultimately increases monotonically as a quadratic function of $t$. [This result demonstrates the manner in which the wave packet associated with a freely moving particle expands; see also equation (4.7.19).]

17. A particle having magnetic dipole moment $\mu_0$ is placed in a uniform magnetic field. Taking the particle's Hamiltonian to be given by equation (4.5.2), write down Heisenberg's equations of motion for the spin axis components and show that these are equivalent to the equation

$$\dot{\sigma}_i = 2\epsilon_{ijk}\,\omega_j\,\sigma_k,$$

where $\omega_j = \mu_0 B_j/\hbar$.

Taking $B_1 = B_2 = 0$ and assuming that, at $t=0$, the base vectors of the representation correspond to the eigenstates $\sigma_3 = \pm 1$, integrate these equations for $\sigma_i$ ($i = 1,2,3$). Deduce that, if $\sigma_0$ is the component of spin axis in the direction determined by polar angles $(\theta, \phi)$, then at time $t$

$$\sigma_0 = \begin{pmatrix} \cos\theta & \exp[-\iota(\phi+2\omega t)]\sin\theta \\ \exp[\iota(\phi+2\omega t)]\sin\theta & -\cos\theta \end{pmatrix},$$

where $\omega = \omega_3$.

If the particle's spin state is specified by the spinor $\{\alpha_+, \alpha_-\}$, show that the direction $(\theta, \phi)$ in which the component of spin axis has the sharp value $+1$ at time $t$ is given by

$$\theta = 2\tan^{-1}\left|\frac{\alpha_-}{\alpha_+}\right|, \quad \phi = \arg\left(\frac{\alpha_-}{\alpha_+}\right) - 2\omega t.$$

(This again demonstrates that the spin axis precesses with angular velocity $2\omega$ about the direction of the field.)

## Chapter 5

1. Defining the operators $\hat{L}_+$, $\hat{L}_-$ by the equations

$$\hat{L}_+ = \hat{L}_x + \iota\hat{L}_y, \quad \hat{L}_- = \hat{L}_x - \iota\hat{L}_y,$$

obtain the results

(a) $[\hat{L}_+, \hat{L}_z] = -\hbar\hat{L}_+$,

(b) $[\hat{L}_-, \hat{L}_z] = \hbar\hat{L}_-$,

(c) $[\hat{L}^2, \hat{L}_+] = [\hat{L}^2, \hat{L}_-] = 0$,

(d) $\hat{L}_+\hat{L}_- = \hat{L}^2 - \hat{L}_z^2 + \hbar\hat{L}_z$,

(e) $\hat{L}_-\hat{L}_+ = \hat{L}^2 - \hat{L}_z^2 - \hbar\hat{L}_z$,

(f) $[\hat{L}_+, \hat{L}_-] = 2\hbar\hat{L}_z$.

2. If $\mathfrak{r}$ is the position vector of a particle in an inertial frame and $\mathfrak{L}$ is its angular momentum about the origin, obtain the results

(a) $\hat{\mathfrak{r}}.\hat{\mathfrak{L}} = \hat{\mathfrak{L}}.\hat{\mathfrak{r}} = 0,$

(b) $[\hat{L}^2, \hat{\mathfrak{r}}] = \iota\hbar(\hat{\mathfrak{r}} \times \hat{\mathfrak{L}} - \hat{\mathfrak{L}} \times \hat{\mathfrak{r}}),$

(c) $[\hat{\mathfrak{L}}, \hat{r}^2] = 0,$

(d) $\hat{\mathfrak{L}} \times \hat{\mathfrak{r}} + \hat{\mathfrak{r}} \times \hat{\mathfrak{L}} = 2\iota\hbar\hat{\mathfrak{r}},$

(e) $[\hat{L}^2, [\hat{L}^2, \hat{\mathfrak{r}}]] = 2\hbar^2(\hat{L}^2\hat{\mathfrak{r}} + \hat{\mathfrak{r}}\hat{L}^2).$

3. If $x_i$, $p_i$, $L_i$ $(i=1,2,3)$ are the components of the position vector, linear momentum and angular momentum of a particle relative to a rectangular inertial frame, prove that

(a) $[\hat{L}_i, \hat{x}_j] = \iota\hbar\epsilon_{ijk}\hat{x}_k,$

(b) $[\hat{L}_i, \hat{p}_j] = \iota\hbar\epsilon_{ijk}\hat{p}_k,$

(c) $\hat{L}_i\hat{p}_i = \hat{p}_i\hat{L}_i = 0,$

(d) $[\hat{L}^2, \hat{\mathfrak{p}}] = \iota\hbar(\hat{\mathfrak{p}} \times \hat{\mathfrak{L}} - \hat{\mathfrak{L}} \times \hat{\mathfrak{p}}),$

(e) $\hat{\mathfrak{L}} \times \hat{\mathfrak{p}} + \hat{\mathfrak{p}} \times \hat{\mathfrak{L}} = 2\iota\hbar\hat{\mathfrak{p}},$

(f) $[\hat{\mathfrak{L}}, \hat{p}^2] = 0,$

where the repeated subscript summation convention is in operation and $\epsilon_{ijk}$ is $+1$ if $(ijk)$ is an even permutation of (123), is $-1$ if $(ijk)$ is an odd permutation and is zero otherwise.

4. Show that, if $\hat{L}_+$ and $\hat{L}_-$ are the operators defined in Ex. 1 (Chap. 5) and if $\psi$ is an eigenvector of $\hat{L}_z$ corresponding to the eigenvalue $k\hbar$, then $\hat{L}_+\psi$, $\hat{L}_-\psi$ are eigenvectors of $\hat{L}_z$ corresponding to the eigenvalues $(k+1)\hbar$, $(k-1)\hbar$ respectively. (Hint: employ the results (a) and (b) from Ex. 1.)

In particular, prove the identities

$$\hat{L}_+ Y_{lk} = -\hbar\sqrt{[(l-k)(l+k+1)]}\, Y_{l,\,k+1},$$
$$\hat{L}_- Y_{lk} = -\hbar\sqrt{[(l+k)(l-k+1)]}\, Y_{l,\,k-1}.$$

(Hint: employ the recurrence relationships (D.10), (D.14) and (D.16) from Appendix D.)

5. If $L_z$ is sharp, prove that $\bar{L}_x = 0$. (Hint: employ the commutation identity $[\hat{L}_y, \hat{L}_z] = \iota\hbar\hat{L}_x$.)

6. $Oxyz$ are rectangular axes and $Oz'$ is in a direction having direction cosines $(l, m, n)$ relative to $Oxyz$. Express $L_{z'}$ in terms of $L_x$, $L_y$, $L_z$ and hence show that, if $L_z$ is sharp,

$$\bar{L}_{z'} = nL_z.$$

7. Write down equation (5.3.12) for the case when

$$V = \tfrac{1}{2}m\omega^2 r^2,$$

i.e. the case of the isotropic harmonic oscillator. Putting

$$x = \alpha r^2, \quad R = x^{\frac{1}{2}l}\exp(-\tfrac{1}{2}x)\,y, \quad \beta = E/\hbar\omega,$$

where $\alpha = m\omega/\hbar$, obtain the equation

$$x\frac{d^2y}{dx^2} + (l+\tfrac{3}{2}-x)\frac{dy}{dx} + \tfrac{1}{2}(\beta - l - \tfrac{3}{2})\,y = 0.$$

Deduce that $E$ is quantised according to the equation

$$E = \hbar\omega(l+2n+\tfrac{3}{2}),$$

and show that the wave function is given by

$$\psi = Ar^l \exp(-\tfrac{1}{2}\alpha r^2)\, L_n^{l+\frac{1}{2}}(\alpha r^2)\, Y_{lk}(\theta,\phi)\exp(-\iota Et/\hbar).$$

8. Write down equation (5.3.12) for the case when

$$V = 0, \quad r < a,$$
$$= \infty, \quad r > a,$$

i.e. the case of a spherically symmetric potential well. Deduce that

$$R = Ar^{-\frac{1}{2}} J_{l+\frac{1}{2}}(\lambda r),$$

where $\lambda^2 = 2mE/\hbar^2$. Hence prove that $E$ is quantised and that the energies of the states of least energy when $l=0, 1, 2$ are approximately $4{\cdot}9\hbar^2/ma^2$, $10{\cdot}6\hbar^2/ma^2$ and $16{\cdot}8\hbar^2/ma^2$ respectively.

9. A particle having mass $m$ and electric charge $e$ moves in an electromagnetic field $(\mathfrak{B}, \mathfrak{E})$ for which the vector potential is $\mathfrak{A}$. If $\mathfrak{v}$ denotes the velocity of the particle and $\mathfrak{p}$ is the linear momentum of the system, show that

(a) $[\hat{v}_x, \hat{v}_y] = \dfrac{\iota\hbar e}{m^2 c}B_z$, etc,

(b) $\hat{\mathfrak{p}}.\hat{\mathfrak{A}} - \hat{\mathfrak{A}}.\hat{\mathfrak{p}} = \dfrac{\hbar}{\iota}\operatorname{div}\mathfrak{A}$.

(Hint: refer to equation (F. 4) of Appendix F.)

10. A spinless particle having mass $m$ and charge $e$ is in motion in a uniform magnetic field of intensity $B$ directed along $Oz$. If the particle is in a

steady state with energy $E$, write down the wave equation for the wave function $\psi$ employing cylindrical polar coordinates $(\rho, \phi, z)$. Assuming that the motion is parallel to the plane $Oxy$ and that $L_z$ takes the sharp value $k\hbar$, show that

$$\psi = R(\rho)\exp[\iota(k\phi - Et/\hbar)],$$

and that $R$ satisfies

$$\frac{d^2R}{d\rho^2}+\frac{1}{\rho}\frac{dR}{d\rho}+\left[\alpha-\left(\beta\rho-\frac{k}{\rho}\right)^2\right]R = 0,$$

where

$$\alpha = 2mE/\hbar^2, \quad \beta = eB/2c\hbar.$$

Putting

$$\xi = \beta\rho^2, \quad R = w\xi^{\frac{1}{2}|k|}\exp(-\tfrac{1}{2}\xi),$$

show that $w$ satisfies the Laguerre-type equation

$$\xi\frac{d^2w}{d\xi^2}+(|k|+1-\xi)\frac{dw}{d\xi}+\left(\frac{\alpha}{4\beta}+\frac{k-|k|-1}{2}\right)w = 0.$$

Deduce that $E$ is quantised according to the equation

$$E = \frac{eB\hbar}{mc}\left(n+\frac{|k|+1-k}{2}\right).$$

11. The $z$-coordinate of a particle is known to be $z_0$ precisely. Show that the product of the standard deviations of the $x$-component of the angular momentum about the origin and the $y$-coordinate, is not less than $\frac{1}{2}\hbar z_0$.

12. If $\mathfrak{L}$ is the orbital angular momentum of an electron and $\mathfrak{J}$ is the total angular momentum including spin, prove that

$$[\hat{J}^2, \hat{\mathfrak{L}}] = -2\iota\hbar\hat{\mathfrak{J}}\times\hat{\mathfrak{L}}-2\hbar^2\hat{\mathfrak{L}}.$$

13. Allowing for spin in the Hamiltonian (5.6.5) by the addition of a term

$$-\frac{e}{mc}\mathfrak{B}.\mathfrak{s},$$

obtain an equation of continuity for the flow of probability and show that

$$\rho = \psi\psi^\dagger,$$

$$\mathfrak{j} = \frac{\iota\hbar}{2m}(\psi\nabla\psi^\dagger-\psi^\dagger\nabla\psi)-\frac{e}{mc}\psi\psi^\dagger\mathfrak{A},$$

where $\psi$ is a $2\times1$ matrix and $\psi^\dagger$ is its conjugate transpose.

14. A spinless particle is moving in a given force field (not necessarily central). $L_z$, $L^2$ are measured and found to take values $k\hbar$, $l(l+1)\hbar^2$ respectively. At the same instant a further compatible observable is measured, so that the particle is prepared in a pure state. Show that the wave function for the particle at this instant is given by

$$\psi = R(r)\, Y_{lk}(\theta, \phi),$$

where $R$ is normalised to satisfy the condition (5.4.17).

15. $P$ is a point having spherical polar coordinates $(r, \theta, \phi)$ relative to a frame $Oxyz$ and $Oz'$ is an axis passing through $P$. The values of $L^2$, $L_{z'}$ for a particle moving in the vicinity of the frame are measured to be $2\hbar^2$, $\hbar$ respectively. Show that the probabilities that an immediate measurement of $L_z$ will yield the results $+\hbar$, 0, $-\hbar$ are $\cos^4 \frac{1}{2}\theta$, $\frac{1}{2}\sin^2\theta$, $\sin^4 \frac{1}{2}\theta$ respectively. If the measurement of $L_{z'}$ yields the value 0 (instead of $\hbar$), show that these probabilities are $\frac{1}{2}\sin^2\theta$, $\cos^2\theta$, $\frac{1}{2}\sin^2\theta$, respectively.

16. Employing as a basis the complete set of eigenstates $\psi_{lkn}$ for the observables $L^2$, $L_z$, $H$, where $H$ is the Hamiltonian for a spinless particle moving in a central field, show that the elements of the matrix representing $L_z$ are given by

$$L_z(lkn, l'k'n') = k\hbar\delta_{ll'}\,\delta_{kk'}\,\delta_{nn'}.$$

Show, also, that the elements of the matrices representing $L_x$, $L_y$, are zero except where $l=l'$, $n=n'$ and $k=k'+1$ or $k'=k+1$, in which cases

$$\begin{aligned} L_x(lkn, lk'n) &= \tfrac{1}{2}\hbar\sqrt{[(l-k+1)(l+k)]}, \quad \text{if } k = k'+1, \\ &= \tfrac{1}{2}\hbar\sqrt{[(l-k'+1)(l+k')]}, \quad \text{if } k' = k+1, \\ L_y(lkn, lk'n) &= -\tfrac{1}{2}i\hbar\sqrt{[(l-k+1)(l+k)]}, \quad \text{if } k = k'+1, \\ &= \tfrac{1}{2}i\hbar\sqrt{[(l-k'+1)(l+k')]}, \quad \text{if } k' = k+1. \end{aligned}$$

(Hint: employ the results of Ex. 4 (Chap. 5).)

17. An atom of an isotope of hydrogen is in its ground state. The charge on the nucleus changes abruptly from $e$ to $2e$ as a consequence of $\beta$-decay. Show that the probability of finding the orbital electron in the $n$th excited state of the new atom after the change is

$$2^9 n^5 \frac{(n-2)^{2n-4}}{(n+2)^{2n+4}}.$$

Deduce that the probability that the orbital electron also escapes from the atom is approximately 2·6%.

## Chapter 6

1. The one-dimensional harmonic oscillator whose Hamiltonian is given by equation (3.10.36), is perturbed by a small uniform field of force of magnitude $\epsilon F$ in the positive $x$ sense. Employing the theory of section 6.1 show that, to the first order in $\epsilon$, the perturbed eigenvectors are

$$\psi^n + \epsilon F(2m\omega^3\hbar)^{-\frac{1}{2}}[n^{\frac{1}{2}}\psi^{n-1} - (n+1)^{\frac{1}{2}}\psi^{n+1}]$$

and, to the second order in $\epsilon$, the perturbed energy levels are

$$E_n - \frac{\epsilon^2 F^2}{2m\omega^2}.$$

Show that this second result is exact.

2. A hydrogen atom in its ground state is placed between the plates of a condenser. A voltage pulse is applied to the condenser so as to produce a homogeneous electric field whose intensity at time $t$ is given by the equations

$$\begin{aligned} E &= 0, & t < 0, \\ &= \epsilon \exp(-t/\tau), & t > 0. \end{aligned}$$

If $\epsilon$ is small, show that to $O(\epsilon^2)$ the probability of finding the atom in the 2S-state ($n=2, l=0, k=0$), after a long time has elapsed, is zero. To the same order of approximation, show that the probabilities of finding it in the 2P-states ($n=2, l=1, k=\pm 1$) are also zero, but that the probability of finding it in the 2P-state ($n=2, l=1, k=0$) is

$$\frac{2^{21}}{3^{12}} \frac{\hbar^8}{m^4 e^{10}}.$$

(Assume $\tau \gg \hbar^3/me^4$.)

## Chapter 7

1. Prove that

$$\boldsymbol{\gamma}_1 = \tfrac{1}{2}\iota[\boldsymbol{\alpha}_x, \boldsymbol{\beta}], \text{ etc.}$$

and deduce that the $\boldsymbol{\gamma}_i$ are all Hermitian. Calculate the canonical forms of the $\boldsymbol{\gamma}_i$.

2. Prove that

$$[\hat{\mathbf{H}}, x] = -\iota c\hbar\boldsymbol{\alpha}_x$$

and deduce that

$$\hat{\mathfrak{r}} = c\mathfrak{A}.$$

Hence show that the eigenvalues of any component of the velocity of a Dirac particle are $\pm c$.

3. Show that the matrix $\boldsymbol{\rho}$ defined by equation (7.8.10) may be expressed in terms of Dirac matrices thus:

$$\boldsymbol{\rho} = \iota \boldsymbol{\alpha}_x \boldsymbol{\alpha}_y \boldsymbol{\alpha}_z.$$

Deduce that

$$\boldsymbol{\rho}^2 = \mathbf{I}, \quad \boldsymbol{\rho\beta} = -\boldsymbol{\beta\rho}.$$

4. $\hat{\mathfrak{a}}$, $\hat{\mathfrak{b}}$ are two vector operators which commute with $\mathfrak{S}$. Show that

(i) $(\mathfrak{S}.\hat{\mathfrak{a}})\,\mathfrak{S} = \hat{\mathfrak{a}}\mathbf{I} + \iota\mathfrak{S}\times\hat{\mathfrak{a}}$,

(ii) $\mathfrak{S}(\mathfrak{S}.\hat{\mathfrak{a}}) = \hat{\mathfrak{a}}\mathbf{I} - \iota\mathfrak{S}\times\hat{\mathfrak{a}}$,

(iii) $(\mathfrak{S}.\hat{\mathfrak{a}})(\mathfrak{S}.\hat{\mathfrak{b}}) = \hat{\mathfrak{a}}.\hat{\mathfrak{b}}\mathbf{I} + \iota\mathfrak{S}.(\hat{\mathfrak{a}}\times\hat{\mathfrak{b}})$,

(iv) $[\mathfrak{S}, \mathfrak{S}.\hat{\mathfrak{a}}] = 2\iota\hat{\mathfrak{a}}\times\mathfrak{S}$.

5. If $\boldsymbol{\psi} = \{\psi_1, \psi_2, \psi_3, \psi_4\}$ is an eigenfunction of $\hat{\mathbf{H}}$ corresponding to the eigenvalue $E$, show that $\boldsymbol{\psi}' = \{-\psi_3, -\psi_4, \psi_1, \psi_2\}$ is an eigenfunction for the eigenvalue $-E$.

6. If $\psi$ is normalised, show that the functions $u$, $v$ defined in section 7.8 must satisfy the condition

$$\int_0^\infty (u^2+v^2)\,dr = \frac{1}{2l-1}.$$

7. Show that, in the eigenstate $J_z = (s+\frac{1}{2})\hbar$, $J^2 = (k^2-\frac{1}{4})\hbar^2$, $K = k\hbar$, $H = E$ of the hydrogen atom, the expected value of $\Sigma_z$ is given by

$$\bar{\Sigma}_z = \frac{2s+1}{2l+1}\left[4l\int_0^\infty u^2\,dr - 1\right],$$

where $l = |k|$.

8. Prove that, if $\hat{\mathfrak{J}} = (\hat{\mathbf{J}}_x, \hat{\mathbf{J}}_y, \hat{\mathbf{J}}_z)$, then

(i) $\hat{\mathfrak{J}}\times\hat{\mathfrak{J}} = \iota\hbar\hat{\mathfrak{J}}$,

(ii) $\hat{L}^2\mathbf{I} = \hat{\mathbf{K}}^2 - \hbar\boldsymbol{\beta}\hat{\mathbf{K}}$,

(iii) $\hat{\mathbf{J}}^2 = \hat{\mathbf{K}}^2 - \frac{1}{4}\hbar^2\mathbf{I}$.

# Bibliography


1. R. BECKER and F. SAUTER, *Electromagnetic Fields and Interactions, Vol. 2, Quantum Theory of Atoms and Radiation*, London, Blackie, 1964.

2 D. BOHM, *Quantum Theory*, New York, Prentice-Hall, 1952.

3 P. A. M. DIRAC, *The Principles of Quantum Mechanics*, Oxford, O.U.P., 1958.

4 A. LANDÉ, *New Foundations of Quantum Mechanics*, Cambridge, C.U.P., 1965.

5 A. MESSIAH, *Quantum Mechanics*, Vols. I and II, North-Holland, Amsterdam, 1962.

6 E. T. WHITTAKER, *A History of the Theories of Aether and Electricity; the Modern Theories*, London, Nelson, 1953.

# Index

A CATALOG OF SELECTED

# DOVER BOOKS

IN SCIENCE AND MATHEMATICS

# Math–Decision Theory, Statistics, Probability

ELEMENTARY DECISION THEORY, Herman Chernoff and Lincoln E. Moses. Clear introduction to statistics and statistical theory covers data processing, probability and random variables, testing hypotheses, much more. Exercises. 364pp. 5⅜ x 8½. 65218-1

STATISTICS MANUAL, Edwin L. Crow et al. Comprehensive, practical collection of classical and modern methods prepared by U.S. Naval Ordnance Test Station. Stress on use. Basics of statistics assumed. 288pp. 5⅜ x 8½. 60599-X

SOME THEORY OF SAMPLING, William Edwards Deming. Analysis of the problems, theory, and design of sampling techniques for social scientists, industrial managers, and others who find statistics important at work. 61 tables. 90 figures. xvii +602pp. 5⅜ x 8½. 64684-X

LINEAR PROGRAMMING AND ECONOMIC ANALYSIS, Robert Dorfman, Paul A. Samuelson and Robert M. Solow. First comprehensive treatment of linear programming in standard economic analysis. Game theory, modern welfare economics, Leontief input-output, more. 525pp. 5⅜ x 8½. 65491-5

PROBABILITY: An Introduction, Samuel Goldberg. Excellent basic text covers set theory, probability theory for finite sample spaces, binomial theorem, much more. 360 problems. Bibliographies. 322pp. 5⅜ x 8½. 65252-1

GAMES AND DECISIONS: Introduction and Critical Survey, R. Duncan Luce and Howard Raiffa. Superb nontechnical introduction to game theory, primarily applied to social sciences. Utility theory, zero-sum games, n-person games, decision-making, much more. Bibliography. 509pp. 5⅜ x 8½. 65943-7

INTRODUCTION TO THE THEORY OF GAMES, J. C. C. McKinsey. This comprehensive overview of the mathematical theory of games illustrates applications to situations involving conflicts of interest, including economic, social, political, and military contexts. Appropriate for advanced undergraduate and graduate courses; advanced calculus a prerequisite. 1952 ed. x+372pp. 5⅜ x 8½. 42811-7

FIFTY CHALLENGING PROBLEMS IN PROBABILITY WITH SOLUTIONS, Frederick Mosteller. Remarkable puzzlers, graded in difficulty, illustrate elementary and advanced aspects of probability. Detailed solutions. 88pp. 5⅜ x 8½. 65355-2

PROBABILITY THEORY: A Concise Course, Y. A. Rozanov. Highly readable, self-contained introduction covers combination of events, dependent events, Bernoulli trials, etc. 148pp. 5⅜ x 8¼. 63544-9

STATISTICAL METHOD FROM THE VIEWPOINT OF QUALITY CONTROL, Walter A. Shewhart. Important text explains regulation of variables, uses of statistical control to achieve quality control in industry, agriculture, other areas. 192pp. 5⅜ x 8½. 65232-7

# Math–Geometry and Topology

ELEMENTARY CONCEPTS OF TOPOLOGY, Paul Alexandroff. Elegant, intuitive approach to topology from set-theoretic topology to Betti groups; how concepts of topology are useful in math and physics. 25 figures. 57pp. 5⅜ x 8½. 60747-X

COMBINATORIAL TOPOLOGY, P. S. Alexandrov. Clearly written, well-organized, three-part text begins by dealing with certain classic problems without using the formal techniques of homology theory and advances to the central concept, the Betti groups. Numerous detailed examples. 654pp. 5⅜ x 8½. 40179-0

EXPERIMENTS IN TOPOLOGY, Stephen Barr. Classic, lively explanation of one of the byways of mathematics. Klein bottles, Moebius strips, projective planes, map coloring, problem of the Koenigsberg bridges, much more, described with clarity and wit. 43 figures. 210pp. 5⅜ x 8½. 25933-1

CONFORMAL MAPPING ON RIEMANN SURFACES, Harvey Cohn. Lucid, insightful book presents ideal coverage of subject. 334 exercises make book perfect for self-study. 55 figures. 352pp. 5⅝ x 8¼. 64025-6

THE GEOMETRY OF RENÉ DESCARTES, René Descartes. The great work founded analytical geometry. Original French text, Descartes's own diagrams, together with definitive Smith-Latham translation. 244pp. 5⅜ x 8½. 60068-8

PRACTICAL CONIC SECTIONS: The Geometric Properties of Ellipses, Parabolas and Hyperbolas, J. W. Downs. This text shows how to create ellipses, parabolas, and hyperbolas. It also presents historical background on their ancient origins and describes the reflective properties and roles of curves in design applications. 1993 ed. 98 figures. xii+100pp. 6½ x 9¼. 42876-1

THE THIRTEEN BOOKS OF EUCLID'S ELEMENTS, translated with introduction and commentary by Thomas L. Heath. Definitive edition. Textual and linguistic notes, mathematical analysis. 2,500 years of critical commentary. Unabridged. 1,414pp. 5⅜ x 8½. Three-vol. set. Vol. I: 60088-2 Vol. II: 60089-0 Vol. III: 60090-4

GEOMETRY OF COMPLEX NUMBERS, Hans Schwerdtfeger. Illuminating, widely praised book on analytic geometry of circles, the Moebius transformation, and two-dimensional non-Euclidean geometries. 200pp. 5⅝ x 8¼. 63830-8

DIFFERENTIAL GEOMETRY, Heinrich W. Guggenheimer. Local differential geometry as an application of advanced calculus and linear algebra. Curvature, transformation groups, surfaces, more. Exercises. 62 figures. 378pp. 5⅜ x 8½. 63433-7

CURVATURE AND HOMOLOGY: Enlarged Edition, Samuel I. Goldberg. Revised edition examines topology of differentiable manifolds; curvature, homology of Riemannian manifolds; compact Lie groups; complex manifolds; curvature, homology of Kaehler manifolds. New Preface. Four new appendixes. 416pp. 5⅜ x 8½. 40207-X

## History of Math

THE WORKS OF ARCHIMEDES, Archimedes (T. L. Heath, ed.). Topics include the famous problems of the ratio of the areas of a cylinder and an inscribed sphere; the measurement of a circle; the properties of conoids, spheroids, and spirals; and the quadrature of the parabola. Informative introduction. clxxxvi+326pp; supplement, 52pp. 5⅜ x 8½. 42084-1

A SHORT ACCOUNT OF THE HISTORY OF MATHEMATICS, W. W. Rouse Ball. One of clearest, most authoritative surveys from the Egyptians and Phoenicians through 19th-century figures such as Grassman, Galois, Riemann. Fourth edition. 522pp. 5⅜ x 8½. 20630-0

THE HISTORY OF THE CALCULUS AND ITS CONCEPTUAL DEVELOPMENT, Carl B. Boyer. Origins in antiquity, medieval contributions, work of Newton, Leibniz, rigorous formulation. Treatment is verbal. 346pp. 5⅜ x 8½. 60509-4

THE HISTORICAL ROOTS OF ELEMENTARY MATHEMATICS, Lucas N. H. Bunt, Phillip S. Jones, and Jack D. Bedient. Fundamental underpinnings of modern arithmetic, algebra, geometry, and number systems derived from ancient civilizations. 320pp. 5⅜ x 8½. 25563-8

A HISTORY OF MATHEMATICAL NOTATIONS, Florian Cajori. This classic study notes the first appearance of a mathematical symbol and its origin, the competition it encountered, its spread among writers in different countries, its rise to popularity, its eventual decline or ultimate survival. Original 1929 two-volume edition presented here in one volume. xxviii+820pp. 5⅜ x 8½. 67766-4

GAMES, GODS & GAMBLING: A History of Probability and Statistical Ideas, F. N. David. Episodes from the lives of Galileo, Fermat, Pascal, and others illustrate this fascinating account of the roots of mathematics. Features thought-provoking references to classics, archaeology, biography, poetry. 1962 edition. 304pp. 5⅜ x 8½. (Available in U.S. only.) 40023-9

OF MEN AND NUMBERS: The Story of the Great Mathematicians, Jane Muir. Fascinating accounts of the lives and accomplishments of history's greatest mathematical minds–Pythagoras, Descartes, Euler, Pascal, Cantor, many more. Anecdotal, illuminating. 30 diagrams. Bibliography. 256pp. 5⅜ x 8½. 28973-7

HISTORY OF MATHEMATICS, David E. Smith. Nontechnical survey from ancient Greece and Orient to late 19th century; evolution of arithmetic, geometry, trigonometry, calculating devices, algebra, the calculus. 362 illustrations. 1,355pp. 5⅜ x 8½. Two-vol. set. Vol. I: 20429-4 Vol. II: 20430-8

A CONCISE HISTORY OF MATHEMATICS, Dirk J. Struik. The best brief history of mathematics. Stresses origins and covers every major figure from ancient Near East to 19th century. 41 illustrations. 195pp. 5⅜ x 8½. 60255-9

# Physics

OPTICAL RESONANCE AND TWO-LEVEL ATOMS, L. Allen and J. H. Eberly. Clear, comprehensive introduction to basic principles behind all quantum optical resonance phenomena. 53 illustrations. Preface. Index. 256pp. 5⅜ x 8½. 65533-4

QUANTUM THEORY, David Bohm. This advanced undergraduate-level text presents the quantum theory in terms of qualitative and imaginative concepts, followed by specific applications worked out in mathematical detail. Preface. Index. 655pp. 5⅜ x 8½. 65969-0

ATOMIC PHYSICS: 8th edition, Max Born. Nobel laureate's lucid treatment of kinetic theory of gases, elementary particles, nuclear atom, wave-corpuscles, atomic structure and spectral lines, much more. Over 40 appendices, bibliography. 495pp. 5⅜ x 8½. 65984-4

A SOPHISTICATE'S PRIMER OF RELATIVITY, P. W. Bridgman. Geared toward readers already acquainted with special relativity, this book transcends the view of theory as a working tool to answer natural questions: What is a frame of reference? What is a "law of nature"? What is the role of the "observer"? Extensive treatment, written in terms accessible to those without a scientific background. 1983 ed. xlviii+172pp. 5⅜ x 8½. 42549-5

AN INTRODUCTION TO HAMILTONIAN OPTICS, H. A. Buchdahl. Detailed account of the Hamiltonian treatment of aberration theory in geometrical optics. Many classes of optical systems defined in terms of the symmetries they possess. Problems with detailed solutions. 1970 edition. xv+360pp. 5⅜ x 8½. 67597-1

PRIMER OF QUANTUM MECHANICS, Marvin Chester. Introductory text examines the classical quantum bead on a track: its state and representations; operator eigenvalues; harmonic oscillator and bound bead in a symmetric force field; and bead in a spherical shell. Other topics include spin, matrices, and the structure of quantum mechanics; the simplest atom; indistinguishable particles; and stationary-state perturbation theory. 1992 ed. xiv+314pp. 6⅛ x 9¼. 42878-8

LECTURES ON QUANTUM MECHANICS, Paul A. M. Dirac. Four concise, brilliant lectures on mathematical methods in quantum mechanics from Nobel Prize–winning quantum pioneer build on idea of visualizing quantum theory through the use of classical mechanics. 96pp. 5⅜ x 8½. 41713-1

THIRTY YEARS THAT SHOOK PHYSICS: The Story of Quantum Theory, George Gamow. Lucid, accessible introduction to influential theory of energy and matter. Careful explanations of Dirac's anti-particles, Bohr's model of the atom, much more. 12 plates. Numerous drawings. 240pp. 5⅜ x 8½. 24895-X

ELECTRONIC STRUCTURE AND THE PROPERTIES OF SOLIDS: The Physics of the Chemical Bond, Walter A. Harrison. Innovative text offers basic understanding of the electronic structure of covalent and ionic solids, simple metals, transition metals and their compounds. Problems. 1980 edition. 582pp. 6⅛ x 9¼. 66021-4